Springer Aerospace Technology

The series explores the technology and the science related to the aircraft and spacecraft including concept, design, assembly, control and maintenance. The topics cover aircraft, missiles, space vehicles, aircraft engines and propulsion units. The volumes of the series present the fundamentals, the applications and the advances in all the fields related to aerospace engineering, including:

- structural analysis,
- aerodynamics,
- aeroelasticity,
- aeroacoustics,
- flight mechanics and dynamics,
- orbital maneuvers,
- avionics,
- systems design,
- materials technology,
- launch technology,
- payload and satellite technology,
- space industry, medicine and biology.

The series' scope includes monographs, professional books, advanced textbooks, as well as selected contributions from specialized conferences and workshops.

The volumes of the series are single-blind peer-reviewed.

To submit a proposal or request further information, please contact: Mr. Pierpaolo Riva at pierpaolo.riva@springer.com (Europe and Americas) Mr. Mengchu Huang at mengchu.huang@springer.com (China)

The series is indexed in Scopus and Compendex

More information about this series at http://www.springer.com/series/8613

Simon Appel · Jaap Wijker

Simulation of Thermoelastic Behaviour of Spacecraft Structures

Fundamentals and Recommendations

Springer

Simon Appel
ATG Europe (Netherlands)
Noordwijk, The Netherlands

Jaap Wijker
Velserbroek, The Netherlands

ISSN 1869-1730 ISSN 1869-1749 (electronic)
Springer Aerospace Technology
ISBN 978-3-030-78998-5 ISBN 978-3-030-78999-2 (eBook)
https://doi.org/10.1007/978-3-030-78999-2

This Springer imprint is published by the registered company Springer Nature Switzerland AG
The registered company address is: Gewerbestrasse 11, 6330 Cham, Switzerland

The authors dedicate this book to their former boss ir. Marinus P. Nieuwenhuizen of the structural department of Fokker Space and Systems BV, The Netherlands.

Foreword

Thousands of spacecraft have been launched into space since the launch of Sputnik in 1957. We have landed men on the moon, visited planets in our Solar System with robotic probes, stationed scientists on the International Space Station, established global telecommunication and navigation systems, and nowadays, we are setting new goals in exploration, and consequently, despite all the reached accomplished, we are facing new, as well as old, challenges. Among them, the design of spacecraft structures subjected not only to mechanical, vibrational and acoustic loading, but also to extreme thermal conditions, requires new analysis techniques, and the consideration of thermal and structural analyses at the same time.

This book covers the topics necessary to the understanding of the thermoelastic behaviour of spacecraft structures, describing the essential steps of the analysis and verification, without missing important aspects such as handling uncertainties in the thermoelastic analysis process. Each topic is presented in the book with the necessary background, the up-to-date perspective and the essential theory–practical connection, thanks to the extensive industrial experience of both authors.

This book will be of immense help to thermal and mechanical engineers who are looking for concrete answers of their problems and, at the same time, to graduate students who would like to acquire knowledge in the fascinating world of the space structures.

October 2020

Chiara Bisagni
Professor of Aerospace Structures
Delft University of Technology
Delft, The Netherlands

Preface

The authors, Jaap Wijker and Simon Appel, have been working together since October 1986. In that month, Simon Appel arrived as a trainee at the Space division of Fokker Aircraft in the group led by Jaap. In those years, the group was running a technology development activity for ESA-ESTEC of which part of it aimed at developing an interface between the most important lumped parameter thermal analyser of those days, SINDA ("Systems Improved Numerical Differencing Analyser") and the finite element code used in the structures section at ESTEC, which was ASKA ("Automatic System for Kinematic Analysis"). In this activity, the foundation was established of the PAT method, the prescribed average temperature method, that you will find explained in this book and its implementation in the SINAS software: SINda—ASka. The mentioned technology development was initiated by Steve Stavrinidis, and the follow-on developments were run under the supervision of Michel Klein.

Jaap with his team members could be called the founding fathers of the PAT method. In the years after, Simon matured the implementation of the method in the SINAS software and the interface with the commonly used finite element tool MSC Nastran. The tool has been used at ESTEC and at Fokker Space and Systems. These were the years that Jaap and Simon worked on several thermoelastic problems.

In 2001, Simon joined the company that is called today ATG Europe and has been supporting ESA-ESTEC since then. An important part of his time has been, and still is, dedicated to thermoelastic problems and further development of the SINAS software. Jaap kept working for Fokker Space as it was called in the meantime, but also started to lecture at the Delft Technical University. Inspired by his students, he got motivated to write several books on spacecraft structures and related structural dynamics. Despite his retirement in 2009, Jaap continued writing books and managed even to obtain his Ph.D. degree.

The never-ending drive of Jaap to share his knowledge and keep on studying made him to contact Simon to join him writing the book that is now in your hands on the subject of thermoelastic simulations. It is the subject with which the relationship between Jaap and Simon started in the 1980s.

Jaap and Simon enjoyed writing this book. They know there is much more to discuss, and it will never be finished, but at some point, the time is there to share the results with you as reader.

Whether you are a graduate student or an experienced senior thermal or mechanical engineer, Jaap and Simon hope you find some useful information in this book. They also invite you to send feedback and suggestions that they may consider for a potential next edition.

Velserbroek, The Netherlands
Voorhout, The Netherlands
May 2021

Jaap Wijker
Simon Appel

Acknowledgements

The authors thank all those supported them with advices and encouragements during preparation of this book.

A special thanks goes to:

- The management and colleagues of ATG Europe and the Mechanical Department of ESA-ESTEC for their moral support and advice
- The colleagues of Engineering Lab of ATG Europe, and specifically Alexander van Oostrum and Alberto Peman, for the inspiring discussions
- Dick Wijker for facilitating the application of scientific software for the preparation of the numerical examples for this book
- Daniele Stramaccioni for his thorough and critical review of the thermal aspects discussed in the book
- MSC Software and In Summa Innovation for their support on the structural aspects of the numerical examples prepared for this book
- ITP for providing an ESATAN-TMS licence for solving the thermal problems that are part of the examples

and many others who contributed with ideas without realising this.

Perspective

How did the Universe originate, what are its fundamental physical laws and what is it made of? How does the Solar System work? What are the conditions for planet formation and the emergence of life within and outside our Solar System? How can we monitor, understand and preserve the delicate ecosystem of our planet and how can we protect life on Earth? These are some of the fundamental questions we, as human beings, are trying to answer since hundreds of years, and these are the questions the European Space Agency (ESA) is addressing within its world leading space science programmes and Earth observation missions.

With our "time-machine" Planck and his astronomy observer companion Herschel, we changed our view of the Universe, by revealing the relic radiation left by the Big Bang to an unprecedented level of accuracy, measuring fluctuations in temperature of a few millionth of degree, hence tracing the birth of stars and the evolution of galaxies throughout time. Gaia, the billion-star surveyor, has been making precise measurements of the positions, motions and characteristics of stars in order to create a three-dimensional map of our Milky Way and explore the past and future evolution of the Galaxy. PLATO will soon hunt Sun–Earth analogue systems in relatively nearby stars, identifying and studying thousands of exoplanetary systems, with emphasis on discovering and characterising Earth-sized planets in the habitable zone of their parent star. Closer to us, the Copernicus Earth observation programme is taking the pulse of our planet and is providing decision-makers with indisputable data in order to understand global changes and intervene effectively to resolve them.

These are some of the most recent missions ESA has developed in an attempt to shed light in the understanding of the Universe and our home, the Earth. They have posed exceptional challenges to the thousands of European engineers and scientists working in industry, academia and in the agency for the development, the verification and the operation of these scientific jewels. Space is probably the most demanding and hostile operational environment in which human engineering products are required to function. Without any repair or maintenance option, space structures shall guarantee years of performances reaching incomparable levels of accuracy and stability of the telescopes and optical instruments on-board. And, this shall be ensured despite the very demanding mechanical loading during launch and separation and the prohibitive temperature environment and fluctuations encountered in-service.

Although more and more the design of our space and Earth exploring satellites is driven by thermomechanical performances, hardly any textbooks or university classes are devoted to this fundamental discipline. Many good references exist addressing spacecraft thermal analysis and control, and also, a vast amount of literature can be found on the different satellite structural engineering subjects. However, these sources only cover the relevant topics within the context of the respective disciplines. In some cases, the interaction with the other subject is briefly discussed, but mainly from the perspective of their own domains. Moreover, the thermal and structural disciplines for the development of spacecraft structures and payloads have been traditionally supported by two distinct entities in most space engineering organisations, and in addition, both disciplines are using different analysis methodologies and associated numerical tools.

Thermomechanical and thermoelastics analyses aim to predict the deformations and the stresses affecting a structure or a component due to temperature fields and variations. In order to have a complete understanding of the problem, the two above-mentioned disciplines need to be addressed in a synergistic and cross-sectorial manner. A structural finite element model for thermoelastic predictions of a structure under a given thermal environment cannot be precisely established without a detailed knowledge of the temperature fields described by the thermal model. Conversely, a thermal engineer needs to have an adequate understanding of what are the temperature results and the resolution required by the structural model for reliable analysis.

The present book aims at capitalising the vast experience of the authors in the field of thermoelastic predictions applied to spacecraft structures and provides a coherent approach to solve practical and real-life problems. While analysing the different modelling and verification objectives of the thermal and structural domains, the authors address the current state-of-the-art approaches and limitations for both analyses and provide a suitable and verified method for addressing both disciplines in a synergistic fashion. This also includes specific numerical tools for transferring results from the structural to the thermal numerical environment and vice versa.

The book is most welcome in the space community and will provide a unique guidance to senior and the younger generation of engineers involved in the structural and thermal analyses of sophisticated spacecraft structures and instruments. It can constitute a sound basis for the building of a dedicated European Cooperation for Space Standardisation (ECSS) standard related to thermoelastic analysis and verification, ultimately leading to more performing space missions, improving our understanding of the Universe and contributing to a better life on our planet.

May 2021

Tommaso Ghidini
Head of the Structures, Mechanisms
and Materials Division
European Space Agency
The Hague, The Netherlands

Contents

Acronyms and Abbreviations

Abbreviations

ASM	Aerospace Specification Metals
AU	Astronomical unit
CNES	Centre National d'Etudes Spatiales
CPPT	Centre-Point Prescribed Temperature method
CSA	Canadian Space Agency
CTE	Coefficient of thermal expansion
DOF	Degree of freedom
ECSS	European Cooperation for Space Standardisation
ESA	European Space Agency
ESTEC	European Space Technology Center
FE	Finite element
FEA	Finite element analysis method
FEM	Finite element model
FoS	Factor of safety
GMM	Geometric mathematical model
HST	Hubble Space Telescope
IDW	Inverse distance weighting
IR	Infrared
JAXA	Japan Aerospace Exploration Agency
KARI	The Korean Aerospace Research Institute
KSASS	Korean Society Aeronautical Space Science
LEOP	Launch early orbit phase
LHS	Latin hypercube sampling
LoS	Line of sight
LPM	Lumped parameter method
MCRT	Monte Carlo ray tracing
MCS	Monte Carlo sampling
MoS	Margin of safety
MPC	Multipoint constraint

NAFEMS	National Agency for Finite Element Methods and Standards
NASA	National Aeronautical and Space Administration
PAT	Prescribed average temperature
PCB	Printed circuit board
PCL	MSC.Patran Command Language
PEM	Point estimates moments
RBE	Rigid body element
RF	Radio frequency
SM	Service mission
SRC	Standardised regression coefficient
STOP	Structural, thermal and optical performance
SWOT	Surface Water Ocean Topography
TMM	Thermal mathematical model
TN	Thermal node
TRP	Temperature reference point

Symbols

A	Area, cross-section (m^2)
a	Radius (m), constant, coefficient, $[A]$-matrix term
$[B]$	Interpolation matrix
b	Radius (m), width (m)
C	Heat capacitance
$[C]$	Conduction matrix
CTE	Coefficient of thermal expansion (n/m/°C, m/m/K)
c_p	Specific heat (J/kg/°C, J/kg/K)
^{o}C	Centigrade (degree Celsius)
E	Energy (W)
$[E]$	Unitary diagonal matrix
$[D]$	Elasticity tensor
E	Young's modulus (Pa)
F	Force (N), view factor
(F_T)	Equivalent thermal load vector
$F(x)$	Function of x
$F(x,\ y)$	Function of $x,\ y$
$\mathcal{F}$	Hottel's total view
f	Frequency (Hz)
G	Shear modulus (Pa), conductance (W/°C,K)
G_{sc}	Solar constant (W/m^2)
G_{ss}	Transformation matrix
GL	Conductor, thermal conductance coefficient (W/°C,K)
GR	Radiative conductor
h	Height, thickness (m), convective heat transfer coefficient (W/m/°CK)

H	Panel height
I	Second moment of area (m^4), integral
J	Joules, Jacobian, thermal functional
K	Kelvin
K	Stiffness matrix
K_c	Conduction matrix
k	Conductivity coefficient (W/m/K, W/m/°C)
$\mathcal{N}$	Normal distribution
l, L	Length (m)
L_{ref}	Reference length (m)
$\mathcal{LN}$	Log normal distribution
m	Discrete mass (kg)
M_T	Equivalent thermal moment (NM)
M	Mass matrix
M_{eff}	Modal effective mass (kg)
N	Number of samples
P	Pressure (Pa)
P_T	Thermal force (N)
q	heat transfer rate
Q_s	Solar constant (W)
Q	Heat flux (W/m^2)
(R_Q)	Heat flow vector
r	Radius (m)
R	Distance (m), resistance (°C/W, K/W), residual, Rayleigh quotient
T	Temperature (°C, K)
T_{ref}	Reference temperature (°C, K)
t	Time (s), thickness (m)
r	Radius (m)
u	Displacement m, stochastic variable
$\mathcal{U}$	Uniform distribution
U	Strain energy
U_m	Matrix of coefficients (MPC equations)
v	Displacement (m)
V	Volume (m^3), coefficient of variation
W	Watts
w	Deflection (m)
x	Coordinate, variable
y	Coordinate, variable
Y	Stochastic variable

Greek Symbols

α	Absorption coefficient (-), absorptivity

α	Coefficient of thermal expansion (m/m/K, m/m/°C)
β	(Thermal) Expandability (m/m/°C), thermal stress modulus (Pa), constant
β_x	Sensitivity index
ε	(Thermal) Emittance (-), emissivity, (engineering) strain (m/m)
δ_{ij}	Kronecker delta
δ	Displacement (m), differential operator, virtual (displacement)
Δ	Difference, evaluated (temperature), prescribed displacement
ζ	Isoparametric coordinate, dummy variable
μ	(Ensemble) average value
η	Bond thickness (m), isoparametric coordinate, dummy variable
λ	Lamé modulus (Pa), constant
Λ	Lagrange multiplier
ν	Poisson's ratio (-)
Π	Potential energy
σ_{ii}	Component stress tensor (Pa)
σ	Boltzmann constant (W/m^2/K^4), standard deviation, constant
θ	Angle
τ	Shear stress (Pa)
ϕ_R	Rigid body mode
Φ	Probability function
Ψ	Trial function, (nodal) shape function
ω^2	Eigenvalue

Chapter 1
Thermoelastic Verification

1.1 The Thermoelastic Problem

The objective of every scientific and Earth observation mission is to do observations of various physical effects in space or on planet surfaces. Also missions to the Sun aim at measuring different physical phenomena. All these observations have in common that these want to measure the distribution of different properties over a planet surface or the sky to capture space as deep possible. Advances in the optical sensors and radar technologies offer the potential to do these observations with increasing resolution allowing to enhance the quality of the information collected from space.

With the benefit of this enhancement in the resolution of the observations, also disturbances of the instruments, such as in the variation in the orientations or other movements, will unfortunately show up in equal "quality" and could reduce the benefit of the higher resolutions. This requires that the stability of the instruments and their supporting structures have to be increased as well to limit the impact of the disturbances.

During so-called operational phases of the mission, when instruments are turned on and are running the observations, there may be several sources of disturbances affecting the quality of the observations. In this book, the focus is on thermoelastic disturbances, which are thermally induced deformations of the instruments and their supporting structure. Since the thermal environment of a spacecraft orbiting around a planet is constantly changing with time, also the induced deformations show this transient behaviour. These deformations can cause that at some moments in time for instance telescopes are not pointing accurately enough at the intended point on the sky or radar apertures are scanning different parts of the Earth's surface than was planned. Also mirrors and lenses inside an instrument can be become not sufficiently aligned or even deformed.

S. Appel and J. Wijker, *Simulation of Thermoelastic Behaviour of Spacecraft Structures*, Springer Aerospace Technology, https://doi.org/10.1007/978-3-030-78999-2_1

In general, it is less complicated to cope with a steady state or static deformation of most of the components. However, the challenge lays in handling the amplitude of the variation of the deformation over time. The amplitude of this variation is one of the properties defining the stability characteristics of a spacecraft. For the performance, it is essential that this variation is kept within determined limits which defines the stability requirement.

Directly linked to the thermally induced deformations are the stresses in the structural components. These may cause damage leading to non-recoverable failure.

The stability of the systems in a spacecraft is based on a design, with the right material combinations and adequate thermal control limiting the temperature excursions of the structural supports and the instruments. The design has to include as well support concepts that can isolate sufficiently the sensitive items from deformations in the underlying structure. A stable design forms the basis for a successful operating spacecraft. The subject of this book is not about designing stable spacecraft and instruments, but on determining the thermally induced deformation and stresses that can be used to verify the stability of the spacecraft and its sub-systems. Basically, the topic of this book, thermoelastic analysis, is to support the verification of the design of spacecraft structures and instruments.

There are many good textbooks available on spacecraft structural verification covering topics like structural vibration and acoustics such as [81, 94]. Likewise in the domain of thermal verification of spacecraft and instruments, quite some literature is available. On the topic of thermoelastic verification, especially for spacecraft, the abundance of textbooks appears not to be present. This book tries therefore to support engineers, ranging from undergraduate to experienced, by sharing suggestions and guidelines for the analysis of thermoelastic problems.

1.2 Structure of This Book

The book starts with some examples showing the challenges of the thermoelastic verification. In Chap. 2, a limited number of spacecraft are discussed illustrating the thermoelastic topic with past and recent spacecraft and one under development.

In Chap. 3, the reader is introduced in the physics of thermoelastic analysis. For most readers, this will be well-known information, but for completeness it is included.

Although still thermal and structural analyses with corresponding model development are done very much in isolation, this book tries to explain the benefits of having the thermal and structural engineers working closely together during the development of their models. This allows for good exchange of modelling needs between the thermal and structural disciplines.

All the simulation work in the end has the single objective to verify the stability and corresponding performance or the strength of the instrument or spacecraft. It is therefore essential for thermal and structural engineers to have a good understanding of what kind of deformation are affecting the performance of the instruments. For this reason, thermoelastic verification shall always be an interaction between the disciplines linked to the performance of the instrument on one side, and the disciplines thermal and structures at the other side.

To reflect the integrated approach of the structures and thermal disciplines, modelling aspects of both disciplines are discussed together in Chap. 4. It discusses the essential interaction of both space thermal analysis and structural analysis for thermoelastic verification.

Chapter 5 aims at introducing the reader into the thermal side of the thermoelastic verification. Space thermal analysis has its own specificities in terms of thermal environment to be simulated and the method that is commonly used. Traditionally, spacecraft thermal analysis has the objective to verify the thermal control design. One of the topics discussed in Chap. 5 is the difference in objective between thermal analysis for thermal control and thermoelastic. Although there are essential elements in common, the differences are very important.

In the end, the structural model has to be used to compute the thermally induced distortions and stresses. Chapter 6 aims at providing recommendations on how to represent the structure in such a way that thermoelastic phenomena are simulated as good as possible. The structural model has to provide the results with which the performance of the structure as part of an instrument or spacecraft can be verified. This can be the thermal stresses with which the structural integrity is checked. Often, it is more challenging to verify the verification of the stability. The objectives of the structural model have to be defined together with the users of the produced data: the analysts verifying the performance of the instruments. Equally, this derives requirements for the data produced by the thermal model.

The thermal and structural models are in general based on two different simulation methods. The thermal model is commonly based on the lumped parameter method, while the structural model is almost always based on the finite element method. Besides the difference in methods behind the models, there is in most cases also the difference in mesh densities. Different methods exist to solve this interface problem and to transfer, or map as it is often called, the temperatures of the thermal model to the structural finite element model. In Chap. 7, different temperature mapping methods are discussed and compared for a small number of cases.

Of the different temperature mapping methods explained in Chap. 7, the prescribed average temperature (PAT) method shows several advantages compared to the other methods presented. Therefore, Chap. 8 is dedicated to the PAT method.

The finite element model, build for the same structural components as the thermal model, in combination with the PAT method offers the possibility to both automate generation and increase the quality of the conductive links (GLs) in the thermal model. Obviously, the increase of quality is always useful, but at the same time having a method suitable for automation is important when models have to be updated to reflect design changes, but also to be able to run efficiently mesh convergence checks as discussed in Chap. 4.

The last chapter of this book, Chap. 10, tries to cover aspects of the uncertainties in the thermoelastic analyses. Various sources of errors or uncertainties are discussed and tries to deal with these uncertainties in two ways. The first one is based on the application of factors of safety. This is a pragmatic way, but unfortunately suffers from the limitations coming with the ambiguous values of the factors of safety. The second approach discussed makes of use of the stochastic nature of the uncertainty of many of the parameters in the models. Stochastic methods are in general infamous due to the fact that these require high amount of computational resources, which would make the thermoelastic process even more heavier in that sense. However, a very efficient method, the Rosenblueth 2k+1 PEM method, is proposed to overcome this problem and allows with reasonable effort to do stochastic assessment of the thermoelastic results.

In all chapters is tried to illustrate the provided suggestions and recommendations with examples.

The appendices complement the book by providing background information on the finite element method for both thermal and structures domain. The system equations are derived for a one-dimensional rod element that is used in many examples throughout the book.

Chapter 2
Occurrence of Thermoelastic Phenomenon in Spacecraft

Abstract Thermoelastic aspects in the development of spacecraft are illustrated with a few examples.

2.1 Introduction

As explained in the introduction (Chap. 1), the thermoelastic topic is becoming more and more an important design driver for the instruments and spacecraft for scientific and Earth observation missions. In this chapter, a few spacecraft are briefly discussed to illustrate how the thermoelastic aspects are dealt with in the development of those spacecraft.

2.2 Hubble Space Telescope

The Hubble Space Telescope (HST), launched 1990, is one of NASA's most productive astronomical observatories. This spacecraft experienced an interesting thermoelastic phenomenon. The initially applied design of the solar arrays was based on a concept of flexible foils with solar cells that were kept under tension by two deployable booms (see Fig. 2.1).

These booms were made from a pair of circularised steel sheets which fit together in a C-shape to form a cylindrical boom. A schematic of the cross section of these booms in deployed configuration is shown in Fig. 2.2. The two C-shape sheets were overlapping, and for that reason, it was assumed that both sheets would have the same temperature. However, after leaving eclipse when one side of the boom came into sunlight and other one in shadow, due to the apparent poor thermal contact between the two blades and the low thermal inertia, a temperature difference between these blades rapidly occurred. It caused the two blades to expand at different rates causing clicking against each other producing small shocks. The effect was a small movement of the solar arrays that was big enough to cause the so-called loss of lock disturbing the pointing of the complete spacecraft every time the spacecraft left eclipse [40].

S. Appel and J. Wijker, *Simulation of Thermoelastic Behaviour of Spacecraft Structures*, Springer Aerospace Technology, https://doi.org/10.1007/978-3-030-78999-2_2

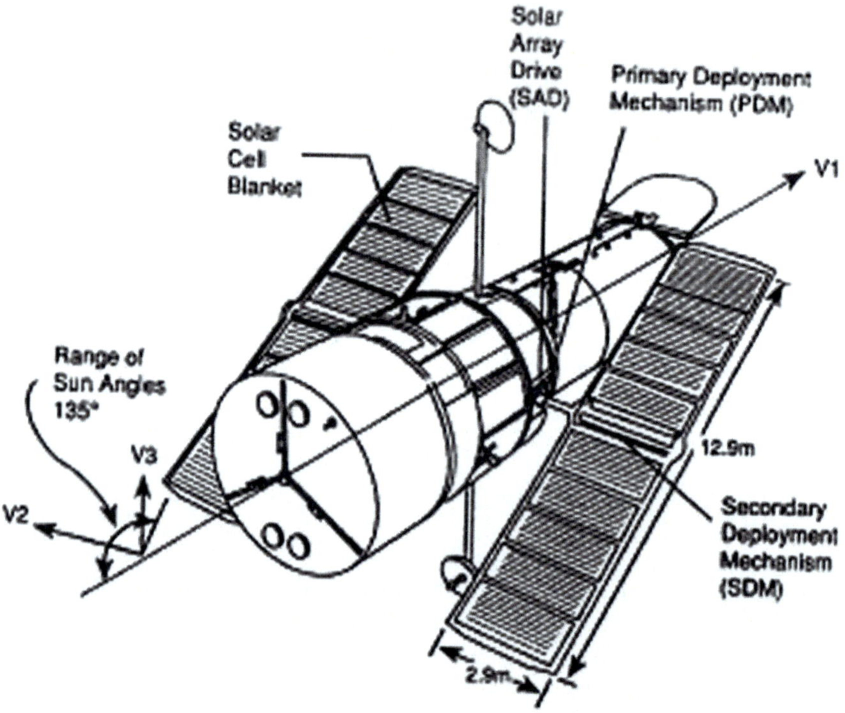

Fig. 2.1 HST on orbit configuration [40]

Fig. 2.2 Schematic cross section of solar array boom with radiative properties of the Hubble Space Telescope [40]

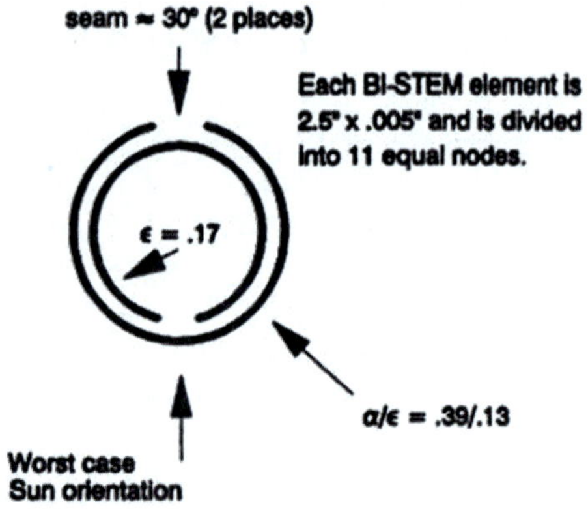

NASA has routinely conducted servicing missions with the space shuttle to refurbish older equipment as well as to replace existing scientific instruments with better, more powerful instruments. During the servicing mission SM3B, one of the major refurbishment efforts, a new rigid-panel solar arrays, was installed as a replacement for the existing flexible-foil solar arrays. This was necessary in order to increase electrical power availability for the new scientific instruments. Before the new solar arrays were installed on HST, an extensive test campaign in the Large Space Simulator (LSS) at ESTEC was run to verify that these new solar arrays will not cause any performance degradations to the observatory. The major concern was that also the new solar arrays were causing pointing loss of lock for the telescope. The tests were successful [40], and the analysis of the signal of the thermomechanical disturbances [30] showed that the new solar arrays were stable enough to prevent the loss of lock pointing problem.

2.3 Korean Observation Satellite

In [64], a thermoelastic pointing error analysis of Korean observation satellites considering seasonal and daily temperature variation was reported by the Korean Society of Aeronautical and Space Sciences (KSASS). Points on the spacecraft were experiencing a temperature excursion up to 200 degrees during on-orbit operation, which caused a disturbance of the coalignment of the star trackers and line of sight (LoS) the payload (see Fig. 2.3) due to thermoelastic deformation.

It is interesting to note that the PAT method (see Chap. 8) was used to map the temperatures computed with the thermal model to the structural finite element model to compute the distortion of the spacecraft and payload. The thermoelastic distortions (see Fig. 2.4) were then used to compute the pointing error.

2.4 Gaia

Another more recent spacecraft is Gaia, which was launched on 19 December 2013 (Fig. 2.5).

The main aim of Gaia is to measure the three-dimensional spatial and velocity distribution of stars and to determine their astrophysical properties, such as surface gravity and effective temperature, to map and understand the formation, structure, and past and future evolution of our Galaxy [37].

Gaia is relying on two fields of view, separated by large angle, the basic angle, on the sky along the scanning circle. The two viewing directions allow to map their images onto a common focal plane such that the observation times can be converted into small-scale angular separations between stars inside each field of view and large-scale separations between objects in the two fields of view [37]. For the performance of the spacecraft, it is important that the variation of the basic angle is kept within limits.

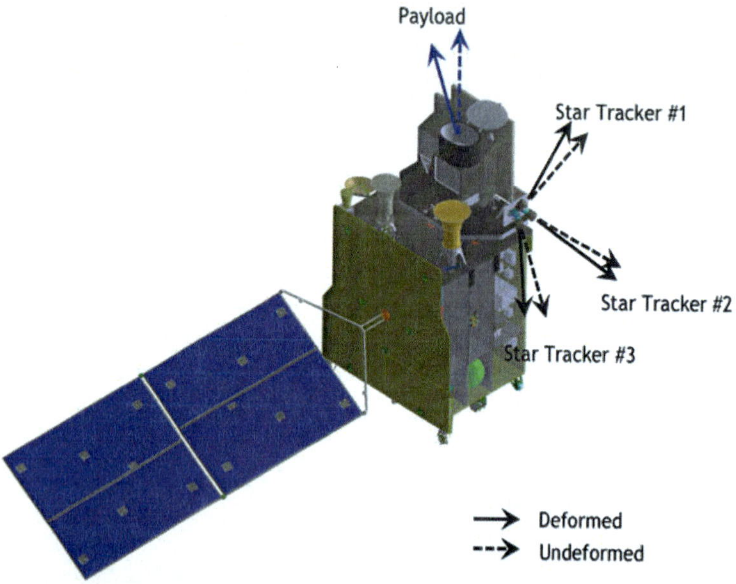

Fig. 2.3 Thermal pointing error of observation satellite (courtesy of KSASS) [64]

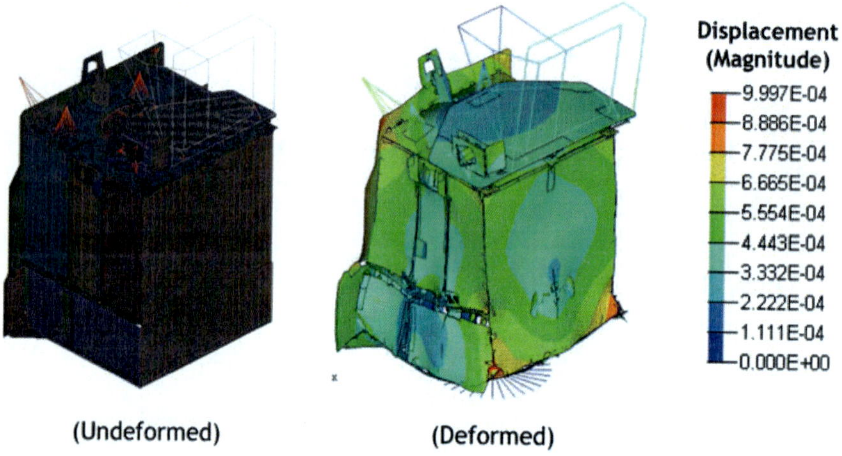

Fig. 2.4 Deformed shape of the satellite due to thermoelastic deformation (courtesy of KSASS) [64]

Fig. 2.5 Artist's impression of the Gaia satellite (courtesy ESA)

It is obvious that during the design and development, the demanding thermoelastic stability requirement got a significant amount of attention. Despite all the efforts during the development, it turned out that the measured fluctuations of the basic angle were two orders of magnitude higher than expected based on prelaunch predictions [37].

A dedicated working group, involving ESA and Airbus DS, was set up and concluded that there is strong evidence of a thermoelastic origin of the basic angle variation. The working group also managed to conclude that measurement of the basic angle was very accurate and would allow the measured basic angle variation to be taken into account in de processing of the data. Frustrating for the working group was that the exact cause of the basic angle variation could not be identified [37].

2.5 Surface Water and Ocean Topography (SWOT)

The Surface Water Ocean Topography (SWOT) (Fig. 2.6) is a partnership between NASA, Centre National d'Etudes Spatiales (CNES) and Canadian Space Agency (CSA), to conduct a comprehensive global survey of water and ocean topography of the surface of the earth, planned for launch in 2022. The main objective of SWOT is

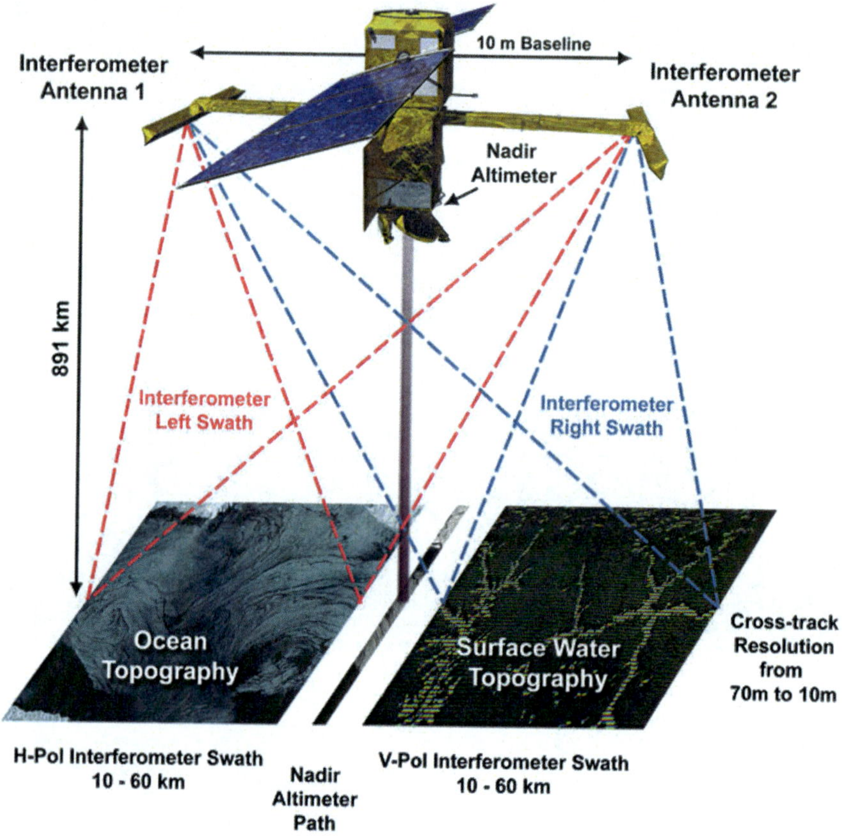

Fig. 2.6 Artist's impression of SWOT (courtesy NASA)

to collect precise measurements of surface water hydrology, observe details of ocean surface topography and circulation and measure how water bodies change over time [91].

The SWOT mission is composed of six optical and RF payloads. The pointing of these instruments towards the earth surface and the quality of signals are all affected by the distortions induced by the temperature field. In [91], the complete verification chain is summarised, starting from the thermal analysis, temperature mapping and thermoelastic distortion analysis. In the paper is also briefly mentioned that so-called STOP analysis has been performed. STOP analysis stands for structural thermal and optical performance analysis, which requires also the assessment of the optical performance of lenses and mirrors that are deformed due to the thermoelastic effects. As is indicated [91] for this assessment, three models are involved for each of the interacting disciplines. The optical design determines the distortion limits for the structural design, which then provides limits in the form of temperature gradients to the thermal design [91].

2.6 PLATO

PLAnetary Transits and Oscillations of stars (PLATO) is the third medium-class mission in ESA's Cosmic Vision programme. The primary goal of PLATO is the detection and characterisation of terrestrial exoplanets around bright solar-type stars, with emphasis on planets orbiting in the habitable zone [10].

In the payload of the PLATO, spacecraft will be equipped with of 26 separate cameras, as shown in Fig. 2.7.

The 26 cameras have to be kept well aligned under varying temperature fields. Because of the demanding stability requirements and the complexity of the structure to support these cameras, the development includes a thermoelastic test campaign on a demonstrator structure containing the mounts of two of the cameras as is shown in Fig. 2.8. The objective of the tests was to identify the best way of modelling the structure to capture as good as possible the thermoelastic responses [53].

In parallel, at ESTEC an end-to-end STOP analysis has been performed on one of the PLATO telescope optical units (TOU) to establish the process for verifying this kind of optics. The process involved very detailed analysis of the impact of the deformation of the six lenses (see Fig. 2.9) due to thermoelastic effects. Besides the deformation of the optical elements, also the change of refraction index of the lens material due to the small variation of temperature over the lenses was analysed. Finally, the impact of the low stresses was included in the predictions of the performance of the optics.

Fig. 2.7 PLATO spacecraft [10]

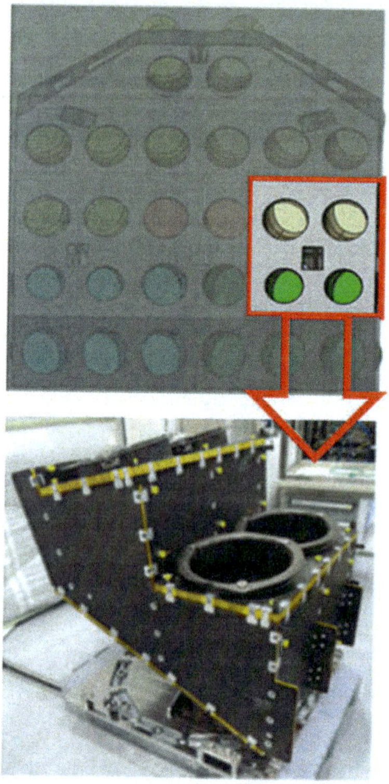

Fig. 2.8 PLATO demonstrator structure for thermoelastic tests [53]

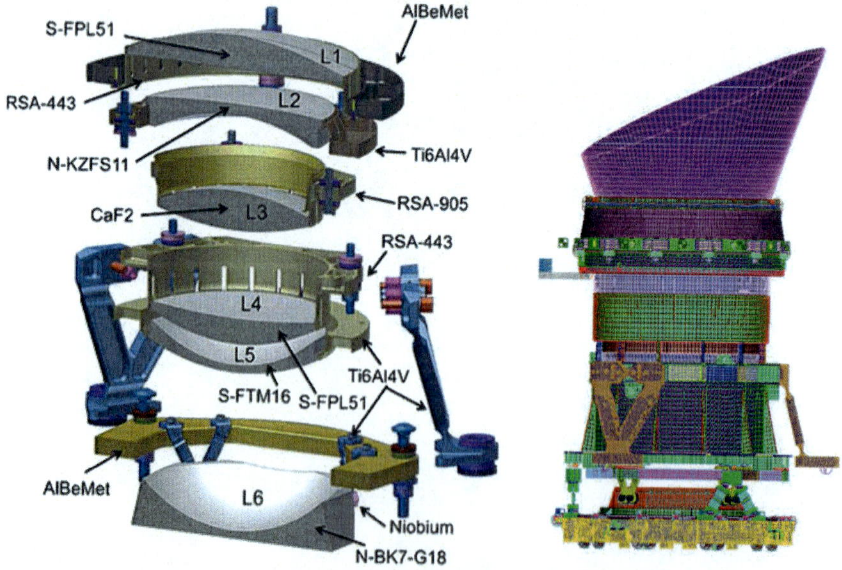

Fig. 2.9 PLATO TOU. Left: overview of lenses. Right the FE model [10]

Interesting to note is the fact that the lenses are foreseen to be manufactured at room temperature in a shape that is predicted to deform at the low operational temperatures (−80 °C) to a shape that is correct for the right optical performance.

Chapter 3
Physics of Thermoelastics

Abstract In this chapter, the main physical parameters involved in thermoelastic effects, the coefficient of thermal expansion (CTE) and Young's modulus, are discussed. The equilibrium and constitutive relations between stress and strain including thermoelastic effects are briefly recapitulated. At the end of the chapter, a summary of governing equilibrium and constitutive equations is given.

3.1 Introduction

Thermoelasticity in the context of this book is considered to be the interaction of temperature, stiffness and thermal expansion in a structure. As a consequence, this leads to what is referred to as thermoelastic responses such as thermal stresses and deformations.

In spacecraft structural design, it is exceptional that plastic deformation is tolerated. With the structural analyses, it has to be demonstrated that materials stay in their linear elastic domain. This is the reason why this book is about thermoelastic problems, rather than the general thermomechanical problems that would include as well plasticity resulting from temperature-induced deformations.

In this chapter is started with the discussion of the physical parameters involved in thermoelastic problems and the corresponding analyses to predict the responses.

The following physical material properties discussed:

- The coefficient of thermal expansion (CTE)
- The Young's modulus
- Constitutive laws of linear thermoelasticity.

The constitutive laws discussed in this chapter are applicable to isotropic materials only. Although the principles are not different for anisotropic and orthotropic material properties, the details on handling this non-isotropy in material models are not discussed here. For this aspect is referred to good textbooks about theory of elasticity, e.g. [86].

S. Appel and J. Wijker, *Simulation of Thermoelastic Behaviour of Spacecraft Structures*, Springer Aerospace Technology, https://doi.org/10.1007/978-3-030-78999-2_3

3.2 Coefficient of Thermal Expansion

The coefficient of thermal expansion (CTE) is a material property that is relevant for the thermoelastic behaviour of a material. In this section is explained how the CTE of a material is derived and how it is used to compute the deformation of a structure with for instance a finite element (FE) model.

For the prediction of thermal expansion, it is essential to define the reference temperatures at which the deformation and thermal stresses are zero. In most cases, this is the temperature at which a component is manufactured or assembled.

Let us consider a bar with length L_{ref} at a temperature T_{ref}. The increase or decrease of the length of the bar is then $\Delta L(T)$ (m)

$$\Delta L(T) = L(T) - L_{ref}(T_{ref}) \tag{3.1}$$

The thermal expandability $\beta(T)$ is defined as

$$\beta(T) = \frac{\Delta L(T)}{L_{ref}(T_{ref})} \tag{3.2}$$

The definition of the thermal expandability β is the same as the definition of the engineering strain ε_e (m/m) due to an external force F (N), thus

$$\varepsilon_e(F) = \frac{\Delta L}{L_{ref}} \tag{3.3}$$

Because of this similarity, the thermal expandability $\beta(T)$ is also referred to as the thermal engineering strain $\varepsilon_t(T)$.

The CTE, $\alpha(T)$, is defined as being the change in engineering strain due to a unit increase of the temperature. In other words, when $\varepsilon_t(T)$ is assumed to be a continuous function, the coefficient of thermal expansion $\alpha(T)$ can be expressed as

$$\alpha(T) = \frac{d\varepsilon_t(T)}{dT}. \tag{3.4}$$

Using this definition of the CTE, the total thermal engineering strain $\varepsilon_t(T)$ can then be obtained with

$$\varepsilon_t(T) = \beta(T) = \int_{T_{ref}}^{T} \alpha(\xi)d\xi. \tag{3.5}$$

When the CTE is constant over the full temperature range of interest, it implies that $\varepsilon_t(T)$ is a linear function of T. In that case, with the help of Eq. (3.5) the one-dimensional elongation can be written as

$$\Delta L = L_{ref}\varepsilon_t(T) = L_{ref}\alpha\Delta T = L_{ref}\alpha(T - T_{ref}). \tag{3.6}$$

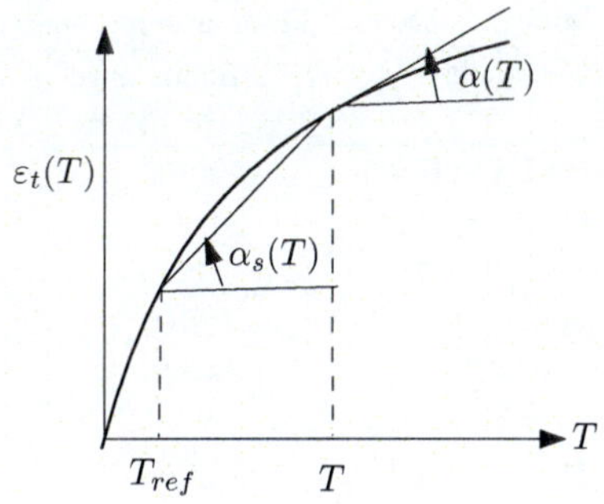

Fig. 3.1 Definition of α_s and α

Although the CTE values of many materials are nearly constant in the temperature range of interest, however, there are occasions where the constant CTE approach is not anymore valid. For instance, Fig. 3.1 presents the total thermal engineering strain $\varepsilon_t(T)$ as a nonlinear function of the temperature. In such a case, the expression in Eq. (3.6) is not any more valid. For those cases, the secant CTE α_s is introduced and defined by

$$\alpha_s(T) = \frac{1}{(T - T_{ref})} \int_{T_{ref}}^{T} \alpha(\xi) d\xi \approx \frac{(\varepsilon_t(T) - \varepsilon_t(T_{ref}))}{(T - T_{ref})}. \tag{3.7}$$

From mathematical point of view, α_s corresponds to the average of α, thus the average slope of the function $\varepsilon_t(T)$ over the temperature interval.

When the secant CTE α_s is used, the temperature-dependent CTE $\alpha(T)$ is linearised between the reference temperature T_{ref} and the current temperature T. The associated thermal deformations and stresses are linearised too. Many FE codes, like MSC Nastran uses the secant CTE $\alpha_s(T)$ over the temperature range T_{ref} to T.

The CTE, α, and the secant CTE, α_s, are illustrated in Fig. 3.1.

The ASM (Aerospace Specification Metals Inc.) database provides for the Al-alloys 6061-T6, 6061-T651 and Ti-alloy Ti-6-4 the secant CTE's, which are presented in Table 3.1

Using Eq. (3.3) with varying temperature $\Delta T = (T - T_{ref})$, the elongation ΔL of the bar is

$$\Delta L = L_{ref}\alpha(T)\Delta T = L_{ref}\alpha_s(T - T_{ref}) \tag{3.8}$$

After a temperature increase or decrease ΔT the new length of the bar is given by

$$L = L_{ref}(1 + \alpha_s \Delta T) \tag{3.9}$$

Table 3.1 Secant CTE for a some commonly used materials

Metal-alloy	Temp. range $(T - T_{ref})$ (°C)	CTE α_s (m/m/°C)	Remark
6061-T6 (T651)	20–100	23.6×10^{-6}	AA; typical average
6061-T6 (T651)	20–300	25.2×10^{-6}	Estimated from trends in similar Al alloys.
Ti-6-4	20–100	8.6×10^{-6}	
Ti-6-4	20–315	9.2×10^{-6}	Average over the temperature range
Ti-6-4	20–650	9.7×10^{-6}	Average over the temperature range

Table 3.2 $\varepsilon_t(T)$ and $\alpha(T)$ titanium alloy

T (°C)	ΔT (°C)	α_s (m/m/°C)	$\varepsilon_t(T)$
20			0
100	20–100	8.6×10^{-6}	6.88×10^{-4}
315	20–315	9.2×10^{-6}	27.14×10^{-4}
650	20–650	9.7×10^{-6}	61.11×10^{-4}

Example $\varepsilon_t(T)$ and $\alpha(T)$ Titanium Alloy

The $\varepsilon_t(T)$ curve is reconstructed by curve fitting from the three given secant CTE's values provided in Table 3.1. At T=20 °C, the thermal strain is set to zero $\varepsilon_t(20) = 0$ m/m/°C. With Eq. (3.6), the values of the thermal strain $\varepsilon_t(T)$ are calculated and are presented in Table 3.2.

The third-order polynomial curve fitting is applied with the function `lscurvefit` in MATLAB using the polynomial.

$$\varepsilon_t(T) = \beta(T) = a_0 + a_1 T + a_2 T^2 + a_3 T^3 \tag{3.10}$$

where T is expressed in °C, the coefficients are $a_0 = -0.16582 \times 10^{-3}$, $a_1 = 0.82288 \times 10^{-5}$, $a_2 = 0.31157 \times 10^{-8}$ and $a_3 = -0.2175 \times 10^{-14}$. The values of the coefficients show that $\varepsilon_t(T)$ is not a linear function.

The result of the curve-fitting process is illustrated in Fig. 3.2.

The CTE $\alpha(T)$ values are obtained by numerical differentiation of the thermal strain $\varepsilon_t(T)$ and shown in Fig. 3.3.

With Eq. (3.7), the secant CTE is calculated and presented in Table 3.3.

This example shows how the secant CTE values can be used to construct the total thermal engineering strain $\varepsilon_t(T)$ function. This function could be considered as the results from material characterisation tests in which the elongation of a material as

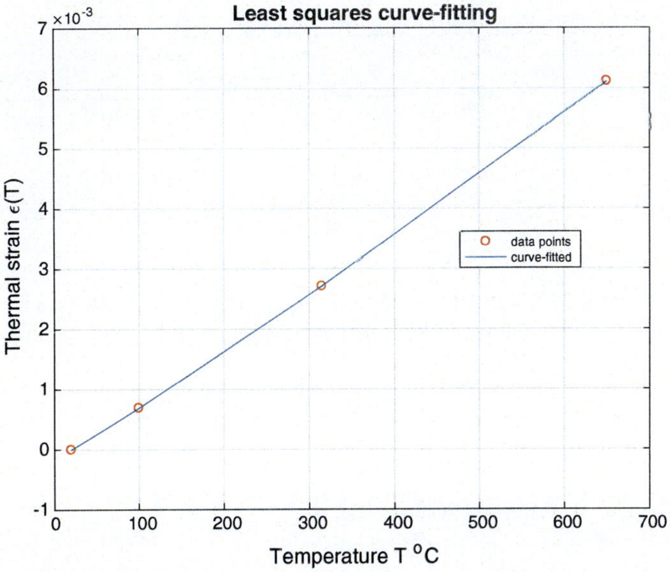

Fig. 3.2 Curve-fitted thermal strain

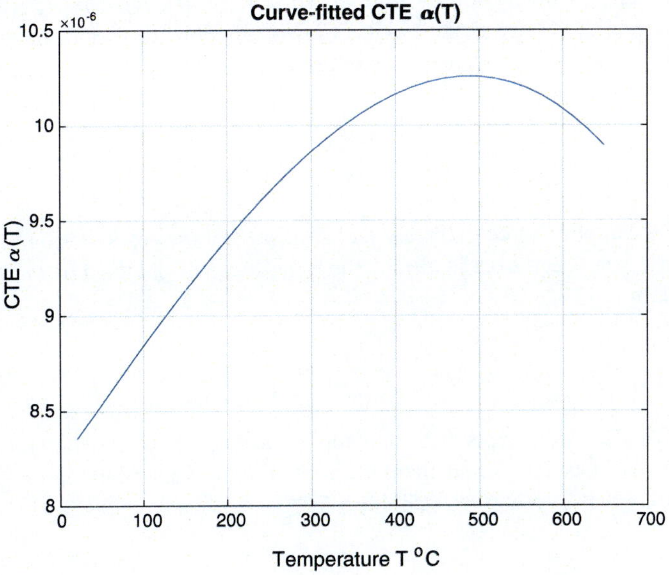

Fig. 3.3 Curve-fitted CTE

Table 3.3 Derived values of $\alpha_s(T)$ by using Eq. (3.7)

T (°C)	ΔT (°C)	α_s (m/m/°C)
100	20–100	8.59×10^{-6}
315	20–315	9.19×10^{-6}
650	20–650	9.69×10^{-6}

a function of temperature is measured. Subsequently, the example then shows how the secant CTE is obtained from the (artificial) data represented by the total thermal engineering strain $\varepsilon_t(T)$ function.

3.3 Young's or Elasticity Modulus

The Young's modulus E (Pa) is a material property and is a measure of stiffness and is the ratio between the one-dimensional stress σ_{xx} (Pa) and the strain ε_{xx} (-) obtained from the linear part of the stress–strain curve. The proportional relation between stress and strain is known as the Hooke's law [29]. The stress and the strain are measured using a special made loaded one-dimensional test item (bar). The Young's modulus is defined by

$$E = \frac{\sigma_{xx}}{\varepsilon_{xx}} = \frac{\frac{F}{A}}{\frac{\Delta L}{L_{ref}}} \tag{3.11}$$

where F is the applied force to the bar and A its area of the cross section.

The elongation ΔL of a bar loaded by the force F can be obtained from Eq. (3.11) and is given by

$$\Delta L = \frac{F L_{ref}}{EA} = \frac{FL}{EA} \tag{3.12}$$

The ASM (Aerospace Specification Metals Inc.) database provides for the Al-alloys 6061-T6, 6061-T651 and titanium alloy Ti-6-4 Young's modulus (at room temperature), which is presented in Table 3.4

Table 3.4 Young's modulus for 6061-T6, 6061-T651 and Ti-6-4 (room temperature)

Metal-alloy	Young's modulus E (GPa)	Remark
6061-T6 (T651)	68.9	AA; typical; average of tension and compression
Ti-6-4	113.8	

In general, the Young's modulus can be a function of temperature; therefore, the values of the Young's moduli presented in Table 3.4 should be accompanied with the temperature range for which these values are applicable.

When a structure that is built from different materials with most likely different values for the Young's modulus and CTE is subjected to a temperature change, these material properties play an important role in the interaction of the components of the structure. Especially the ratio of the stiffness values of the parts is an important factor in the interaction between the parts.

When one part is significantly stiffer than the other part to which it attached, the thermoelastic deformation of the stiffer part will to a large extent be imposed on the less stiff part. The following example illustrates this.

Effect of relative stiffness in thermoelastic problems

This example illustrates the interaction of relative stiffness between two components of a structure. In this example, a fictitious schematic optical bench is considered. In Fig. 3.4, a schematic representation is given with the applied material properties and dimensions. The Young's modulus of the "optical bench" is varied to determine the effect of the relative stiffness. The structure is modelled with beam elements and the structure is subjected to a temperature increase of 20 °C.

The support material has a CTE of 25 × $10^{-6} m/m/$°C, while the bench material has a CTE of 4 × $10^{-6} m/m/$°C. The temperature increase of 20 °C makes that the both the support and the bench want to expand. Due to the lower CTE of the bench, it wants to expand less than the support.

When the optical bench has a very small stiffness compared to the support, the support is able to impose its deformation on the bench (see the top left picture in

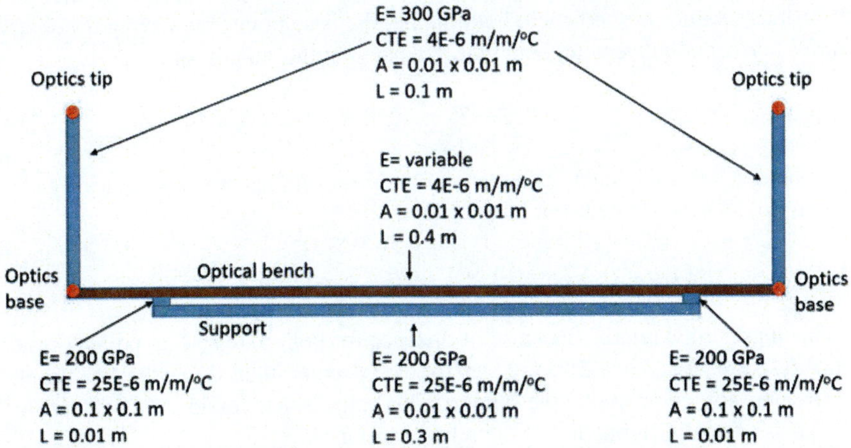

Fig. 3.4 Schematic model of a fictitious optical bench

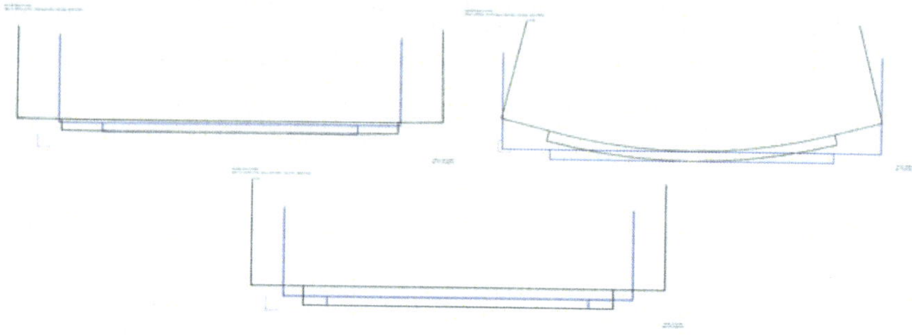

Fig. 3.5 Deformation in case of top left: soft optical bench; top right: stiffness values of the same order of magnitude; bottom: significant stiffer optical bench

Fig. 3.5). When the stiffness of the bench is gradually increased to a level that is of the same order of magnitude as that of the support, the bench becomes able to resist the deformation that the support would like to impose. As a consequence, due to the relative offset of the axes of the two beams, the system is bending (see top right picture in Fig. 3.5). Increasing the stiffness of the bench further to a level that it becomes orders of magnitude stiffer than the support, the opposite of the first extreme situation occurs: the deformation of the bench is imposed on the support (see bottom picture in Fig. 3.5).

Besides through the deformed shapes of Fig. 3.5, also graphically the transition of one extreme to the other can be made visible. In Fig. 3.4, the tips and the base points of the "optics" are marked with red dots for reference in this text. During bending of the system, the relative horizontal movement of these two tip points is amplified by the bending deformation of the system. The relative horizontal movement of the base points of the "optics" is (for a linear model) not affected by the bending effect, as can also be seen in the top right picture of Fig. 3.5. In Fig. 3.6 is shown that for low stiffness values of the bench, both the tip and base points are having the same relative horizontal movement, being the extension of the support of

$$\Delta L_{sup} = L_{sup}\alpha_{sup}\Delta T = 0.3 \times 25 \times 10^{-6} \times 20 = 1.5 \times 10^{-4}\ \mathrm{m} \tag{3.13}$$

To be exact also the length extension of the pieces of the bench overhanging the support points needs to be taken into account:

$$\Delta L_{bo} = L_{bo}\alpha_{bench}\Delta T = 0.1 \times 4 \times 10^{-6} \times 20 = 8 \times 10^{-6}\ \mathrm{m} \tag{3.14}$$

This makes total relative increase of distance for both base and tip points of the optics $\Delta L_{sup} + \Delta L_{ob} = 1.58 \times 10^{-4}$ m for the extreme situation in which the bench is extremely soft compared to the support. This value is also found at the left side of Fig. 3.6 as a negative reduction of the relative distance.

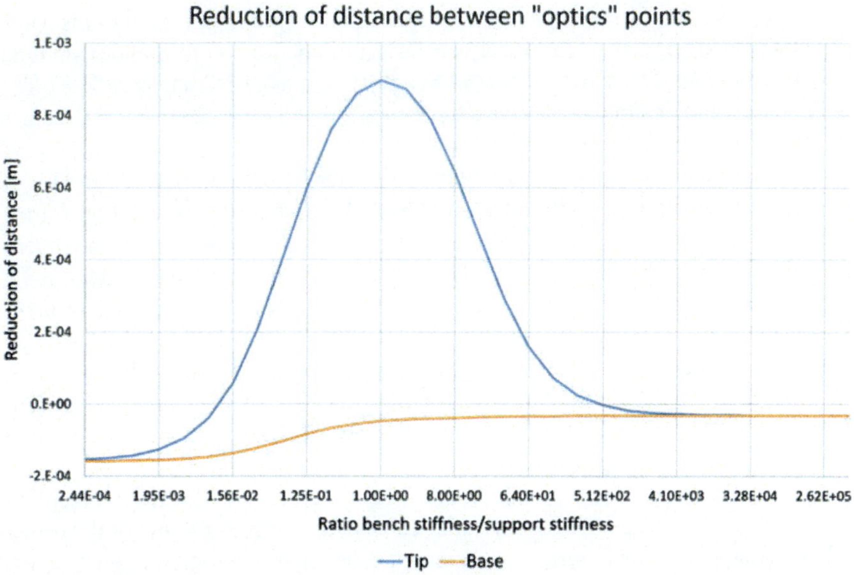

Fig. 3.6 Effect of stiffness ratios on relative responses

The other extreme case in which the bench is orders of magnitude stiffer than the support is found at right side of the graph in Fig. 3.6. The coinciding relative movement of both tip and base is for that case driven by the extension of the bench:

$$\Delta L_{bench} = L_{bench}\alpha_{bench}\Delta T = 0.4 \times 4 \times 10^{-6} \times 20 = 3.2 \times 10^{-5}\ \text{m} \tag{3.15}$$

In between the two extreme cases, the stiffness values of the support and the bench are challenging each other (sometimes called "thermal fight"). In this situation, the difference in CTE is driving the bending deformation. This bending deformation can be observed in Fig. 3.6 from the high level of relative displacement for the optics tip points. Note that the horizontal axis of Fig. 3.6 is using a logarithmic scale to allow better visualising the range in which the interaction between the stiffness of two structural elements is operational.

In this example is illustrated what the thermoelastic consequences are of the ratio of the Young's modulus of two structural components. From this example can be deducted that for an optical bench, it is preferred to have a high stiffness relative to the supporting structure to limit the amount of deformation that support can induce in the optical bench. In this example, only the Young's modulus has been varied. Especially the bending stiffness can be increased by changing the cross section of the optical bench.

As for most structural design tasks, material choices are important. For instance, supports of optical elements (optical benches) thermoelastic deformations shall be as low as possible. This can be achieved by selecting a material that has a low CTE. At the same time, it should be prevented that parts attached to the support structure are able to impose a strong deformation on it. This is one of the reasons why often for optical benches the material Zerodur is selected. This material has an almost zero CTE value (see Table 3.5) and a Young's modulus value that is a bit higher than for an Al-alloy. Another often used material for optical benches is silicon carbide (SiC). This material has also a low CTE, although a bit higher than Zerodur, but a much higher value for the Young's modulus. Table 3.5 lists the properties of these materials.

The key properties of a material needed for optical benches are a high specific stiffness (E/ρ) (Nm/kg) combined with a high thermoelastic stability coefficient (E/α) (N/m^2), where E is the Young's modulus, ρ (kg/m^3) is the density, α is the coefficient of thermal expansion (CTE).

Although the material properties themselves are important, also the geometrical dimensions and shape provide an essential contribution to the stiffness of the bench and the resistance against imposed deformation from interfacing structural elements.

Euclid telescope

The Euclid payload comprises of a 1.2 m Korsch telescope designed to provide a large field of view. The telescope is built on a truss hexa-pod concept: six struts are connecting the secondary mirror (M2), mounted on a frame through spiders to the primary mirror (M1) optical bench. The optical bench, made of SiC [41], provides also interface points to the service module. An implementation of this concept is shown in Fig. 3.7.

The deflection at the centre of a circular plate simply supported along the circumference is given by [19]

$$w = \frac{6a^2 M_T}{Et^3} \tag{3.16}$$

in which a is the radius, M_T is the constant thermal bending moment due to a thermal gradient across the thickness t, and E is Young's modulus of the circular plate. The thermal bending moment M_T is obtained by the following expression

$$M_T = \alpha E \int_{-\frac{t}{2}}^{\frac{t}{2}} \Delta T(y) y dy \tag{3.17}$$

where α is the CTE of the plate material, $T(y)$ is the temperature as a function of the through the thickness coordinate y. $\Delta T(y) = T(y) - T_{ref}$ is the temperature elevation above the reference temperature T_{ref}. The CTE α and E are assumed to be independent of the temperature T.

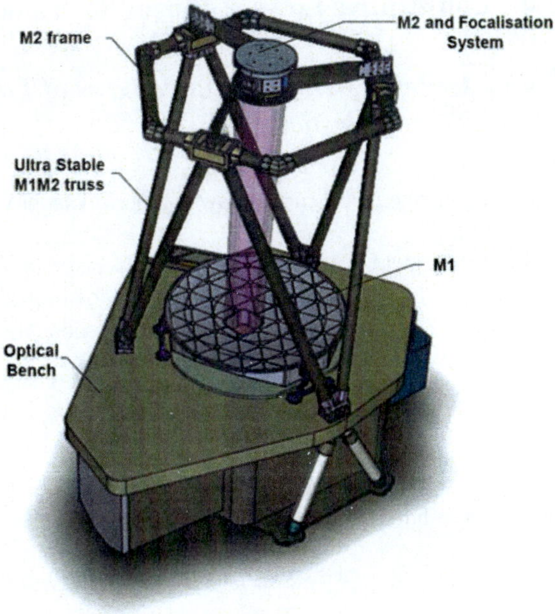

Fig. 3.7 Euclid telescope on optical bench

If Young's modulus E is constant over the circular plate, the deflection Eq. (3.16) can be simplified to

$$w = \frac{6a^2\alpha \int_{-\frac{t}{2}}^{\frac{t}{2}} \Delta T(y) y dy}{t^3}. \tag{3.18}$$

The deflection w will go down when α and $\int_{-\frac{t}{2}}^{\frac{t}{2}} \Delta T(y) y dy$ are small and the plate thickness t high. The deflection w will increase with an increasing radius a. Young's modulus is involved when E is dependent of y (i.e. a sandwich plate). A high thermal conductivity reduces thermal gradients.

Typical values for commonly used ceramic materials are found in Table 3.5.

Table 3.5 Typical material properties for SiC and Zerodur

Material	Young's modulus E (GPa)	Conductivity (W/m°C)	CTE ($\times 10^{-6}$ m/m/°C)	Remark
Silicon carbide	410	120	4.0	Room temperature[a]
Zerodur	90.3	1.46	0.02	At 0–50 °C[b]

[a]Accuratus Corporation, 35 Howard Street, Phillipsburg, NJ 08865, USA
[b]SCHOTT North America, Inc., 2 International Drive, Suite 105 Rye Brook, NY 10573, USA

3.4 Constitutive Laws of Linear Thermoelasticity

3.4.1 General 3-D Constitutive Laws of Linear Thermoelasticity

This section, about constitutive laws of thermoelsticity, is based on associated chapters in [44, 89].

In the classical theory of linear thermoelasticity, the components of the strain tensor $\varepsilon_{ij},\ i, j = 1, 2, 3\ (x, y, z)$ are a superposition of the components of the elastic tensor $\varepsilon^e_{ij},\ i, j = 1, 2, 3\ (x, y, z)$ and the components of the strain tensor ε^T_{ij} due to temperature variations

$$\varepsilon_{ij} = \varepsilon^e_{ij} + \varepsilon^T_{ij}, \tag{3.19}$$

The thermal strain for an isotropic material due to the temperature change ΔT can be formulated as

$$\varepsilon^T_{ij} = \alpha(T - T_{ref})\delta_{ij} = \alpha \Delta T \delta_{ij}, \tag{3.20}$$

where α is the CTE of the material and δ_{ij} is the Kronecker delta function ($\delta_{ij} = 1, i = j, \delta_{ij} = 0, i \neq j$). There are no thermal shear strains, but only strains along all three perpendicular directions with equal values resulting in a change of volume.

The elastic strain tensor ε^e_{ij} is linearly proportional to the stress tensor σ_{ij}

$$\varepsilon^e_{ij} = \frac{1}{2G}\left(\sigma_{ij} - \frac{\nu}{1+\nu}\sigma_{kk}\delta_{ij}\right) \tag{3.21}$$

where G is the shear modulus and ν is Poisson's ratio. In accordance with the Einstein summation convention $\sigma_{kk} = \sigma_{11} + \sigma_{22} + \sigma_{33} = \sigma_{xx} + \sigma_{yy} + \sigma_{zz}$.

Equation (3.21) is known as the constitutive law or Hooke's law.

For an isotropic material, the following relations can be written [57]

$$G = \frac{E}{2(1+\nu)}, \quad \nu = \frac{E}{2G} - 1. \tag{3.22}$$

The Aerospace Specification Metals (ASM) database provides for the Al-alloys 6061-T6 and 6061-T651 numerical values for Young's modulus, shear modulus and Poisson's ratio, which are presented in Table 3.6.

The total strain tensor ε_{ij} is the sum of Eqs. (3.20) and (3.21)

$$\varepsilon_{ij} = \frac{1}{2G}\left(\sigma_{ij} - \frac{\nu}{1+\nu}\sigma_{kk}\delta_{ij}\right) + \alpha \Delta T \delta_{ij} \tag{3.23}$$

Table 3.6 Young's modulus, Poisson's ratio and shear modulus for 6061-T6 and 6060-T651 (room temperature)

Al-alloy 6061-T6 (T651)	Numerical value	Remark
Young's modulus E (GPa)	68.9	AA; typical; average of tension and compression
Shear modulus G (GPa)	26	Estimated from similar Al alloys
Poisson's ratio ν (−)	0.33	Estimated from similar Al alloys

Equation (3.23) is called the constitutive law of the linear thermoelasticity. Solving Eq (3.23) for the stress tensor σ_{ij} gives

$$\sigma_{ij} = 2G\left[\varepsilon_{ij} + \frac{\nu}{1-2\nu}\left(\varepsilon_{kk} - \frac{1+\nu}{\nu}\alpha\Delta T\right)\delta_{ij}\right] \tag{3.24}$$

In the following sections, five simple examples will illustrate thermoelastic problems (thermal stresses) in structural parts included:

- Thermal stress in a rod
- Thermal stresses in a circular plate
- Thermal stresses in bonded joint
- Thermal axisymmetric stress in long pipe.

3.4.2 1-D Stress–Strain Relation

The 1-D stress–strain relation (e.g. in a rod) can be easily derived from Eqs. (3.23) and (3.24), leading to

$$\epsilon_{xx} = \frac{1}{E}\sigma_{xx} + \alpha\Delta T, \tag{3.25}$$

and

$$\sigma_{xx} = E\left(\varepsilon_{xx} - \alpha\Delta T\right). \tag{3.26}$$

In the absence of body forces, the equilibrium of stress is [57]

$$\frac{d\sigma_{xx}}{dx} = 0 \tag{3.27}$$

thus Eq. (3.27) can be written with displacement $u(x)$ and strain $du/dx = \varepsilon_{xx}$

$$\frac{d^2u}{dx^2} = \alpha\frac{dT}{dx} \tag{3.28}$$

Example: Sinusoidal temperature distribution in rod

A rod with length $L = 1$ m is fixed at both ends ($x = 0, x = L$). The elongation of the rod is $u(x)$. The rod is made of Al-alloy with Young's modulus $E = 70$ GPa and CTE $\alpha = 24 \times 10^{-6}$ m/m/°C. The temperature distribution is

$$\Delta T(x) = T(x) - T_{ref} = T_o \sin\left(\frac{\pi x}{L}\right), \tag{3.29}$$

where $T_o = 50$ °C.

The second derivative of the axial displacement with respect to the axial coordinate is

$$\frac{d^2u}{dx^2} = \alpha T_o \frac{\pi}{L} \cos\left(\frac{\pi x}{L}\right) \tag{3.30}$$

The solution of $u(x)$ is

$$u(x) = A + Bx - \frac{\alpha L T_o}{\pi} \cos\left(\frac{\pi x}{L}\right) \tag{3.31}$$

The average displacement u_{avg} is

$$u_{avg} = \frac{1}{L} \int_o^L u(x)dx = 0 \tag{3.32}$$

After the introduction of the boundary conditions, the the solution for $u(x)$ results in

$$u(x) = \frac{\alpha L T_o}{\pi} \left[1 - \frac{2x}{L} - \cos\left(\frac{\pi x}{L}\right)\right] \tag{3.33}$$

leading to the strain ε_{xx} as

$$\varepsilon_{xx} = \frac{\alpha L T_o}{\pi} \left[-\frac{2}{L} + \frac{\pi}{L} \sin\left(\frac{\pi x}{L}\right)\right] \tag{3.34}$$

Subsequently, the constant stress σ_{xx} can now be calculated

$$\sigma_{xx} = E\left(\varepsilon_{xx} - \alpha \Delta T\right) = -\frac{2}{\pi} \alpha E T_o \tag{3.35}$$

resulting in the internal force F

$$F = A\sigma_{xx} = -\frac{2}{\pi} \alpha A E T_o \tag{3.36}$$

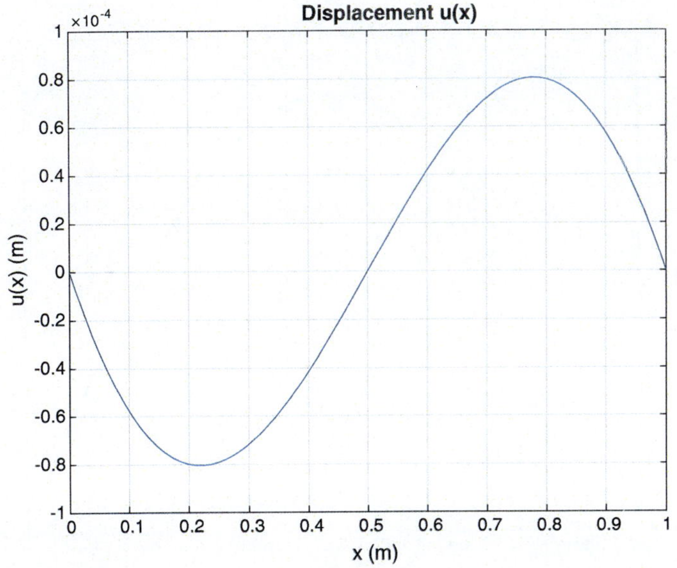

Fig. 3.8 Displacement $u(x)$

The stress $\sigma_{xx} = -53.4761$ MPa.

The numerical results of the displacement $u(x)$ and the strain ε_{xx} are graphically presented in Figs. 3.8 and 3.9.

3.4.3 *Plane Stress State*

The plane stress state is a simplification of a 3-D stress state to 2-D elastic problems. The 2-D stress components are functions of the two coordinates (x, y), and the transverse stresses are zero.

$$\begin{aligned} \sigma_{xx} &= \sigma_{xx}(x, y), \ \sigma_{yy} = \sigma_{yy}(x, y), \ \sigma_{zz} = 0, \\ \sigma_{xy} &= \tau_{xy}(x, y) = \tau_{xy}(x, y), \ \tau_{yz} = \tau_{yz} = 0 \end{aligned} \tag{3.37}$$

The strain–stress relations for the plane stress state can be derived from Eq. (3.23) in combination with the relation in Eq. (3.22)

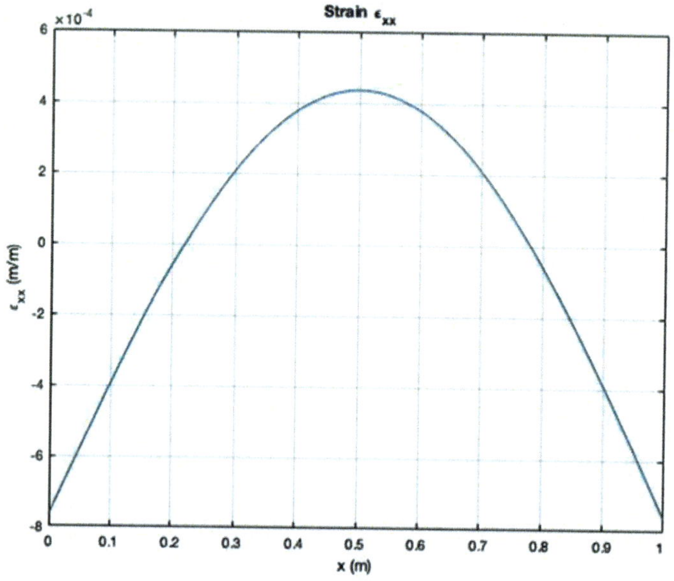

Fig. 3.9 Strain ε_{xx}

$$\begin{aligned} \epsilon_{xx} &= \frac{1}{E}(\sigma_{xx} - \nu\sigma_{yy}) + \alpha\Delta T, \\ \epsilon_{yy} &= \frac{1}{E}(\sigma_{yy} - \nu\sigma_{xx}) + \alpha\Delta T \\ \epsilon_{xy} &= \frac{1}{2G}\tau_{xy} \end{aligned} \tag{3.38}$$

and the stress–strain relations Eq. (3.24)

$$\begin{aligned} \sigma_{xx} &= \frac{E}{1-\nu^2}\left[(\varepsilon_{xx} + \nu\varepsilon_{yy}) - \alpha(1+\nu)\Delta T\right], \\ \sigma_{yy} &= \frac{E}{1-\nu^2}\left[(\varepsilon_{yy} + \nu\varepsilon_{xx}) - \alpha(1+\nu)\Delta T\right], \\ \tau_{xy} &= \frac{E}{1+\nu}\varepsilon_{xy} \end{aligned} \tag{3.39}$$

Example: thermal stresses in thin circular disc

Considered is a thin disc with a radius b and an internal concentric hole with a radius a. The disc is subjected to an axisymmetric temperature field allowing the temperature elevation to be written as

$$\Delta T(r) = T(r) - T_{ref}. \tag{3.40}$$

Assumed is that the temperature is constant through the thickness of the of the disc. As a consequence, the stress and displacement do not vary over the thickness.

Both the internal and external edges are not constrained. Due to axisymmetry, the tangential displacement $u_\theta = 0$. The expressions for the radial displacement u_r, the radial stress σ_{rr} and tangential stress $\sigma_{\theta\theta}$ are derived in many textbooks, such as [44, 73, 90]. The radial displacement u_r is given by (with $\sigma_{rr}(a) = \sigma_{rr}(b) = 0$)

$$u_r = \frac{(1+\nu)\alpha}{r}\int_a^r \Delta T r dr + \frac{(1+\nu)a^2 + (1-\nu)r^2}{(b^2 - a^2)r}\alpha \int_a^b \Delta T r dr \tag{3.41}$$

and the radial stress σ_{rr} and tangential stress $\sigma_{\theta\theta}$ can be obtained by the following expressions

$$\begin{aligned} \sigma_{rr} &= -\frac{E\alpha}{r^2}\int_a^r \Delta T r dr + \frac{E\alpha}{(b^2 - a^2)}\left(1 - \frac{a^2}{r^2}\right)\int_a^b \Delta T r dr, \\ \sigma_{\theta\theta} &= \frac{E\alpha}{r^2}\int_a^r \Delta T r dr - E\alpha\Delta T + \frac{E\alpha}{(b^2 - a^2)}\left(1 + \frac{a^2}{r^2}\right)\int_a^b \Delta T r dr \end{aligned} \tag{3.42}$$

The temperature distribution from radii $a = 0.25$ m to $b = 0.5$ m of the disc is

$$\Delta T(r) = \Delta T_a \left(1 - \frac{r-a}{b-a}\right) \tag{3.43}$$

with $\Delta T_a = 20\,°\text{C}$. When the Young's modulus is taken to be $E = 70$ GPa, Poisson's ratio is $\nu = 0.33$ and the CTE is $\alpha = 24 \times 10^{-6}$ m/m/°C, the displacement u_r and the stresses $\sigma_{rr}, \sigma_{\theta\theta}$ as a function of the radial coordinate become as shown in Fig. 3.10. The tangential stress $\sigma_{\theta\theta}$, also called hoop-stress, is more significant than the radial stress σ_{rr}.

3.4.4 Plane Strain State

The plain strain state is an other 2-D simplification of the 3-D stress state in which strains are only present in the 2-D plane [89]. In terms of strains, this means

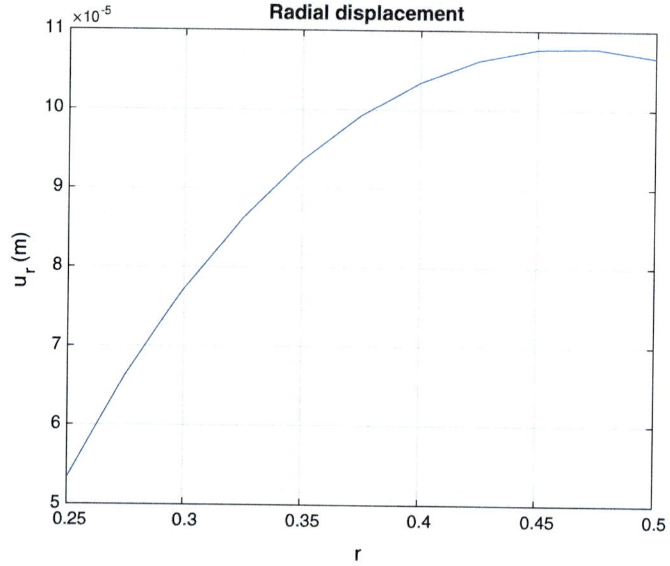

(a) Radial displacement

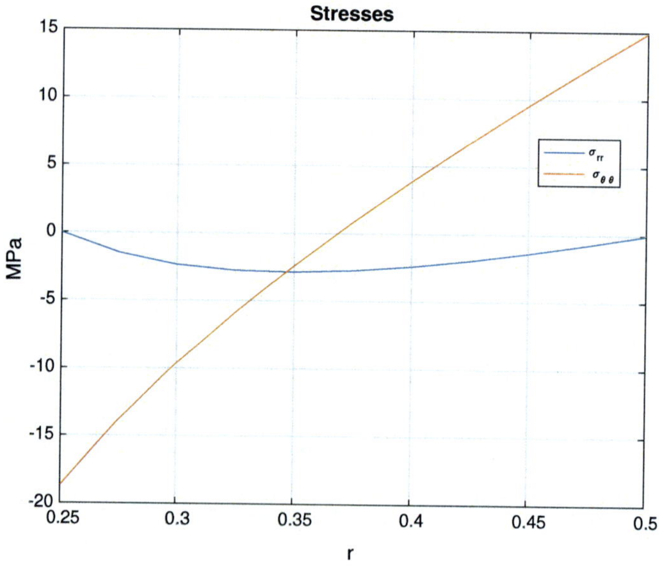

(b) Radial and tangential stress

Fig. 3.10 Radial displacement and stresses

$$\varepsilon_{xx} = \varepsilon_{xx}(x, y),\ \varepsilon_{yy} = \varepsilon_{yy}(x, y),\ \varepsilon_{zz} = 0, \\ \varepsilon_{xy} = \varepsilon_{xy}(x, y) = \varepsilon_{xy}(x, y),\ \varepsilon_{yz} = \varepsilon_{yz} = 0 \tag{3.44}$$

It must be noted that although the strain components $\varepsilon_{zz} = \varepsilon_{yz} = \varepsilon_{yz} = 0$, the corresponding stress components are in general not equal to zero.

$$\sigma_{zz} \neq 0,\ \tau_{yz} \neq 0,\ \tau_{yz} \neq 0 \tag{3.45}$$

The plain strain state may occur in very thick plates. In Eq. (3.24), the material constants E, ν and α are to be replaced by constants E', ν' and α', where

$$E' = \frac{E}{1-\nu^2},\ \nu' = \frac{\nu}{(1-\nu)},\ \alpha' = (1+\nu)\alpha \tag{3.46}$$

Another example of thermoelasticity is the shear stress in a bonded joint exposed to an elevated temperature [24].

Thermal shear stress in bonded joint

Two elastic layers are joined by a joint layer of adhesive bonding material. This is illustrated in Fig. 3.11.

The shear stress τ is given by [24]

$$\tau(x) = \frac{(\alpha_1 - \alpha_2)\Delta T G \sinh \beta x}{\beta \eta \cosh \beta l} \tag{3.47}$$

where β is

$$\beta^2 = \frac{G}{\eta}\left(\frac{1}{E_1 t_1} + \frac{1}{E_2 t_2}\right) \tag{3.48}$$

In these expressions, E_1 and E_2 are the Young's moduli of layers 1 and 2, respectively. G is the shear modulus of the joint layer material. t_1 and t_2 are the thickness values

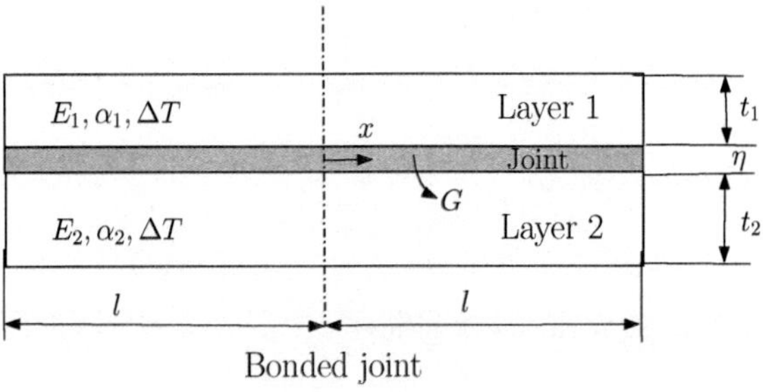

Fig. 3.11 Two elastic layers joined in a bonded joint

of the layers 1 and 2. η is the thickness of the joint layer. The coordinate x is positive in the direction indicated in Fig. 3.11.

Physically, the shear stress is zero at the center, and in theory it increases gradually to a maximum at the free edge. The value of this maximum shear stress is

$$\tau_{max} = \frac{(\alpha_1 - \alpha_2)\Delta T G \tanh \beta l}{\beta \eta} \tag{3.49}$$

When the material properties as provided in Table 3.7 are used, the distribution of the shear stress $\tau(x)$ in the joint layer will be as given in Fig. 3.12.

In this example, the effect of the free edges at both ends of the strip is ignored. At the free edges, no shear stress can be present. In reality, the high shear stress near

Table 3.7 Material properties and geometry of bonded joint (taken from [24])

Parameter	Dimension	Layer-1	Layer-2	Joint
E	GPa	117	275	
G	GPa			1.23
α	m/m/°C	16×10^{-6}	6.5×10^{-6}	
t	mm	1.57	1.52	
η	mm			0.051
l	mm	25.05	25.05	25.05
ΔT	°C	100	100	

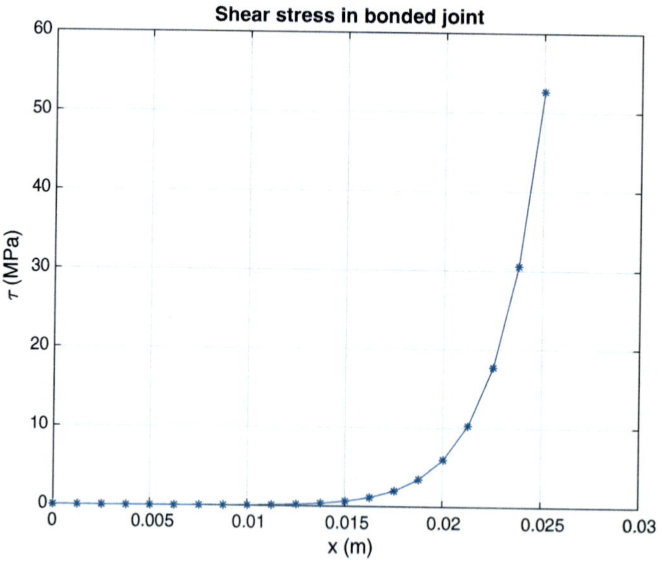

Fig. 3.12 Shear stress in joint layer

the free edge, presented in Fig. 3.12, is in a short distance from this edge reduced to zero shear and converted into a tensile stress normal to the bonded layer surface. This tensile stress causes the joint to open when the joint material fails. This tensile stress is often referred to as "peel-stress".

The next example is an axisymmetric stress state in a pipe.

Axisymmetric thermal stresses

The stresses for a thick-walled pipe occur due to a linear temperature variation through the wall. The temperature difference between the inner and outer surface is

$$\Delta T = T_i - T_o, \tag{3.50}$$

with T_i the temperature at the inner pipe surface and T_o the temperature at the outer surface. The pipe and dimensions are shown Fig. 3.13. In this example is assumed that $T_{ref} = 0$ (°C). This example is partly based on [18].

Expressions for the plain strain radial, circumference and axial stress can be found in [44].

With $a = r_o/r_i$, the radial stress σ_r (Pa) is given by

$$\sigma_r = \frac{\alpha E \Delta T}{2(1-\nu)\ln a}\left[-2\ln\left(\frac{r_o}{r}\right) + \frac{\left(\frac{r_o}{r}\right)^2 - 1}{a^2 - 1}\ln a\right] \tag{3.51}$$

Fig. 3.13 Schematic diagram of axisymmetric pipe section

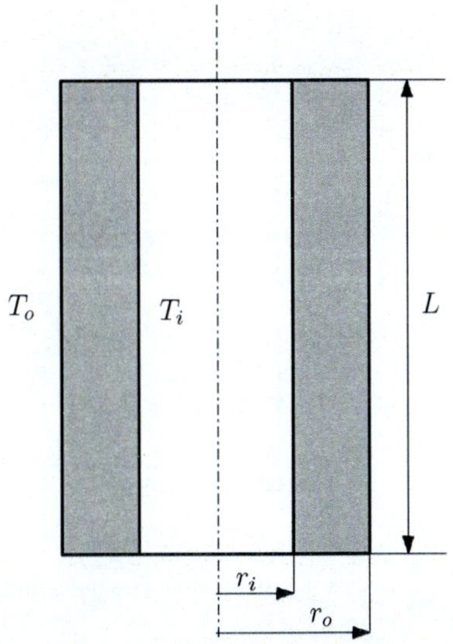

the circumferential stress σ_θ (Pa) is given by

$$\sigma_\theta = \frac{\alpha E \Delta T}{2(1-\nu)\ln a}\left[1 - \ln\left(\frac{r_o}{r}\right) + \frac{\left(\frac{r_o}{r}\right)^2 + 1}{a^2 - 1}\ln a\right] \tag{3.52}$$

and the axial stress σ_z (Pa) is

$$\sigma_z = \frac{\alpha E \Delta T}{2(1-\nu)\ln a}\left[1 - 2\ln\left(\frac{r_o}{r}\right) - \frac{2\ln a}{a^2 - 1}\right] \tag{3.53}$$

The von Mises stress σ_{vM} (Pa) can be obtained by [76]

$$\sigma_{vM} = \frac{1}{\sqrt{2}}\sqrt{(\sigma_r - \sigma_\theta)^2 + (\sigma_r - \sigma_z)^2 + (\sigma_\theta - \sigma_z)^2}, \tag{3.54}$$

The pipe made up of HK40 material is considered with a pipe length $L = 100$ mm, inner radius $r_i = 25$ mm and outer radius $r_o = 50$ mm. The material properties are Young's modulus $E = 138$ GPa, Poisson's ratio $\nu = 0.313$, CTE $\alpha = 15 \times 10^{-6}$ m/m/°C. The inside temperature $T_i = 500$ °C and the outside temperature $T_o = 420$ °C, the thermal gradient $\Delta T = T_1 - T_o = 80$°C.

The Eqs. (3.51), (3.52), (3.53) and (3.54) are programmed into a MATLAB script. The stresses σ_r, σ_θ, σ_z and σ_{vm} are shown in Fig. 3.14.

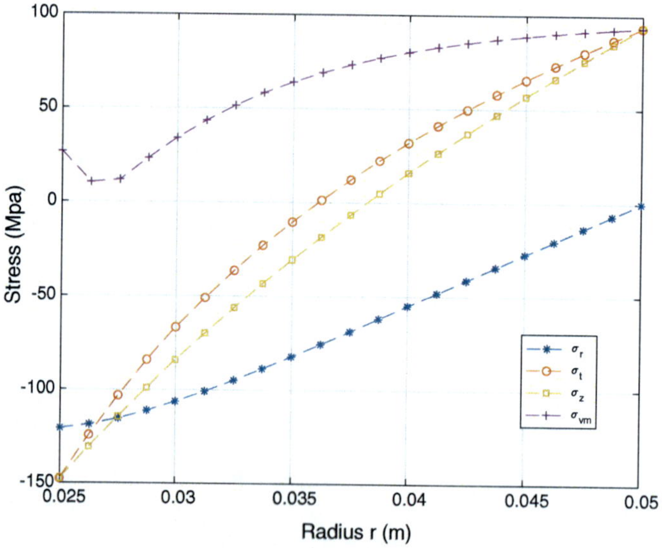

Fig. 3.14 Variation of radial, circumferential, axial and von Mises stresses

3.5 Summary Governing Equilibrium and Constitutive Equations

This section recalls the governing equilibrium and constitutive equations. The matrix notation is used for the expressions [22].

3.5.1 *Equilibrium*

In absence of body forces, the equilibrium of the body is expressed, in Cartesian coordinate system, as

$$\delta^T(\sigma) = 0, \tag{3.55}$$

where (σ) is the vector of stress components, with a tensile normal stress regarded as positive

$$(\sigma) = \begin{pmatrix} \sigma_{xx} \\ \sigma_{yy} \\ \sigma_{zz} \\ \tau_{xy} \\ \tau_{yz} \\ \tau_{zx} \end{pmatrix}. \tag{3.56}$$

The differential operator δ is given by

$$\delta^T = \begin{bmatrix} \frac{\partial}{\partial x} & 0 & 0 & \frac{\partial}{\partial y} & 0 & \frac{\partial}{\partial z} \\ 0 & \frac{\partial}{\partial y} & 0 & \frac{\partial}{\partial x} & \frac{\partial}{\partial z} & 0 \\ 0 & 0 & \frac{\partial}{\partial z} & 0 & \frac{\partial}{\partial y} & \frac{\partial}{\partial x} \end{bmatrix}. \tag{3.57}$$

3.5.2 *Strain–Displacement Relations*

With Eq. (3.57), the relation between displacements and strain can be expressed in matrix form as

$$(\varepsilon) = \delta(u), \tag{3.58}$$

where (ε) is the vector of strain components given by

$$(\varepsilon) = \begin{pmatrix} \varepsilon_{xx} \\ \varepsilon_{yy} \\ \varepsilon_{zz} \\ \varepsilon_{xy} \\ \varepsilon_{yz} \\ \varepsilon_{zx} \end{pmatrix} . \tag{3.59}$$

and the vector of displacements components (u) is

$$(u) = \begin{pmatrix} u_x \\ u_y \\ u_z \end{pmatrix} . \tag{3.60}$$

In the thermoelastic context, it is relevant to note that deformation, in the form of a change of displacement with the position in the structure, leads to strains in the material based on the definition of Eq. (3.58).

3.5.3 Constitutive Law

For the case of thermoelastic deformations, Hooke's law for an isotropic material [22] can be written as

$$(\sigma) = [D]\,(\varepsilon) - \beta \Delta T\,(a)\,, \tag{3.61}$$

in which $\Delta T = T - T_{ref}$ and the vector (a) is

$$(a) = \begin{pmatrix} 1 \\ 1 \\ 1 \\ 0 \\ 0 \\ 0 \end{pmatrix} \tag{3.62}$$

T_{ref} represents the uniform initial (reference) temperature and T the current temperature of the material. The matrix $[D]$ is the matrix of elastic constants, given by

$$[D] = \begin{bmatrix} \lambda + 2G & \lambda & \lambda & 0 & 0 & 0 \\ \lambda & \lambda + 2G & \lambda & 0 & 0 & 0 \\ \lambda & \lambda & \lambda + 2G & 0 & 0 & 0 \\ 0 & 0 & 0 & G & 0 & 0 \\ 0 & 0 & 0 & 0 & G & 0 \\ 0 & 0 & 0 & 0 & 0 & G \end{bmatrix} , \tag{3.63}$$

with λ and G, the Lamé constants, and elastic shear modulus of the material, respectively. The Lamé modulus can be expressed as follows

$$\lambda = \frac{\nu E}{(1+\nu)(1-2\nu)} \tag{3.64}$$

and the thermal stress modulus β is given by

$$\beta = \frac{E\alpha}{(1-2\nu)} \tag{3.65}$$

where E is the Young's modulus, ν the Poisson's ratio and α the CTE.

From Eq. (3.61) can be noted that, through the definition of the vector (a), a temperature change in the material cannot induce shear stresses.

Problems

3.1 This problem about the calculation of the CTE is partly based on an example given in [44]. The thermal expandability $\beta(T)$ is given in Table 3.8.

Perform the following assignments:

- A third-order polynomial curve fitting for $\beta(T)$, Eq. (3.10).
- Calculate with aid of numerical differentiation the CTE $\alpha(T)$.
- Which material had been measured?

3.2 A heat pipe had been stress free mounted into a spacecraft and supported by three supports. At the ends, the supports are fixed and in the mid, a gliding support is

Table 3.8 Thermal expandability $\beta(T)$

Working temperature (°C)	Thermal expandability $\beta(T) \times 10^3$ (m/m)
25	0
150	1.5305
200	2.1913
250	2.8787
300	3.6047
350	4.0555
400	4.9153
450	5.9718
500	6.7963
550	7.6192
600	8.4817

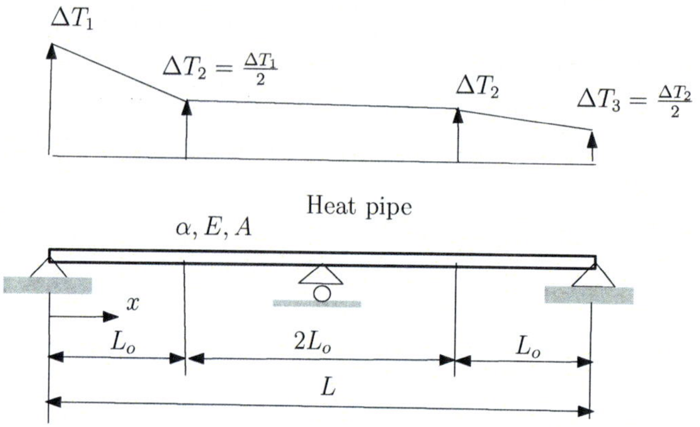

Fig. 3.15 heat pipe configuration

provided to increase the natural frequency. In orbit, the heat pipe shows a piece-wise linear temperature elevation. The heat pipe system and temperature distribution are shown in Fig. 3.15.

Calculate the force build up in the heat pipe.

3.3 A thin walled tube supports an equipment platform of a satellite and is simply supported at both ends. The Euler buckling critical stress σ_{crit} is given by [29]

$$\sigma_{crit} = \frac{\pi^2 EI}{AL^2} \tag{3.66}$$

where L is the length of the tube, $I = \pi d^3 t/8$ (m^4) is the second moment of area and the area $A = \pi dt$ (m^2). Young's modulus is $E = 70 \times 10^9$ Pa. The diameter of the tube is $d = 30$ mm, and the wall thickness is $t = 0.5$ mm. The length of the tube is $L = 75$ cm. The tube is shown in Fig. 3.16. During launch, the tube will encounter a temperate increase ΔT. The CTE is $\alpha = 24 \times 10^{-6}$ m/m/°C. Calculate the allowable ΔT when the tube during launch is also loaded by a compression force $F = 2500$ N. Verify the equations.

3.4 This is a classical strength of materials problem. A cantilevered beam (optical bench) is supported by a rod at point B. The thermoelastic interaction between two

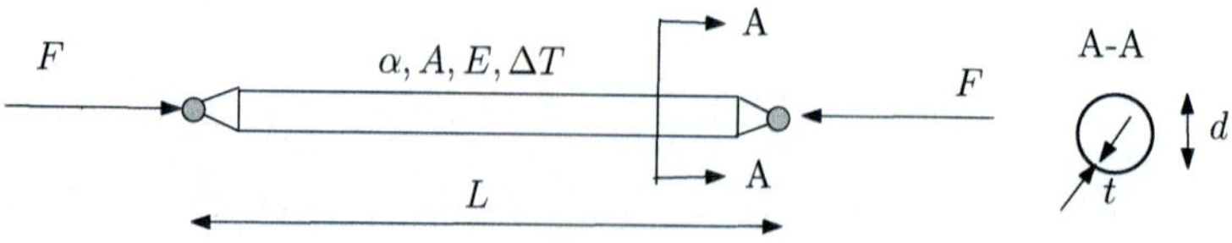

Fig. 3.16 Supporting tube

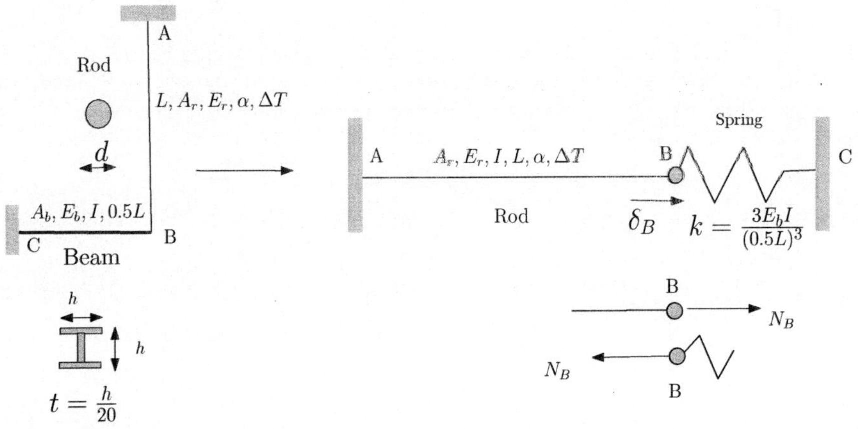

Fig. 3.17 Heated/cooled rod connected to a spring

structural parts (beam and rod) is investigated when the rod is exposed to a uniform temperature decrease ΔT. The beam is replaced by an equivalent spring. The rod is connected to the spring, equivalent to the bending stiffness of the beam, with spring stiffness k. This system and all geometrical and physical parameters are shown in Fig. 3.17. Calculate the interface force at point B. Young's modulus of the rod is E_r and the beam E_b. Assume the second moment of area of the beam is $I = \gamma^2 A_b$, where γ is the radius of gyration and A_b the cross section of the beam. Verify precisely all subsequent assumptions and expressions.

The displacement δ_B of the rod and spring are

$$\begin{aligned} \delta_B &= \frac{N_B L}{E_r A_r} + \alpha \Delta T L \\ \delta_B &= -\frac{N_B}{k} = -\frac{N_B (0.5L)^3}{3E_b I} = -\frac{N_B (0.5L)^3}{3E_b \gamma^2 A_b} \end{aligned} \tag{3.67}$$

Equating both right-hand sides of the equations, the following equation is obtained

$$\frac{N_B L}{E_r A_r} + \alpha \Delta T L + \frac{N_B (0.5L)^3}{3E_b \gamma^2 A_b} = 0 \tag{3.68}$$

The solution for N_B is

$$N_B = -\frac{24 \alpha A_b A_r \Delta T \gamma^2 E_r E_b}{(24 \gamma^2 E_b A_b + A_r E_r L^2)} \tag{3.69}$$

When the ratio $E_r/E_b \rightarrow 0$, the internal force $N_B \rightarrow 0$.

The tensile stress σ_{rod} in the rod can be easily calculated

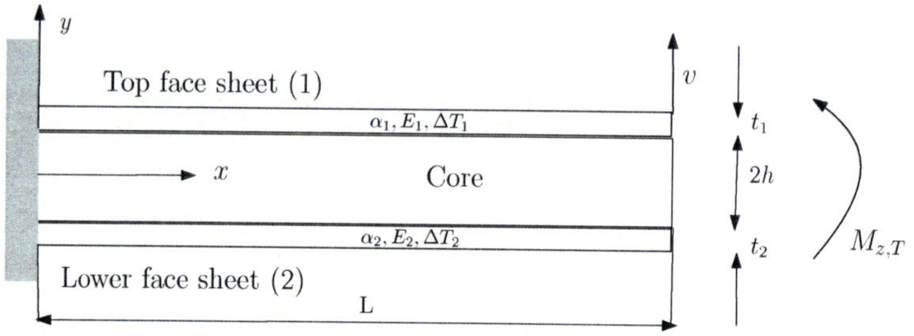

Fig. 3.18 Sandwich beam

$$\sigma_{rod} = \frac{N_B}{A_r} - E_r \alpha \Delta T \tag{3.70}$$

The stiffer part (beam) of the structural system dictates the structural behaviour of that system.

3.5 Sandwich Cantilevered Beam

A cantilevered sandwich beam is shown in Fig. 3.18 as well as the design variables. The temperatures ΔT_1 and ΔT_2 are constant over the length of the beam and constant over the thickness of the face sheets. The core has a very low Young's modulus in length direction but is very large in the direction normal to the two face sheets. Assumed is that there is no shear deformation.

In the following, the relations between the thermoelastic bending moment and deformations are derived. Equation (3.71) can be considered as starting point. Reproduce these relations. It is recommended to consult the mentioned references for background information.

The deflection v is calculated by applying the classical beam bending theory with the Myosotis method [29]. The equivalent thermoelastic bending moment is calculated using following expression [19, 89]

$$M_{z,T} = -\int_A \alpha E \Delta T(y) y dy = -bh[\alpha_1 E_1 \Delta T_1 t_1 - \alpha_2 E_2 \Delta T_2 t_2] \tag{3.71}$$

assuming $t_1, t_2 \ll h$ and the width of the beam is b (m). When ignoring the stiffness of the core, the equivalent bending stiffness EI_{eq} of the sandwich beam is

$$EI_{eq} = bh^2[E_1 t_1 + E_2 t_2] \tag{3.72}$$

The deflection v at the tip of the sandwich beam can be calculated with

$$v = \frac{M_{z,T} L^2}{2EI_{eq}} = -\left(\frac{L^2}{2h}\right) \frac{\alpha_1 E_1 \Delta T_1 t_1 - \alpha_2 E_2 \Delta T_2 t_2}{E_1 t_1 + E_2 t_2} \tag{3.73}$$

When v_1 and ϕ_1 are constrained, the deflection $v_2 = v$ and rotation ϕ_2 at the tip of the beam can be easily calculated with

$$\frac{EI_{eq}}{L}\begin{bmatrix} 12 & -6L \\ -6L & 4L^2 \end{bmatrix}\begin{pmatrix} v_2 \\ \phi_2 \end{pmatrix} = \begin{pmatrix} 0 \\ M_{z,T} \end{pmatrix}. \tag{3.74}$$

3.6 With reference to the example about the axisymmetric thermoelastic stress in the hollow pipe (page 35), it is asked to create a FE model (see Fig. 3.13) and calculate the axisymmetric thermal stresses with your own FEA package. (hint: plane strain boundary conditions)

Chapter 4
Modelling for Thermoelastic

Abstract The physics of thermoelastic involves both thermal and structural phenomena that require to be handled together. Consequently, thermoelastic analysis has to be considered as an integral process including both related disciplines with close interaction of the models. This interaction imposes an important and often overlooked constraint: Making decisions about the importance of a temperature field with only the knowledge of a temperature field is in general not possible and always requires the computation of the structural thermoelastic response.

4.1 Introduction

The objective of thermoelastic analysis is in general twofold: the traditional first objective is to verify the strength of the structure under consideration. The second objective is mostly relevant when deformations of the structure have the potential to degrade the functioning of an instrument or a collection of instruments. The second objective is thus to determine deformation responses as input to the performance verification of instruments. The corresponding thermoelastic modelling and analysis are involving different disciplines with associated models. Quite some information has to flow from one analysis step to the other. Figure 4.1 gives in general terms an overview of the whole analysis process.

The temperature fields are computed through the thermal analysis with the thermal model. The temperature fields are represented by the thermal node temperatures.

The temperature mapping process converts the temperature fields, produced by the thermal model, to corresponding entities in the structural finite element model, FE nodes and elements. This mapping process considers both the practical data format conversion problem and the conversion required to bridge the difference in solution methods between the thermal analysis tools, in most cases the lumped parameter method, and the structural analysis tools for which mostly the finite element method (FE method) is used. One should realise that in most cases, the temperature fields have an important transient behaviour and are therefore evolving over time. Therefore, the temperature mapping process may need to be applied to a significant number of instantaneous temperature fields in order to capture their evolution in time.

S. Appel and J. Wijker, *Simulation of Thermoelastic Behaviour of Spacecraft Structures*, Springer Aerospace Technology, https://doi.org/10.1007/978-3-030-78999-2_4

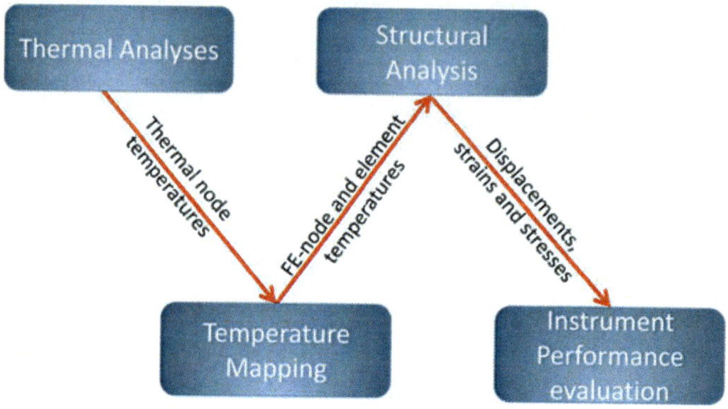

Fig. 4.1 Thermoelastic analysis process

The temperature fields form the loading for the structure that is simulated by the finite element model. The structural analysis produces the thermoelastic responses in the form of displacements, strains, stresses and forces. In many cases, the deformations, i.e. the displacements, being three linear translations along with three angular rotations, are most relevant for the performance of the instrument or spacecraft.

The deformations, in the form of displacements, are therefore then provided for evaluation of the impact on the performance of the instrument or spacecraft. This may require optical, radio frequency (RF) or other type of assessments depending on the type of instruments.

In the following sections, the thermoelastic modelling and simulation process is described and, where available, complemented with suggestions and points of attention. The main theme of the suggestions is that the thermal and structural analysis needs to be considered as one integral analysis process. The importance of a temperature field for the performance of an instrument or a spacecraft can only be identified based on the structural thermoelastic responses and in general not solely on the temperature fields.

4.2 What Is a Thermal Gradient?

There is a general consensus that thermal gradients are important for the thermoelastic response of a structure. The thermal gradient is the change of temperature per unit length along the three different spatial directions. This implies that the thermal gradient is a vector quantity that can mathematically be formulated as

$$(\text{grad } T) = (\nabla T) = \left(\frac{\partial T}{\partial x}, \frac{\partial T}{\partial y}, \frac{\partial T}{\partial z} \right). \tag{4.1}$$

Formally, the thermal gradient as defined in Eq. (4.1) has to be called "temperature gradient", since it is the spatial derivative of the temperature field. In literature, the terms "thermal gradient" and "temperature gradient" are both used and are both using the definion of Eq. (4.1). In this book, both terms are used as well. Because the gradient mathematically refers to change per unit spatial coordinate, in this book the thermal or temperature gradient is also referred to spatial thermal gradient or spatial temperature gradient.

It turns out at various places in engineering documents also the temperature difference between two locations, without any mention of distance between these locations, is referred to as temperature gradient.

$$\Delta T = T_1 - T_2. \tag{4.2}$$

Although the temperature difference between two points may be sometimes a relevant quantity, it is not correct to call this temperature gradient.

When a contour plot of a temperature field is evaluated, the magnitude of the vector components of the spatial thermal gradient can be related to the density of the iso-temperature lines. The closer the iso-temperature lines are along a certain direction, the higher the spatial thermal gradient is along that direction. The normal vector to the iso-temperature lines indicate the direction of the spatial thermal gradient.

The following examples aim at showing the importance of the spatial thermal gradient for thermoelastic stresses and deformations.

Sandwich panel with small ΔT through the thickness

Considered is a simply supported square sandwich plate with the dimension of $L = H = 1$ m and a total thickness of $h = 20$ mm. The plate has aluminium face sheets of 1 mm thick and has an aluminium core. The face sheets have an $E = 70$ GPa, a Poisson's ratio $\nu = 0.33$ and a CTE $\alpha = 24 \times 10^{-6}$ m/m/K.

The honeycomb material is in principle an orthotropic material of which only the out-of-plane shear properties are modelled: $G_{31} = 0.186$ GPa and $G_{23} = 0.09$ GPa, with 1 and 2 the two in-plane material directions.

The sandwich plate is modelled with MSC Nastran using CQUAD4 elements, and plate properties are implemented in the model with the PCOMP entry.

A temperature difference of two degrees has been applied with the temperature at the top two degrees lower than the temperature at the bottom. With the z-axis representing the normal on the plate surface with a positive direction pointing from the bottom to top side of the plate and assuming a linear temperature variation through the thickness, the temperature gradient is then

$$\frac{dT}{dz} = \frac{T_{top} - T_{bot}}{h} = \frac{-2}{0.020} = -100\text{K/m} \tag{4.3}$$

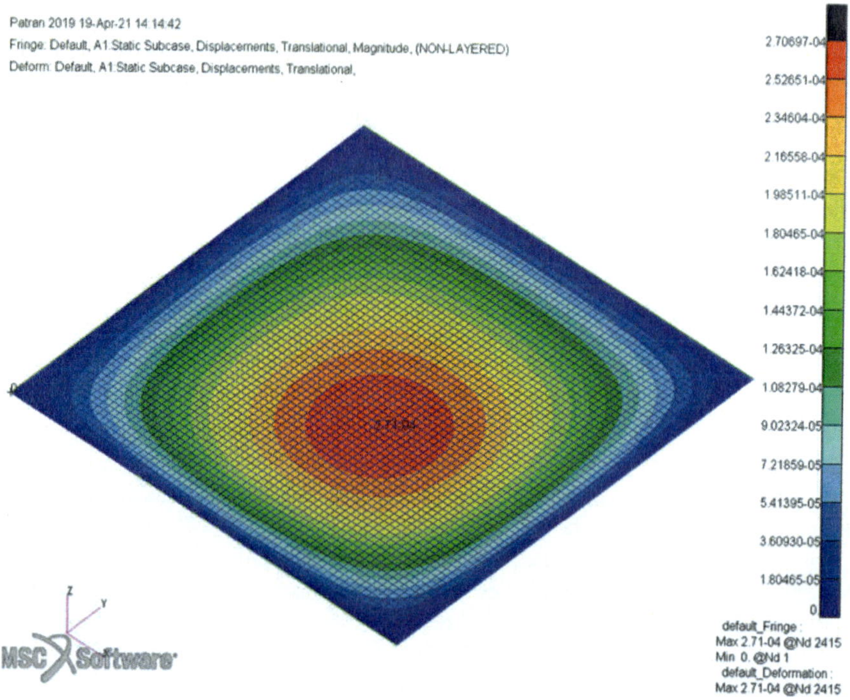

Fig. 4.2 Out-of-plane displacement due to a through thickness temperature gradient

This temperature gradient has been applied with the TEMPP1 entries in on the plate elements.

Due to this temperature gradient through the thickness of the sandwich bending deformation is imposed on the plate, leading to out-of-plane displacements. These are visualised in Fig. 4.2.

This example illustrates that a small temperature difference can lead to a significant temperature gradient, which on its turn is able to induce an important deformation.

Temperature gradient

Considered is a square plate with $L = H = 1$ m and thickness $t = 5$ mm. The plate and boundary temperatures are shown in Fig. 4.3. The conductivity $k = 205$ W/m/°C.

The other material properties are Young's modulus $E = 70$ GPa, the Poisson's ratio $\nu = 0.33$ and the CTE is $\alpha = 24 \times 10^{-6}$ m/m/°C.

The internal temperature field in the plate is calculated with the FEA method and shown in Fig. 4.4.

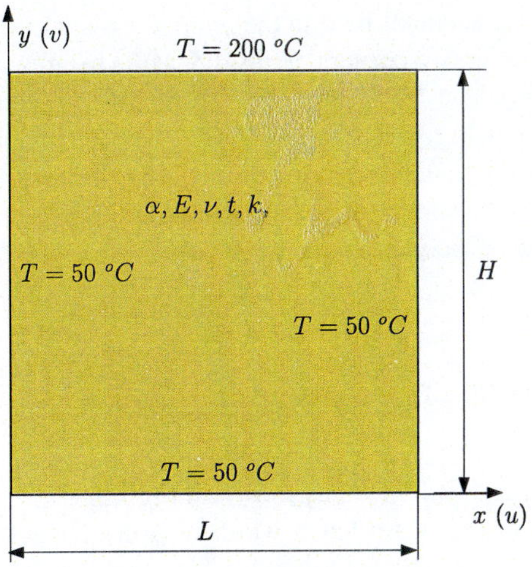

Fig. 4.3 Square plate with prescribed boundary temperatures

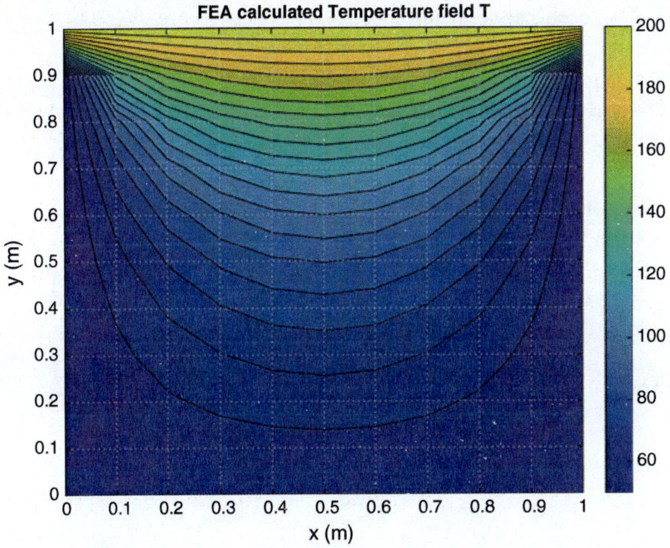

Fig. 4.4 Temperature field in plate

Especially in the top left and right corners of Fig. 4.4, high-temperature gradients are observed. In those corners, a high-temperature difference at a short distance exists, due to the prescribed temperatures at the top boundary (200 °C) and side boundaries (50 °C). In the two top corners, these two boundaries join. The high-temperature gradients can be deducted from the high density of iso-temperature lines at those locations.

To calculate the displacement field in the plate, the displacement in y-direction v is constrained in the lower two corners in Fig. 4.3, at the locations $(x, y) = (0.0, 0.0)$ and $(x, y) = (1.0, 0.0)$. The displacement in x-direction u is constrained in the middle of the bottom edge in Fig. 4.3, at the location $(x, y) = (0.5, 0.0)$. The constraints are almost symmetric. The displacement results are presented in Fig. 4.5. Also the displacement field shows strong gradients. With Eq. (3.58), this can be related to high strain levels with subsequent high stress levels.

Thermoelastic deflection of sandwich beam

Considered is the sandwich beam as shown in Fig. 4.6.

The cross section of the beam has a width of b(m) and a height of h(m). The length of the beam is L(m). The face sheets of the sandwich have a thickness t(m), with $\frac{t}{h} \ll 1$.

The material of the face sheets has a Young's modulus E(Pa) and a CTE α(m/m/°C). The core between the face sheets is very stiff in y-direction, normal to the face sheets, and the stiffness in axial x-direction is assumed negligible and therefore not providing a contribution to the axial and bending stiffness. Only the face sheets are providing axial stiffness.

The total area of the cross section of the face sheets is $A = 2bt(\text{m}^2)$, and the corresponding second moment of area is $I = \frac{1}{2}bh^2t(\text{m}^4)$.

The top face sheet of the beam has a constant temperature increase of ΔT_{top}(°C), and at the bottom face sheet has the constant temperature increase is ΔT_{bot}(°C).

The equivalent thermoelastic bending moment M_T and thermoelastic axial force P_T are given by [19]

$$M_T = -\int_A \alpha(y)E(y)\Delta T(y)y dA = -\alpha bt E\frac{h}{2}[\Delta T_{top} - \Delta T_{bot}] \tag{4.4}$$

and

$$P_T = \int_A \alpha(y)E(y)\Delta T(y)dA = \alpha bt E[\Delta T_{top} + \Delta T_{bot}], \tag{4.5}$$

The curvature $\frac{d^2w(x)}{dx^2}$ can be calculated with

$$\frac{d^2w(x)}{dx^2} = -\frac{M_T}{EI} = -\frac{\alpha}{h}[\Delta T_{top} - \Delta T_{bot}] \tag{4.6}$$

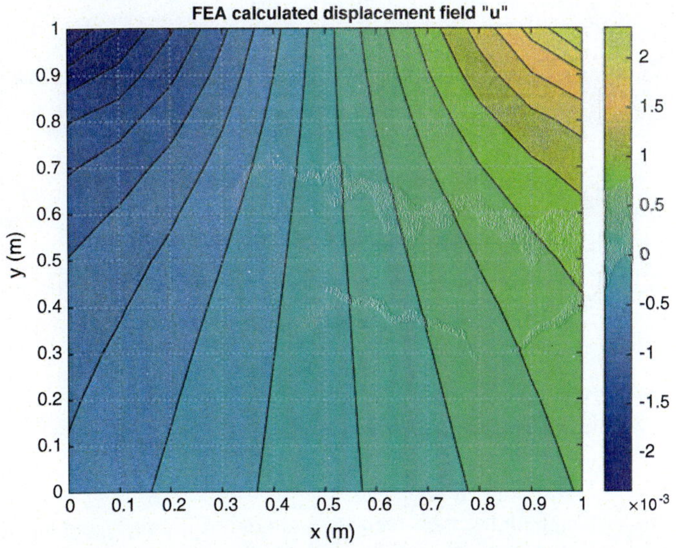

(a) Displacement "u" in x-direction

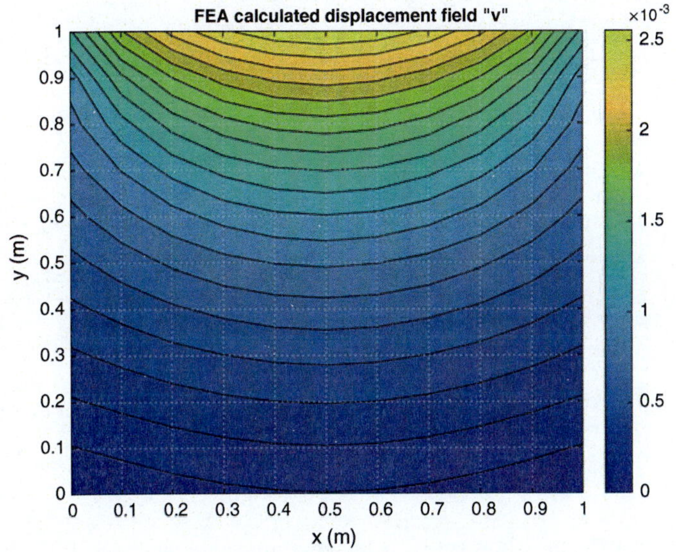

(b) Displacement "v" in y-direction

Fig. 4.5 Displacement fields in plate

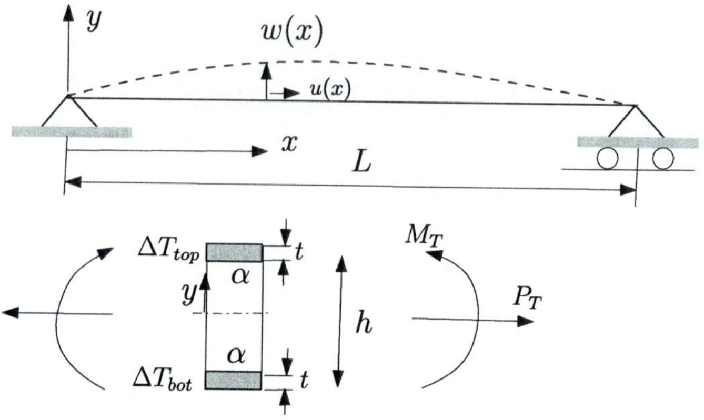

Fig. 4.6 Sandwich beam subjected to a temperature difference

It can be noted that both the thermoelastic moment of Eq. (4.4) and the curvature of the beam from Eq. (4.6) are proportional to the temperature gradient

$$\frac{dT}{dy} = \frac{\Delta T_{top} - \Delta T_{bot}}{h} \tag{4.7}$$

The axial elongation of the beam $u(x)$(m)

$$u(x) = \frac{P_T x}{EA} = \frac{\alpha x}{2}[\Delta T_{top} + \Delta T_{bot}] \tag{4.8}$$

The vertical displacement $w(L/2)$(m) at the mid-span of the sandwich beam is given by

$$w\left(\frac{L}{2}\right) = \frac{\alpha L^2[\Delta T_{top} - \Delta T_{bot}]}{8h} \tag{4.9}$$

The constant axial stress $\sigma_{xx}(y = \pm h/2)$(Pa) in the face sheets can be obtained with

$$\sigma_{xx}(y) = -\alpha E \Delta T(y) + \frac{P_t}{A} - \frac{M_T y}{I}, \ y = \pm\frac{h}{2} \tag{4.10}$$

This simple example shows how a temperature gradient induces the bending deformation of a beam.

4.3 What to Model?

All verifications, by simulation or test, are done with the purpose to check compliance to requirements that in one way or the other are all linked to the performance of the spacecraft or its instruments. Some requirements can be verified by a single engineering discipline in isolation, such as stiffness requirements [31]. Requirements for which a thermoelastic verification is needed obviously require at least the involvement of the two disciplines thermal and structures. These two disciplines can do the job together when the verification has the objective to check only the strength of the structure under thermoelastic environments.

In most other cases, where deformation of the structure due to thermal environment is affecting the performance of an instrument or spacecraft, also other disciplines are needed to verify the impact of the thermoelastic deformation. When the structure is supporting a radio frequency or microwave antenna or if the structure forms the support of a mirror assembly of an optical instrument, then clearly the involvement of the experts of those domains is required.

As mentioned above, the needs for thermoelastic verification originate from the performance requirements of the instrument or spacecraft. The modelling requirements for thermoelastic simulation, like for any discipline, are strongly linked to the requirement that has to be verified. In other words: Which essential physical behaviour do you want to quantify? Or: Is the required physics well captured by the (mathematical) model?

To answer these questions, the thermal and structural engineer together have to consult the specialist, who is responsible for the performance of that subsystem, to get a good understanding of the nature of the requirements. This engineer can tell which types of deformation impact the performance. Also can this engineer indicate which information he or she needs to quantify the impact on the performance of the instrument. This consultation leads to the requirements for the thermal and structural models allowing these to capture the essential physics and have the models producing the right information for the performance evaluation. It might be obvious, but this exchange between the different disciplines is best to be done as one of the first things before the structural and thermal engineer start with the development of their models.

Above-described approach is also referred to as a systems engineering approach [39], in which all the disciplines required for the development of the instrument as a subsystem are as good as possible interacting and considered jointly.

4.4 Structural and Thermal Modelling for Thermoelastic: An Integrated Process

Traditionally in many space engineering companies, there is a thermal department and a structures department. In both these departments, engineers are doing a great job within their engineering domain. For successful thermoelastic verification, it is required that the colleagues from these two engineering disciplines are developing their models closely together, assuring the required accuracy and compatibility of the models. This enables the conditions to develop a thermal model that is able to provide the temperature data needed by the structures engineer for the thermoelastic response simulation, such as:

- temperature fields with sufficient resolution in space and time to capture well enough the temperature variation to be imposed on the structural model
- temperature data for features (such as electronic boxes and interface brackets) in the thermal model that exist in the structure and that are responsible for local temperature variations potentially relevant for distortions of the structure
- temperature fields on stiff components of the structure that have the potential to impose deformation on the surrounding components (e.g. stiffeners and longerons).

The benefit of the close interaction between the two disciplines also works in the other direction. Knowledge of the specificities of the thermal problem, the corresponding model and resulting temperature field allows the structural engineer to adapt the structural model such that it can capture the detailed information of the temperature field.

An important side benefit of developing the two types of models together is to establish from the start a good correspondence between the thermal model and the structural model. As is explained in Chap. 7, a good correspondence will ease the temperature data transfer from the thermal model to the structural model. As a consequence, efforts for subsequent design and modelling iterations are significantly reduced. In this respect, the use of CAD models as single geometry source for both the thermal and structural can be of great help.

To conclude this section, it is good to note that the principle of thermoelastic verification for spacecraft applications relies on the assumption that deformations of the structure have no effect on the heat flow mechanisms in the structure and thus no effect on the temperature field. This allows the thermal model and structural model to be run in sequence, and no feedback is needed from the structural analysis results to the thermal model. A typical situation that would make this assumption to fail is when due to the thermoelastic deformation contact states between parts in the structure are changed. This means that parts that are first in contact are not any more in contact causing an interruption of a (local) heat flow path affecting the resulting temperature field. In those cases, feedback between structural and thermal solutions is needed to solve the problem often in an iterative way. In this book is assumed that this coupling between the thermal and the structural solution is not present and therefore the thermal and structural analysis can be run in sequence without feedback.

4.5 Integrated Model Convergence Checks

Although every engineer's education included the essentials of model discretisation convergence checks, many engineers are not frequently checking the convergence of their models. Even outside the thermoelastic context, it is a good habbit to do this. With a convergence check is meant the verification of the effect on the analysis results due to changes in the model and solver parameters. A typical aspect to verify is whether a change of mesh resolution has an effect on an important outcome of the model. If changes in the values of the results, especially those responses driving the performance, are less than a predefined magnitude, the model can be declared converged.

Convergence checks can of course be done for each discipline in isolation. The convergence checks on the thermal model verify whether changes to a thermal response (temperatures, heat fluxes, ...) as a result of changes in for instance the thermal mesh are below a certain tolerance. Equally with the convergence checks on the structural model, it is verified that further refinements in the finite element model do not change the structural responses (displacements, stresses, ...) more than a tolerance agreed for the structural responses.

It is important to note that convergence checking exercises are only meaningful, when components and features in the structure, that could influence the thermoelastic response, are represented in the model. Only when a complex stiff bracket or an electronic box is presented with its geometry in both the structural and thermal models, a mesh convergence exercise can guide to the right or adequate mesh resolution for these components and the surrounding parts in the model.

To complement, also the basic geometry of thermal and mechanical load application areas must be implemented in the mesh to form a useful starting point for the mesh convergence exercise.

Although the convergence checks on the thermal and structural models in isolation are important and even essential, these are not sufficient in the context of thermoelastic analysis. Since thermoelastic analysis is a multi-step process (see Fig. 4.1), convergence checks need to be executed considering the full thermoelastic analysis chain: thermal analysis, temperature mapping and structural thermoelastic response analysis. For this convergence check, it has to be verified whether refinements in the thermal model, modifications in the mapping process or refinements in the structural finite element model result into changes in the computed performance. When further refinements of the modelling parameters of all the models do not lead to changes in the performance of the instrument bigger than the agreed tolerance, only then it can be claimed that convergence of the whole analysis chain has been reached.

It is obvious that these convergence checks require a high computational effort. In addition, it requires a high level of organisation of the analysis process. In Sect. 4.7, recommendations are provided that can ease the process. Essential for a successful and efficient convergence exercise is to have a parametric set-up of the models already from the first versions. Although not every analysis tool has built in parametrisation features included in its input format, basic scripting will allow for easy changing important parameters in an automated way.

Example: CYLSAT

In this example, the effect of the mesh resolution in the thermal model on the final pointing angle variation is examined.

Considered is a simplified abstract spacecraft with the shape of a cylinder of 2 m high and a diameter of 1 m. The spacecraft outer cylindrical structure is made of an aluminium shell with a thickness of 2 mm. The outer surface has a coating with an infrared emissivity $\varepsilon = 0.7$ and a solar absorptivity $\alpha = 0.3$.

The geometry of the spacecraft is basically built from two disc surfaces, the top and bottom lids, and a cylindrical surface. The surfaces of the discs and cylinder are fully conductively and mechanically coupled at their respective edges. In Fig. 4.7, the thermal geometrical model and structural finite element model are shown. In the following the mesh density of the thermal model is going to be varied. The presented geometric thermal model is one of the models considered for this exercise.

The cylindrical spacecraft is put in an orbit around the earth at 800 km above the earth surface. The orbit and orientation of the spacecraft relative to the sun are shown in Fig. 4.8. The +Y axis of the spacecraft is nadir pointing during the orbit.

Six different versions of the thermal model are created with a mesh resolution ranging from 4 by 4 nodes on each surface up to 18 by 18 nodes on each surface (see Fig. 4.9). The structural FE mesh was kept the same and was combined with each of the six thermal models. Each of these sets of models went through the following steps:

Colour

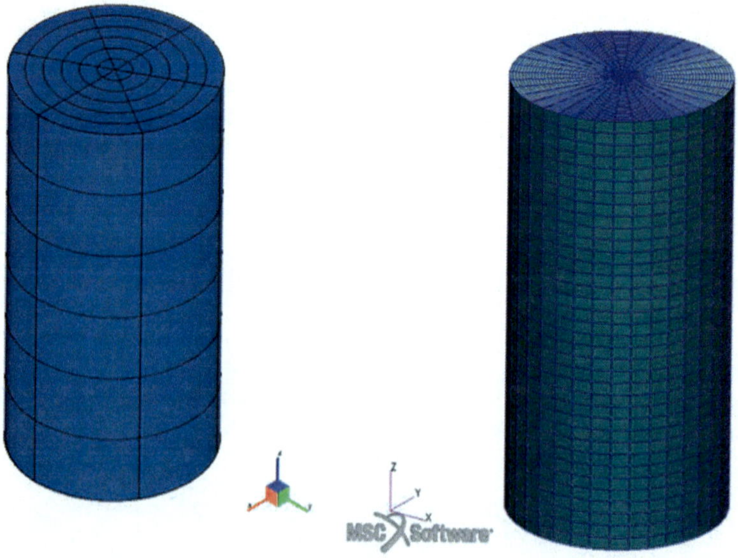

Fig. 4.7 ESATAN-TMS (left) and MSC Nastran FE model (right) of the cylindrical satellite

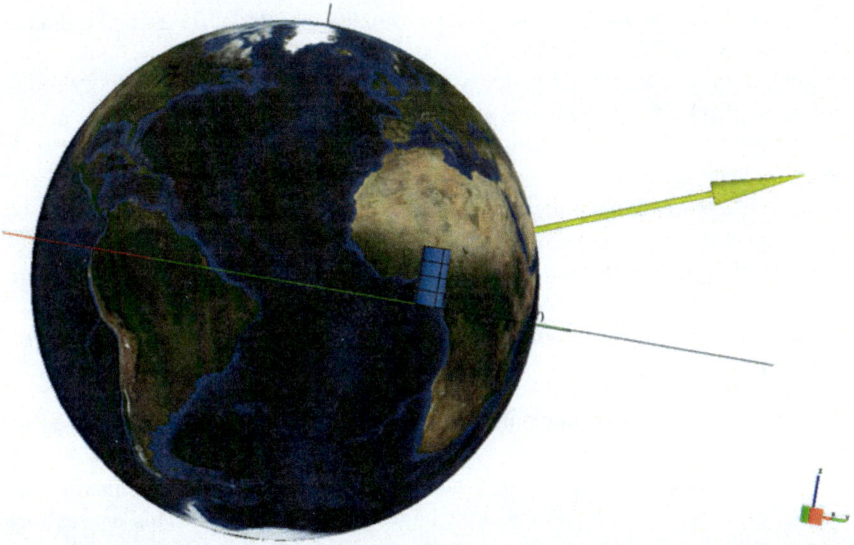

Fig. 4.8 Cylindrical spacecraft in an orbit around the earth modelled with ESATAN-TMS [49]

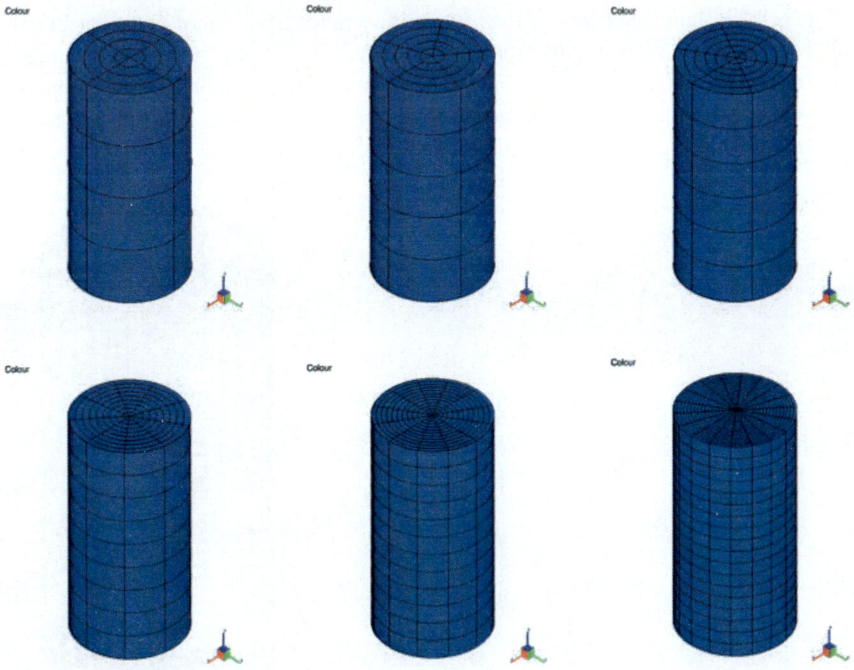

Fig. 4.9 Six thermal models with six different mesh resolutions

- Computation of the radiative analysis to compute the orbital fluxes and radiative couplings with ESATAN-TMS
- Computation of the thermal node temperatures with ESATAN-TMS. Four orbits are simulated in the thermal analysis resulting into 321 temperature fields at 321 orbital positions in the four orbits.
- Transfer of temperature fields produced by the thermal model in the form of thermal node temperatures to the finite element model as finite element node temperatures. This step is done for all the 321 temperature fields with SINAS (PAT method [9]).
- Computation of the deformation fields of the structure due to the 321 temperature fields. This simulation is done with MSC Nastran.
- Finally, the variation of the pointing angle is computed for all the deformed shapes of the structure.

The above steps are an implementation of the schematic process visualisation presented in Fig. 4.1.

While the spacecraft is orbiting around the earth, the sun is illuminating the spacecraft at every position in the orbit under a different angle. This causes that the temperature at each point on the spacecraft is constantly changing with time. To illustrate this, four nodes on the finite element model are considered around the middle section of the cylinder as indicated in Fig. 4.10.

Figure 4.11 shows nicely how the temperature fields at different positions at the wall of the spacecraft vary differently with time. This figure has been produced from the data based on the thermal model with mesh resolution of 18 by 18 thermal nodes per surface, i.e. the thermal model with the highest mesh resolution. Note that

Fig. 4.10 Four nodes in the mid-section of the spacecraft for extraction of temperatures (±X and ±Y)

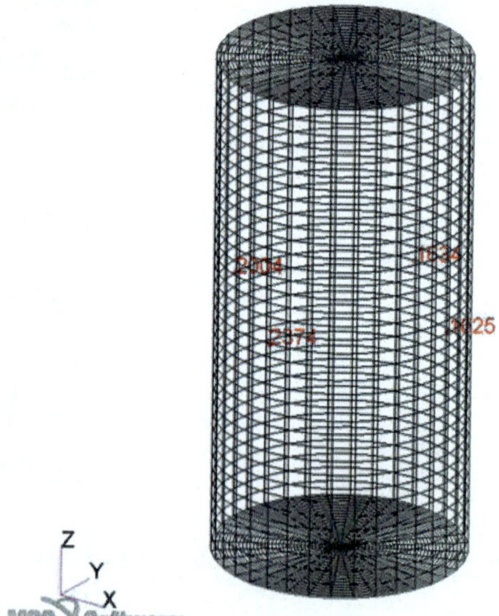

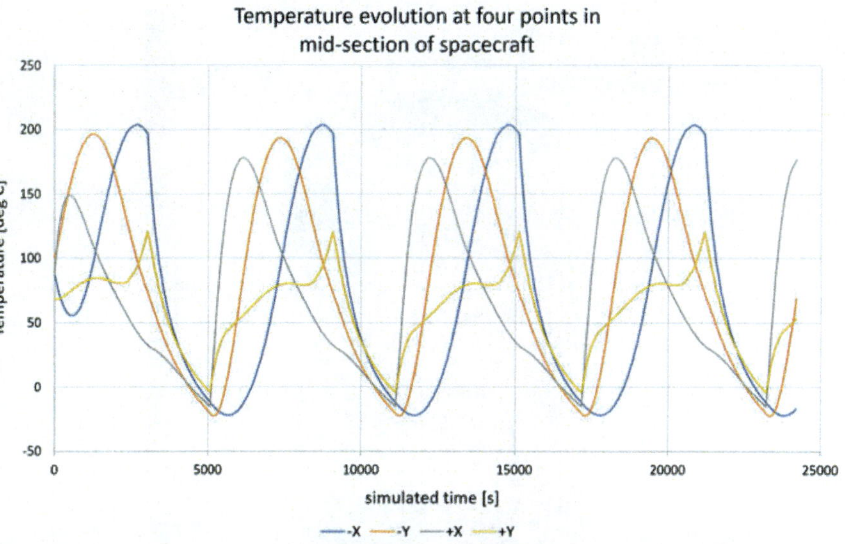

Fig. 4.11 Temperature evolution at four nodes in the mid-section of the spacecraft (±X and ±Y)

three of the four curves have more or less the same shape. However, the curve of the temperature at the +Y side, i.e. nadir side, is quite different. As will be seen in the remaining of this example, a different mesh resolution will give a different temperature distribution, indicating that also the curves of Fig. 4.11 are expected to be dependent on the thermal mesh resolution.

Due to the variation of temperature fields with time, the thermally induced deformation also changes with time. In Fig. 4.12, the temperature field at a specific moment in the transient is plotted on the FE mesh of the structural model with the corresponding thermally induced deformation.

These deformations can be determined for each moment a temperature field is provided by the thermal analysis. This makes it possible to derive from this deformation at each of these moments the performance impact that can for instance be presented as curves showing this impact as a function of time.

Imagine now an antenna mounted on the cylindrical spacecraft that has its main axis pointing normal to the earth (nadir pointing). This antenna is considered to be mounted on just two points. The vector from one point to the other is parallel to the Y-axis of the model. The two points are indicated in Fig. 4.13 together with the pointing vector representing the axis of the antenna.

The angle of rotation of the antenna axis can be approximated by the difference in displacement of the two points (nodes) at the beginning and end of axis, since the two points are one metre apart. Two angles can be distinguished. The most obvious angle is the one in the YZ plane. This angle variation is caused by global bending of the cylinder (see also Fig. 4.12). Angle variation in the XY plane is less obvious. It is caused by torsion of the cylinder. Since deformations are computed for all

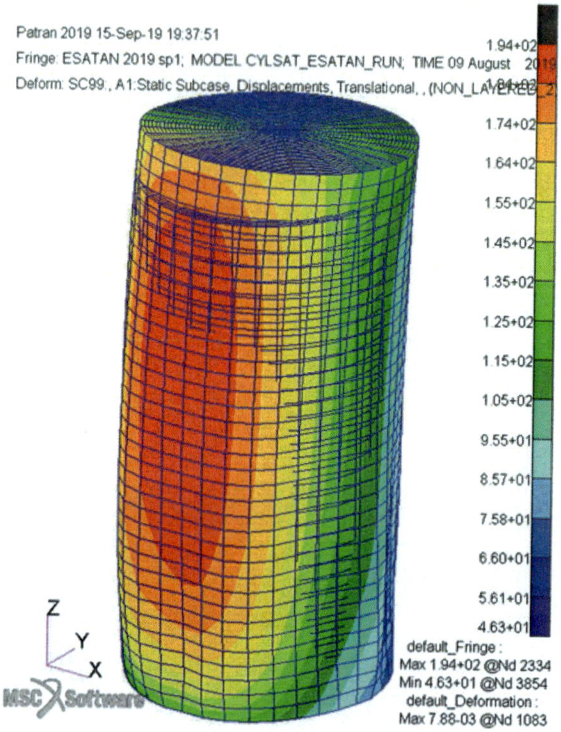

Fig. 4.12 Snapshot of the temperature field with corresponding deformation field

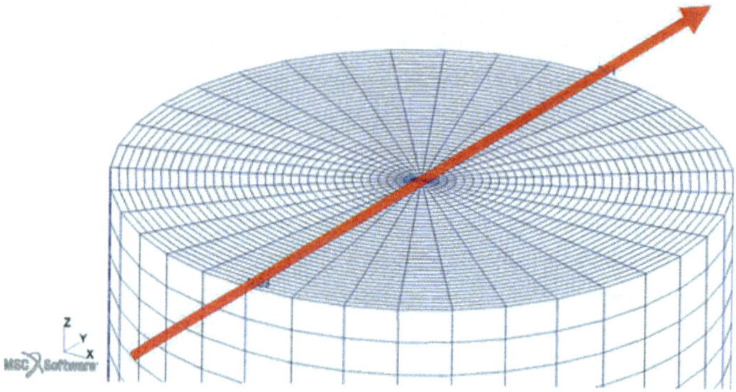

Fig. 4.13 Axis of antenna running in Y-axis trough two points (nodes)

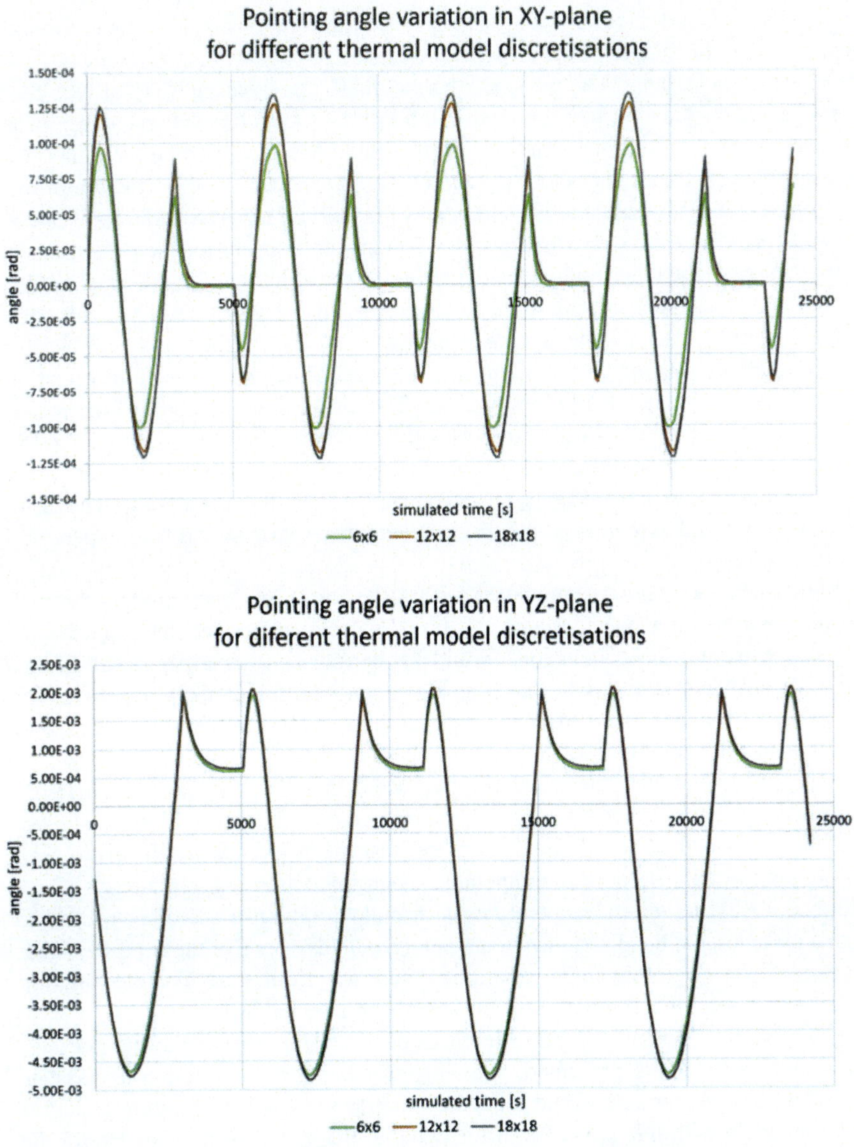

Fig. 4.14 Variation of the pointing angle in XY and YZ planes over time for different thermal mesh resolutions

temperature fields at time increments for four orbits and for six different thermal model resolutions, the results produced with these six thermal models can now be compared.

In the Fig. 4.14, variations of the two angles are presented. Note that for clarity of the graphs, only the curves for the thermal mesh resolution 6×6, 12×12 and

18×18 are plotted. Note as well that the resulting rotations in YZ-plane are one order of magnitude higher than those in XY-plane.

Now the question is: "Did we reach thermal mesh convergence?". To answer this question, the curves have to be studied. From Fig. 4.14, the observation can be made that the curves representing variations of rotation in YZ-plane for the three thermal mesh resolution are all very close to each other. Apparently, the deformation of the structure can already with a low resolution of the thermal mesh be well represented. When is imagined that only rotations in YZ-plane are relevant for the performance of the structure, then one could consider to limit the resolution of the thermal model to 6×6. In most cases, it is not just a single performance that is relevant, in this case rotations in XY-plane also need to be evaluated.

Although the rotations in XY-plane are one order of magnitude less, the effect of thermal resolution can clearly be observed. The differences in height of the peaks between 6×6 and 12×12 are clearly larger than between 12×12 and 18×18. Depending on the required accuracy, one could decide to further refine the thermal mesh. In such a case, it needs to be checked whether the finite element mesh not also needs to be refined to remain compatible with the thermal mesh and the temperature mapping method.

Figure 4.14 shows how the predicted pointing performance depends on the thermal mesh resolution. Now one could ask: "Is it really necessary to have the full analysis chain done to come this conclusion? Would the thermal analysis results not show this also?" To answer this, Fig. 4.11 that contains the mid-section temperature curves for the 18×18 thermal mesh is extended with same curves for the 6×6 thermal mesh. These results are combined in Fig. 4.15.

Figure 4.15 shows that there is little difference in the curves for these two extreme mesh resolutions. Based on this information, one would be very tempted to be already satisfied with just the 6×6 thermal mesh. Because the difference in deformation can only come from a difference in temperature field, this difference is not easy to spot with curves that represent temperatures at discrete points. It is therefore needed to have a close look at the difference in temperature fields at one of the time steps that show the highest pointing angle variation. This difference in the XY angle variation is the largest near one of the peaks, for example at about 6428 s.

In Fig. 4.16, temperature field differences between the extreme thermal meshes, 6×6 and 18×18, are presented. Knowing that the XY plane rotations indicate torsion of the cylinder, one would expect to spot that there are temperature gradients under an angle with the main axis of the cylinder that could be responsible for the torsion. However, the temperature plots do not appear to allow this as an obvious observation. This case should therefore be considered as an example in which the thermoelastic effect can not easily be even estimated from the temperature field only. Also here it is recommend to wait with conclusions on the importance of temperature fields till the results of thermoelastic structural analysis with the temperature field applied to the FE model are evaluated.

The objective of this example is to show the importance of mesh convergence. The mesh convergence exercise is performed on the thermal mesh. The example shows that thermal mesh convergence cannot be just based on thermal analysis results

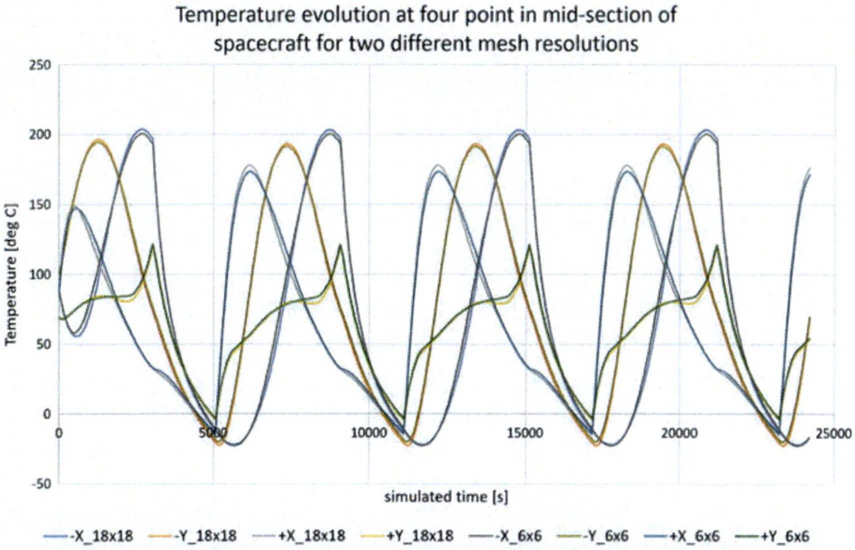

Fig. 4.15 Temperature evolution at four nodes in the mid-section of the spacecraft (±X and ±Y) for 9 × 9 and 18 × 18 thermal mesh resolutions

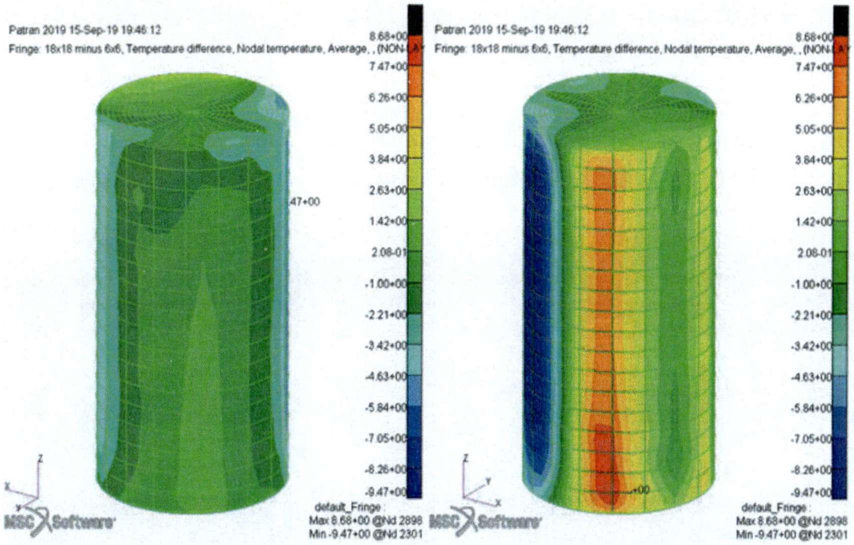

Fig. 4.16 Temperature field differences between 6 × 6 and 18 × 18 thermal mesh

only and has to rely on demonstration of the convergence of the final instrument performance, being in this case the pointing angle variation. The whole analysis process as shown in Fig. 4.1 shall be run through for each thermal mesh variation.

4.6 Modelling Features

4.6.1 Features and How These Are Commonly Modelled

In the previous section, the focus was on mesh resolution. The CYLSAT example showed the importance of end-to-end mesh convergence checking. There are however other aspects that require attention in a way different from what most space structural and thermal engineers are commonly used to.

For instance, boxes with electronics often have a high stiffness that justifies for dynamic analysis to limit the representation in a structural model of a spacecraft to a lumped mass in the centre of gravity of the box connected through rigid elements to a spacecraft panel. A typical representation is provided in Fig. 4.17. In a system level thermal model, similar simplifications are applied. Very often only a few thermal nodes, and sometimes even only a single thermal node, are spent on the representation of the often box shaped item.

The question always is: "How bad is this simplified representation?" or "Which effects do we ignore?". The question that is not often asked is: "Which non-physical effects do we introduce?" Ultimately, the question is what this simplified modelling means for the performance prediction of an instrument on the same or neighbouring

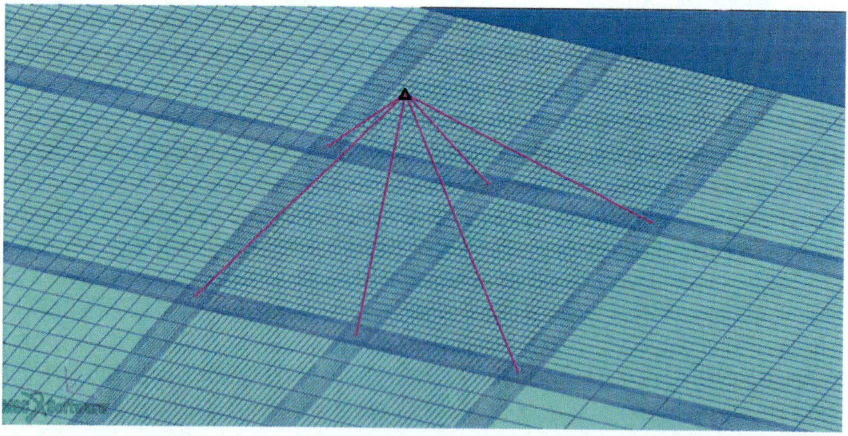

Fig. 4.17 Simplified modelling of electronic box with a rigid element and a lumped mass, typical for dynamic simulations

panel of the electronic box. Are the effects of the electronic box quite localised? Or is there an important influence on the deformation field of the panel in areas of the panel away from the box and affecting the orientation of pointing sensitive instruments sharing the same panel?

Under the pressure of project schedules, there is often the tendency to find a quick, and therefore often possibly not so solid, justification that in many cases has as well a flavour of wishful thinking. In most cases, the result of a quick assessment comes down to a statement like "The modelling is conservative". This is then often the outcome of an examination of a limited number of specific cases, which are conveniently generalised to all other cases.

Although not discussed in great detail in this book, a similar discussion can be held on the representation of joints or interfaces between the various items in a structure. For instance, the representation of cleats, that often form the mechanical interfaces between two panels, is worth a thorough investigation on their contribution to the thermoelastic responses. Rigid elements and springs are extensively used for simplified representations of joints. As is explained in Sect. 6.7.4, the modelling of joints turns out to be less critical for predictions of responses to external loads that do not have imposed deformations involved. Further attention is needed for the modelling of these joints when possible eccentric load transfer is causing bending and thus curvature of a panel. In essence, this all comes down to capturing well all the stiffness components as is essential for the interaction of the structural part and the level of deformation these can impose on each other. So also the modelling of the joints between the structural parts is important for capturing well the influence of these joints on the tilting of pointing sensitive instruments.

4.6.2 Assessment of a Box on a Plate

This section makes use of the detailed parametric exercise, presented in Appendix A, on the effects of an electronics box on the deformation of a sandwich panel. The main outcomes of the exercise are presented here together with some modelling suggestions.

For the purpose of gaining understanding of the effect of an electronic box on the deformation of a sandwich panel relatively high-resolution thermal and structural models are made. The reason for applying this potentially excessive resolution is to exclude as good as possible effects due to too low resolution from the discussion.

For the preparation of Appendix A, several box and plate configurations are analysed. The variations between the configurations include change of dimensions of the box, level of heat dissipation and position of the box on the panel. Of one of these configurations, the structural finite element model and the corresponding geometrical thermal model are presented in Figs. 4.18 and 4.19, respectively.

The panel is a stiff sandwich panel of 50 mm core height and aluminium face sheets. The panel is clamped along all the edges.

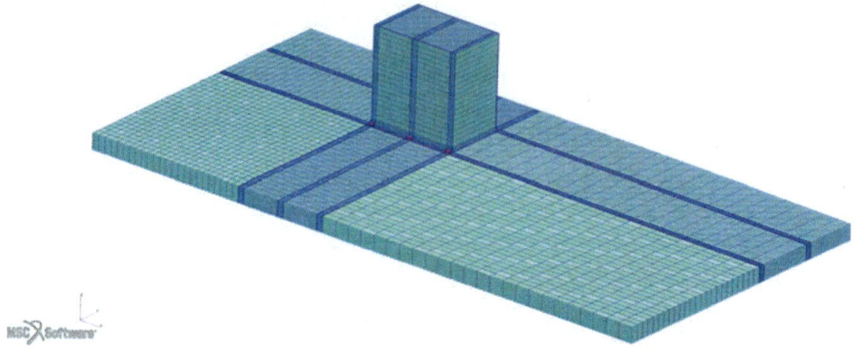

Fig. 4.18 Detailed finite element model of an electronics box on a sandwich plate

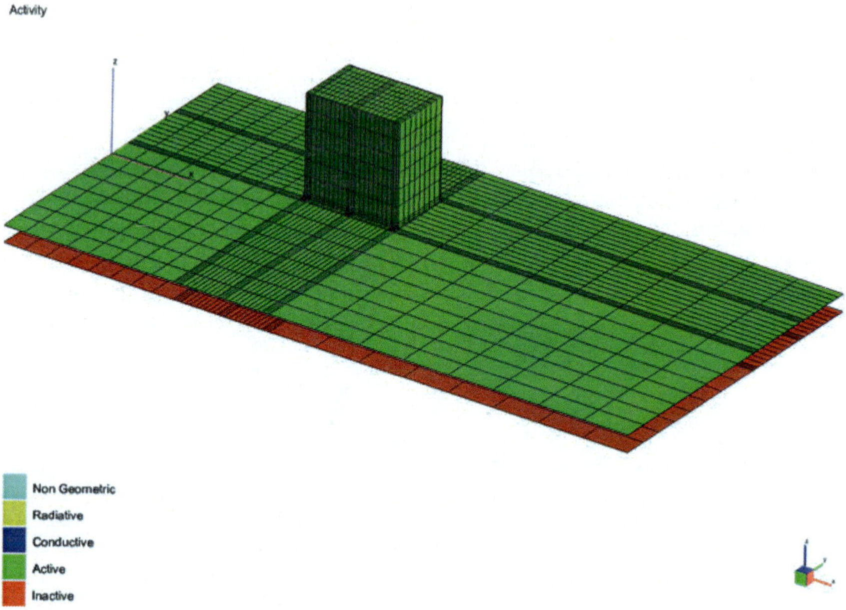

Fig. 4.19 Detailed thermal model of an electronics box on a sandwich plate

The thermal environment is the only loading source for the structure. It consists of a heat dissipation in the box and a radiative boundary condition simulating the thermal environment of an internal panel surrounded by other panels at 20 °C.

Details of these models can be found Appendix A.

For all the models analysed, a steady-state thermal analysis has been run, the temperatures from the thermal model have been mapped on the structural finite element model, and the distortions of the box and the sandwich panel have been

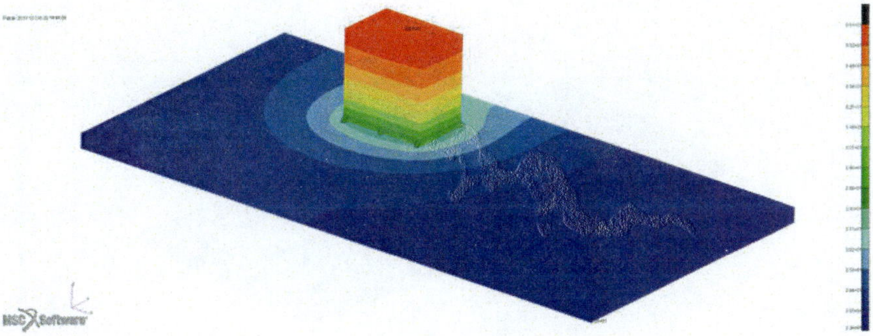

Fig. 4.20 Temperature field induced by the dissipation of the electronics box

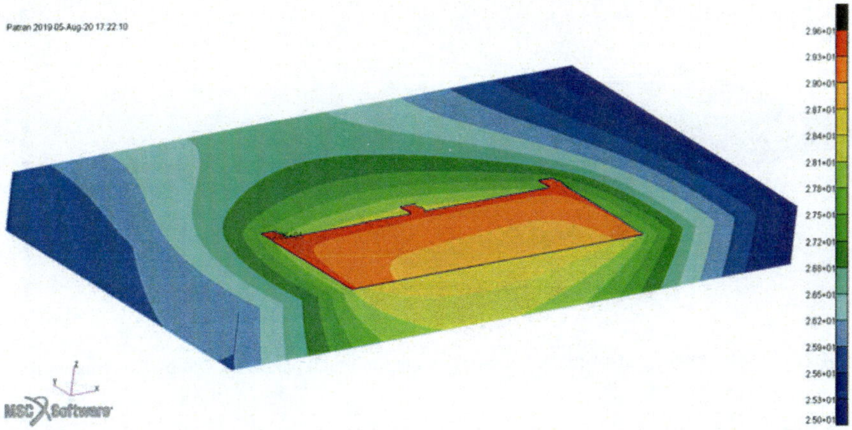

Fig. 4.21 Temperature field induced by the dissipation of the electronics box through the cross section of the sandwich panel

computed. The temperature field mapped on the structural finite element model is presented in Fig. 4.20.

Figure 4.20 shows clearly a temperature gradient along the height of the box. Except for the direct surrounding of the box, the thermal gradient over the surface of the sandwich panel looks quite mild. This may raise the impression that thermoelastic effects are limited to the region close to the box. Interesting is the observation that the temperature field through the thickness of the sandwich right under the box, as presented in Fig. 4.21 is rather uniform. From thermal control perspective, this observation is often a justification for not making the effort to implement thermal nodes at both sides of the sandwich and it is definitely not considered to apply several thermal nodes through the thickness, as is done in this model. Due to the simulated uniform contact resistance, the base plate temperature is a bit higher than the temperature of the sandwich skin.

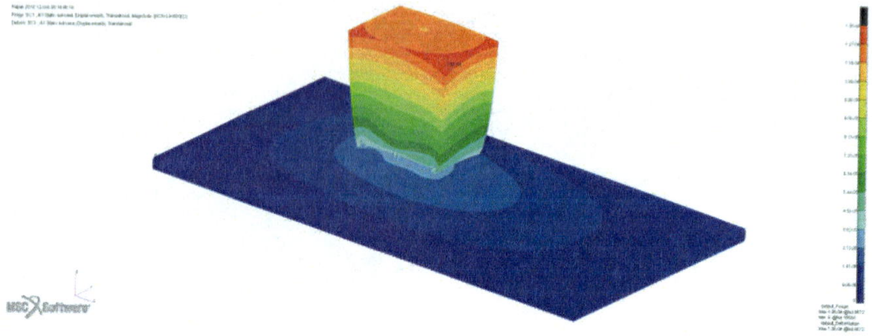

Fig. 4.22 Displacement induced by the temperature field

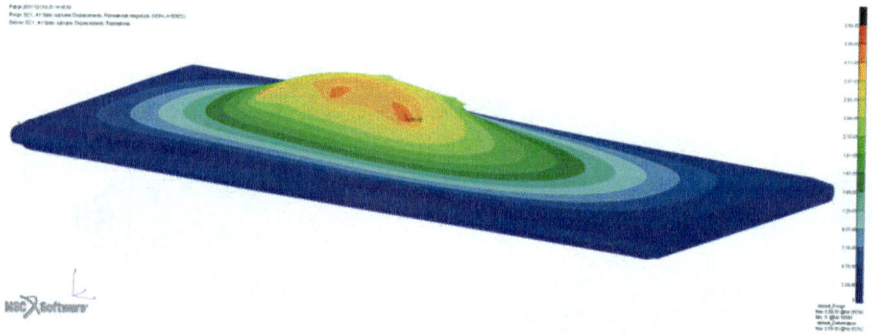

Fig. 4.23 Displacement (out-of-plane bending) on the sandwich panel induced by the temperature field

All the pictures of the temperature fields raise the expectation that the electronic box is only inducing locally a distortion of the sandwich panel. However, Fig. 4.22 and especially Fig. 4.23 show that the deformation due to this dominantly local heating is spreading over the whole panel. This is an other example that it is not trivial to deduct from a picture of the temperature field the severity of the thermoelastic responses.

In Appendix A is examined if and how the interface planes of instruments, sharing the same panel as the electronic box, are tilted due to the distortion of the panel induced by the heat dissipation in the electronic box. The conclusion is that even under mild thermal environments, the tilting induced by equipment can be important. This importance depends of course on the sensitivity of the instrument for rotations of the interface plane. However, the values for the tilting angles, that were obtained with the examples in the Appendix with only a single box, are of the same order of magnitude as the alignment angle budgets of instruments on recent scientific and Earth observation spacecraft. This stresses the importance of paying the right attention to the thermoelastic modelling of dissipating units, especially when these units are located on the same panel as the instrument or in the vicinity of instrument panels. Understanding the influence of the many details, such as contact conduc-

Table 4.1 Delta temperature between top and base panel for three box configurations at position X = 35 cm and Y = 30 cm, Q = 30 W

Box dimensions (h × w × d) [cm]	Top panel max temperatures [°C]	Baseplate min temperatures [°C]	ΔT [°C]	$\Delta T/\Delta h$ [°C/cm]	Overall max tilting [arc s]
30 × 20 × 30	36.9	29.6	7.3	0.24	7.7
15 × 20 × 30	36.1	30.7	5.4	0.36	8.7
7.5 × 20 × 30	35.2	31.2	4.0	0.53	9.4

tance, on the temperature field in the box and subsequent deformation is essential. In this example is not elaborated on the uncertainties and sensitivities of different parameters. Methods to deal with these uncertainties are discussed in Chap. 10.

Based on above observations, especially the observations obtained from the computed temperature fields, it is not obvious how the deformation of the panel is introduced by the electronics box. Appendix A explains how different factors are examined that could be responsible. Different box and panel configurations with different box dimensions are compared. In Table 4.1, different values are collected for three box configurations at the same location of the panel.

The table shows that the box with the smallest height introduces the highest level of tilting of the interface planes of the imaginary instruments. The observed temperature levels at the top panel and base plate for the different configurations are very much similar and do not explain the difference in the level of induced deformation into the sandwich panel. The temperature at the top panel is showing even an opposite trend. What turns out to the be discriminating factor is the temperature gradient in the box along the vertical direction. The gradient was approximated by ΔT, being the temperature difference between the top panel and the base plate, divided by height of the box, Δh.

The box with the lowest height in the example, i.e. the box with dimensions 7.5 × 20 × 30, produced the largest deformations in the sandwich panel. The middle size box produced deformation between those of the smallest and the biggest box.

The results summarised in Table 4.1 help explaining that the temperature gradient along the height of the box is the driving factor behind the deformation of the panel. As the table shows, the highest temperature gradient corresponds to the box that is inducing the largest deformation (max tilting). The smallest deformations are produced by the box with the lowest gradient. Consistently, the box producing deformation in between the largest and the smallest deformation values has a thermal gradient that is also in between the largest and smallest gradient.

As is also explained in Appendix A, the gradient in the box is inducing a curvature in the shape of the box. This curved shape is well illustrated by Fig. 4.24. The top part of the box is warmer than the lower part of the box. As a result, the top part

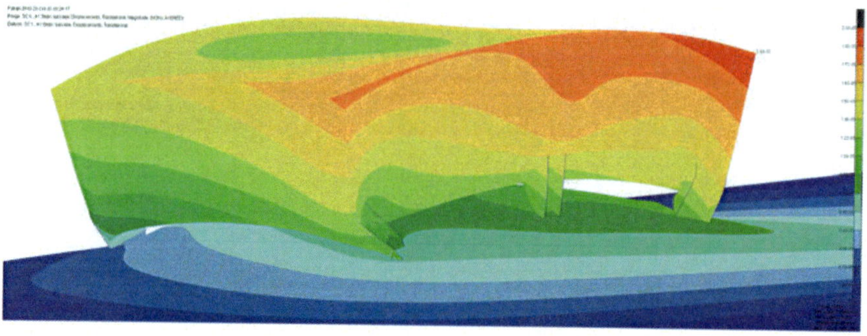

Fig. 4.24 Curvature in the shape of the box due to the thermal gradient along the height of the box

Table 4.2 Thermal gradients over the height of two box configurations at position X = 35 cm and Y = 30 cm with 30 W heat dissipation

Box dimensions (h × w × d) [cm]	Q total [W]	Top panel max temperatures [°C]	Base panel min temperatures [°C]	ΔT [°C]	$\Delta T/\Delta h$ [°C/cm]
30 × 20 × 30	30	36.9	29.6	7.3	0.24
15 × 10 × 15	30	46.1	36.5	9.6	0.64

is expanding more than the lower part, which results then into a curvature that is proportional to the gradient. The same effect is illustrated in the example "Thermoelastic deflection of sandwich beam" in Sect. 4.2. The expressions in this example show nicely showing this proportional relations between the temperature gradient and the induced bending moment. Due to the high stiffness of the box, compared to the sandwich panel, the box is able to induce its deformation into the panel.

Above results appear to consistently explain that the thermal gradient is the driving factor for the level of deformation in the sandwich panel. The comparison is however limited to three boxes with the same base plate dimensions. Therefore, Appendix A includes as well the comparison with a box that has all dimensions of the largest box divided by two, leading to a box that has 1/8th of the volume of the largest box and a quarter of the size of the base plate.

Table 4.2 shows that the small box with dimensions 15 × 10 × 15 has a significantly larger thermal gradient than the large box. Nevertheless, the large box is injecting a larger deformation into the panel than the small box. Despite the higher thermal gradient and the corresponding higher relative deformation of the small box itself, it is not able to cause more deformation of the panel than the larger box. This points to an other factor of importance besides the gradient. The size of the footprint of the box on the sandwich panel is in this case the main difference between the two boxes. The larger box has due its larger footprint a larger area of the panel on which it can impose its curvature. Although it has not been checked by the exercise from

Appendix A, an other cause, also related to the size of the box, could potentially be that the stiffness of the large box is higher due to its larger height and could then better resist the reaction of the panel. For cases analysed, this argument is falsified with the information in Table 4.1, where the effect of the higher stiffness due the larger height was not stronger then the effect of the higher thermal gradient for the box with the smallest height and all with the same footprint. From this assessment is then concluded that the driving difference between the two boxes in Table 4.2 is the size of the footprint that makes that the larger box is able to inject a higher deformation in the panel than the small box, despite the higher thermal gradient in the small box.

This assessment nicely shows how the internally induced deformation due to a thermal gradient is interacting with the stiffness of the different components. The stiffness of the sandwich tries to resist the bending deformation that the box tries to impose through its thermally induced curvature. The deformation of the panel causes through its bending stiffness a reaction bending moment to the box. With this is concluded that the stiffness representation of the structural items involved is essential for the structural modelling of thermoelastic problems. Again is referred to Sect. 6.7.4 where with a small simple example the importance of stiffness representation is explained.

The discussion above mainly focussed on the structural stiffness aspects of the box, which is as explained an important factor in the interaction of the structural items involved. In this discussion should not be ignored at one side the required ability of the structural model to capture the temperature field with its strong gradient causing the deformation and at the other side the requirement for the thermal model the ability to simulate the thermal gradient. To put it simple: if the thermal model is too coarse and therefore not producing or under estimating the thermal gradients or if the structural mesh is too coarse to capture a well-represented spatial temperature field by the thermal model, a good representation of the stiffness will not lead to an adequate prediction of the thermoelastic response.

4.6.3 Simplifying Feature Modelling: Preserve the Physics

The exercise from Appendix A as was discussed in the above was using a high mesh density for both the structural and thermal model in order not spoil the discussion with possible effect coming from too coarse meshes. Such high mesh density could in many cases not be feasible for modelling each unit in system-level thermoelastic models (thermal and structural) of a full spacecraft. Although the sizes of models that are manageable are increasing strongly due to ongoing performance increases of computational hardware, it is often not needed to apply such a high mesh resolutions. In most cases, the mesh resolution of these items can be much lower. On the other hand, a higher than needed mesh resolution would not have a negative impact on the reliability of the predictions. Besides the choice of the mesh resolution, also has to be identified which physical elements in the design play a relevant role in the thermoelastic behaviour, and thus need to be kept in the simplified versions of the models.

Assuming that simplifications can be applied to the modelling, the question then is: To what level can the modelling of features be simplified? The answer may look both simple and not surprising: Also a simplified model shall be able to represent the relevant physical behaviour.

This looks like an easy answer, but what is then the "relevant physical behaviour"? For answering this question, an exercise of the type that is presented in Appendix A is needed. It is needed, without being biased by distortion of results due to limited mesh resolutions, to understand which effect a feature has on its neighbouring structures and vice versa. This allows to capture the understanding of the mechanisms that need to be represented in the models. It turns out that even experienced senior engineers are surprised to see effects from these exercises that they did not expect and without doing this kind of experiments, never would have bothered to take into account in their modelling.

In the following, the "box on plate problem" is taken as an example. The suggested approach for generating a simplified model as is described below can equally be applied to other heat or non-heat dissipating items.

Back to the box on plate problem: Two driving physical phenomena can be identified from the previous section. These phenomena are:

- a strong thermal gradient along the height of the box causing the box to deform into a curved shape that it tries to impose on the panel
- high bending stiffness of the box that makes it possible that the box can resist the stiffness of the panel and therefore impose the curvature deformation on the panel.

Whatever simplifications are going to be made to the box modelling, these two essential phenomena should be captured.

For the generation of a simplified version of the box models for thermoelastic analyses, two scenarios can be imagined. The first one is applicable to early design phases of a project in which in many cases detailed models of equipment are not yet available. In this stage, important conceptual design decisions are taken. It is therefore required that the predictions that form the basis for these decisions are sufficiently accurate.

The second scenario is valid when the project is developed into a more mature stage. Detailed models of various sub-systems become available, so also the models of the various equipment. These detailed models are often too detailed for integration in system-level analyses.

In the scenario of the early design phase of the spacecraft, a rough design concept of the electronic boxes has to be assumed. It could also be that it is foreseen to re-use an equipment design from a previous spacecraft, which is happening quite often. When a detailed design can be reused, then the simplification approach can be similar to the one recommended for the second scenario.

When only a thermal and mechanical design concept is available in the early stage of the project, this design concept should form the basis of the simple model to be integrated in the system-level model. It is recommended to identify the structural elements that provide the stiffness of the box. Also try to prevent underestimating the stiffness of possible joints that are included in the concept. Make sure that any internal structure that is contributing to the stiffness of the box is taken into account. These considerations are relevant for both the thermal and structural models. The thermal model should provide temperatures for the stiff structural elements. At the same time, the thermal model shall simulate well the internal heat dissipation and the way the heat is injected in the structural components with high stiffness. This may require some simplified conductive or radiative modelling of the internal of the box. For the thermal model, it is also important to approximate well the thermal inertia distribution over the model.

One should realise that decisions to include or not combinations of structural elements in both the thermal and structural models is an iterative process. This process is supported with analyses in which it is recommended to set the resolution of both the thermal and structural meshes at a generous level, such that decisions on including or not certain items of the box are not biased by the limitations induced by a too low mesh resolution.

Once it is decided which items are important to be included, then the mesh resolution can incrementally be reduced as long as the thermoelastic results do not show changes per increment larger than a beforehand set value.

In the second scenario, the supplier of the box has already matured the design and a detailed version of the thermal and structural models exist. In this scenario, the simplification starts with removing structural elements from the model that are not driving the thermoelastic response of the box. In most cases, these are not elements with the higher stiffness, such as side panels with high in-plane stiffness. Again this removing of items may require and iterative process in which the effect of removing combinations of structural elements is investigated. Also here a generous mesh resolution shall be preserved for all structural elements in both the thermal and structural models. Once it is clear which items are relevant to be maintained in the simplified models, the mesh resolution can incrementally be reduced to a level at which further reduction of the meshes of both models causes thermoelastic response to change more than a beforehand set value.

The above approach should then lead to an adequate and "light" as possible representation of the electronics box for implementation in a system-level model. As indicated before, this approach is not just applicable to electronic boxes, but also to other dissipating and non-dissipating items.

4.7 Need for Automation of the Analysis Chain

As might have become clear from previous example, running a thermoelastic analyses involves a number of analysis steps (see also Fig. 4.1):

- Thermal analysis
- Temperature mapping

- Thermoelastic response analysis
- Instrument performance analysis.

Obviously, a large amount of data have to be transferred and manipulated from one step to the other. In this book, the focus is on the thermal and structural aspects only, but the last step in the list above, instrument performance analysis, can also require important data transformation and computational intensive analyses. For example, the optical performance analysis requires the deformation of the optical surfaces of the lenses and mirrors. Potentially also the temperature field inside a lens may be required to include the effect of change of refraction index due to change of temperature.

Also Tse et al. [91] worded well their lessons learned from the structural, thermal and optical performance (STOP) analysis for the SWOT mission that the bottleneck in productivity during the analysis flow is the data exchange between different mathematical models and analysis. Well functioning and robust software interfacing between the different tools are important. Tse et al. [91] also noted that data manipulation is often done by one-time-use codes (MATLAB, Python, etc.) or by hand.

For the optical performance impact evaluation, the SigFit tool [3] that integrates mechanical analysis with optical analysis is a very useful exception of the above-mentioned ad hoc one-time use tools for transferring data from one software package used by one discipline to the next software package used by the next discipline in the analysis chain. SigFit can transfer the different mechanical analysis results of finite element simulations into different optical analysis tools so that predictions of optical performance degradation due to mechanical disturbances can be determined.

An essential part of the work is the preparation of the correspondence between the thermal model and the structural finite element model. Although this can be very well automated, it might still require several checks by the engineer, especially when the thermal and structural mesh boundaries are not well aligned. In the CYLSAT example, the correspondence was set up through a PATRAN Command Language (PCL) program in MSC Patran. This was quite convenient, since the objective of the example was to show the effect of thermal mesh resolution. This required to build the correspondence for each of the six thermal models with the finite element model. The resolution of the latter stayed the same. Without automation, it would have been doubtful if the example could have been prepared before the text of this book was finished. The example of the box mounted on the sandwich plate in 32 positions and 4 different box configurations in Sect. 4.6 could only be run thanks to proper automation of the analysis chain.

A typical example where a well-implemented automation of the thermoelastic analysis process pays off is the scanning for the worst-case temperature fields out of the many temperature fields coming from the transient thermal analysis cases. For each time step for which thermal analysis results are available, a fully automated process of temperature mapping, distortion analysis and instrument performance impact calculation removes important constraints on the amount of cases that can be examined. Without automation, it is only feasible to evaluate a limited selection of temperature fields. This selection can only be based on (in)famous engineering judgement. This is not a trivial task, and the justification for the selection of temperature cases often suffers from shortcomings.

Each of the steps in the thermoelastic analysis process requires quite some simulations and produces a lot of data. Handling these large amounts of data is only possible if a proper automation process is in place.

A frequently heard reason for not doing this automation is that "too much data is also not good". In the opinion of the authors of this book, this is a statement of weakness. Real engineering judgement can show its value, when it is based on as much as possible good-quality data and offers the potential to increase knowledge of the behaviour of the structure and allows for better motivated engineering choices.

An other important motivation for automation of the analysis chain is the fact that any analysis cycle in a project needs to be repeated several times. Per cycle several iterations have to be performed. Knowing this, one should realise that investments in setting up at the start of a project an automated infrastructure for running analysis chains is easily returned within the same project. As soon as thermoelastic verification is an important part of the analysis cycle, the pay-off of automation effort is even greater. In addition, the uncertainties can be removed of not knowing the quantified effect of the cases that are not analysed, but "assumed" to have a minor effect.

Besides investment in the development of scripts or tools to run all the different steps nicely sequentially, there might also be a need to invest in computer hardware and licences. These investments have to be balanced with the objective of the project and importance of the required accuracy and reliability of the predictions. Which portion of the project costs are really taken by these automation investments? How many man hours are saved by removing the endless discussions on the need for running or not running cases? How much effort can be saved to convince the customer that the verification is complete by just running these extra cases thanks to the proper infrastructure?

4.8 Summary and Recommendations

In this section, the various topics discussed in this chapter are tried to be summarised into recommendations or points of attention for preparing thermoelastic models. These recommendations are mainly focussing on thermoelastic distortions causing performance degradations of instruments in payloads. The recommendations in this section are all aiming to obtain reliable predictions that are as accurate as possible. The possible effort one needs to put on these recommendations may be driven by the stability and accuracy requirements defined for the system to be simulated.

4.8.1 *Which Deformations Cause Degradation of Performance of Instruments?*

Basic knowledge of the instrument or combination of instruments in a payload is required to understand which deformations cause problems for the performance of the instruments or the payload. The optical or RF engineer can help the thermal and structural engineer to gain this understanding.

Common effects that reduce the performance are:

- Deformation of optical elements, such as lenses and mirrors, or RF elements reflectors, bending of RF panels
- Deformation of the supporting structure causing changes in the relative position and relative orientation of optical or RF elements within the instrument
- Change of relative orientation of different instruments within a payload

In many cases, these distortions can be tolerated up to a certain level. The smaller the value that can be accepted, the more accurate the predictions have to be.

When is known which effects may impact the performance of the instruments, the next step of the preparation of the thermoelastic modelling can start.

4.8.2 *Which Mechanisms Can Make the Degradation of Performance of Instruments Happen?*

For the deformation, that could reduce the performance of the instrument, it is important to identify mechanisms that are potentially responsible. Based on the knowledge of these mechanisms, the thermal and structural model, can be designed to capture these effects.

One could consider two sources of thermoelastic disturbances for an item:

- External sources, introduced by the thermoelastic deformation of the (external) interfaces of an item
- Internal sources, introduced by the thermoelastic deformation of the item itself.

These sources can be considered recursively from the top-level system, being the spacecraft platform, down to optical or RF elements in the instruments in the payload. For instance, the deformation of a mirror in an optical instrument may be affected by the temperature field in the mirror itself, but also by the thermoelastic deformation of the chassis of the instrument on which the mirror is mounted. The deformation of the chassis of the instrument on its turn is caused by its own temperature fields and the thermoelastic deformation of the payload. Finally, the deformation of the payload may have contributions from the deformations of the platform that are passed through the platform to payload interface.

When assessing possible mechanisms, it is therefore important not to look only at the instrument itself, but also at the supporting structure. So called iso-static mounts, for instance in the form of bi-pods and blades, have the objective to let the instrument behave as good as possible as a rigid body and filter out a significant part of the deformation of the supporting structure. This objective will be met in most cases to a large extent, but never completely. It is therefore relevant to identify as well which modes of deformation of the supporting structure may have the potential to deform the instrument.

Since the thermoelastic effects consist of a complex interaction of stiffness, CTE and temperature, it is not trivial to identify beforehand which components are mostly responsible for the thermoelastic responses that are causing the performance impacts on the instruments. Also one should not forget that different modes of deformation may occur at different moments in time due to changing temperature fields.

In general, structural elements with a high stiffness have the highest potential to introduce thermoelastic deformation to the items that are supported by these structural elements. Stiff elements are able to impose deformation on or constrain deformation of other structural elements.

In the above is referred to a recursive contribution of deformations starting even from the platform down to the smallest optical element in an instrument. This indicates the potential that deformation can cumulate along the chain of interfaces. For that reason often star trackers are mounted on the payload to provide a reference for the instruments in the payload and in this way shortcut large deformations coming from the platform. All deformations and movements are then to be related to the orientation of the star trackers, which on their turn are also subject to thermoelastic deformation.

By putting the reference for orientation of an instrument, such as the star trackers, as close as possible to the instrument can filter out to a large extent the deformations from sources that are not on the path with the reference (in this case the path between star tracker and instrument). Basically, the cumulation of deformation is cut short. However, depending on the design of the interfaces on the path to other disturbances, there may still be relevant levels passing through these interfaces. The importance of these disturbances is of course all relative to the stability requirements of the instruments.

4.8.3 What Is Needed to Simulate the Thermoelastic Mechanisms?

Behaviour that the model can not represent will not be observed in the results.

This statement is essential to any type of modelling for what ever discipline. It may sound obvious, but is quite often not taken into account when the validity of results are discussed. This must be kept as a golden rule for modelling decisions. This also implies that effects, of which the importance is not known for a specific application, shall at least be verified once. Care must be taken that heritage or experience are not causing biased modelling decisions.

In the following sub-sections, a number aspects are brought to the attention. This must not be considered as a conclusive list, but should rather be considered as a starting point of becoming aware of potential sources of deformation that may not always be relevant for simulations in other structural analysis domains.

This section may have some overlap with the Sect. 4.8.4, although the current section intends to try to address the aspects from a different point of view.

4.8.3.1 Stiff Elements

Elements in the structure with a high stiffness, compared to the surrounding structural components, have a high potential to impose their thermoelastic deformation on interfacing structural elements. Also the other way around applies: if the thermal control system manages to keep the temperature of these stiff items stable, this stiffness can support the objective to limit variation in deformation of an instrument.

Since these stiff elements can dominate the thermoelastic response of a structure, these items require a high amount of attention in terms of level of detail in which the temperature field is captured. High stiffness often goes hand in hand with high thermal conductance. This may cause that the temperature variation within these stiff elements to be mild, which could be a motivation to limit the thermal mesh resolution. It is however recommended to first confirm the effect in terms of thermoelastic responses of the thermal mesh resolution reduction before applying it to the full analysis campaign.

Capturing the temperature field well is one thing, but this only makes sense when this is accompanied with a corresponding quality of the representation of the stiffness of these important structural elements.

4.8.3.2 A-Symmetric or Eccentric Attachments Points

Longitudinal or in-plane deformation of structural elements may, when this deformation is relative to a constraint at an offset, cause bending and therefore rotation that on its turn can cause a translation normal to the axis or the plane of the structural element. This orthogonal translation can be at a point at a distance of the rotational centre. This is illustrated with the following "off-setted beam" example.

Sandwich panels may be modelled with two-dimensional shell elements. These two-dimensional elements often lack the possibility to specify loads or constraint to just one of the face sheets. In the case of a one-sided insert (i.e. with attachment to one of the face sheets only), it is not possible to simulate the load introduced via this insert at just one of the face sheets. The load will always act at the two face sheets simultaneously. Bending due to these kind of eccentric configurations effect will not be observed in the results. This aspect is explained in more detail in Sect. 6.4.5.

There is also a high likelihood that the model overestimates the stiffness of a sandwich panel, because two-dimensional elements cause both face sheets always to respond simultaneously to applied loads, irrespective of the fact the load is intended to be applied to one face sheet only. For this kind of problems, it is recommended to represent sandwich panels as three-dimensional structures, in which the core material is modelled with solid elements and the face sheets with shell elements.

The relevance of stiffness of joints is discussed in more detail in the next section.

Off-setted beam

Considered is a pinned–pinned beam with offset as presented in Fig. 4.25 that shows the beam with dimensions, loads and properties.

A is the area of the cross section and I the second moment of area of the beam. The offsets with length h are infinitely stiff.

The beam has a constant temperature elevation ΔT. Due to the offset, the thermal expansion of the beam will introduce a bending moment with a displacement $w(0)$.

The displacement $w(0)$ can be calculated using the classical Myosotis equations [29].

The displacement $w(0)$ at C is given by

$$w(0) = \frac{\alpha \Delta T A L^2 h}{2(Ah^2 + I)} \tag{4.11}$$

and the rotation ϕ is

$$\phi = -\frac{\alpha \Delta T A L h}{Ah^2 + I} \tag{4.12}$$

The maximum absolute values of $w(0)_{max}$ and ϕ will be obtained when $h = \sqrt{I/A}$(radius of gyration).

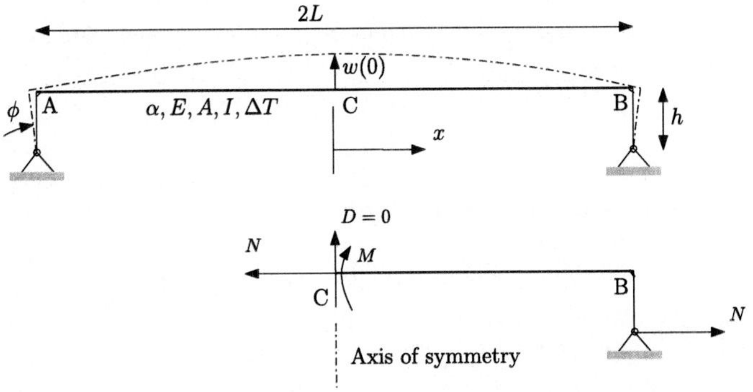

Fig. 4.25 Offsett beam

Table 4.3 Analysis results

Parameter	Dimension	Value
$w(0)$	mm	6.0000
ϕ	Rad	0.0120
M	Nm	−131.25
N	N	−5250

The normal force N is

$$N = -\frac{\alpha \Delta T A E I}{A h^2 + I} \tag{4.13}$$

The bending moment M can be calculated with

$$M = -\frac{\alpha \Delta T A E I h}{A h^2 + I} \tag{4.14}$$

Considered is now that the offset beam is a sandwich strip for which half the length $L = 1$ m, width $b = 25$ cm, and two identical face sheets with a thickness $t = 0.5$ mm. The face sheets are made of Al-alloy with a Young's modulus $E = 70$ GPa and CTE $\alpha = 24 \times 10^{-6}$ m/m°C. The height of the core $h_c = 50$ mm. No shear deformation of the core is taken into account. The temperature in both face sheets is the same $\Delta T = 25$ °C. The offset $h = 0.5\ h_c = 25$ mm represents a constrained face sheet at the edge of the sandwich strip. The analysis results are presented in Table 4.3.

Suppose the offset is varying from $h = 0$ mm to $h = 100$ mm. The corresponding variation of the displacement $w(0)$ is shown in Fig. 4.26.

The maximum value of $w(0)$ occurs at $h = \sqrt{I/A} = 0.5 h_c$

This example illustrates the significant deformation normal to the beam axis due to a support that is located at an offset from the neutral axis of the beam. A similar effect can be observed with a sandwich panel of which attachments are all made via only one of the face sheets.

4.8.3.3 Stiffness Representation of Joints

At various locations in this book, the importance of correct stiffness representation in the structural model is highlighted. For example, Sect. 6.7.4 explains how sensitive the thermoelastic responses for changes in stiffness of joints are, compared to the low sensitivity of for instance eigenfrequencies. Also the example "Effect of relative stiffness in thermoelastic problems" in Sect. 3.3 illustrates the importance of correct representation of stiffness on the predicted thermoelastic responses. Especially Sect.

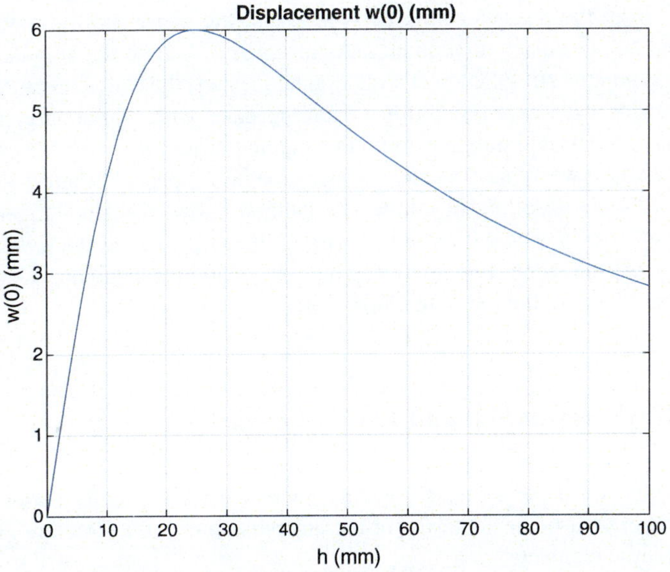

Fig. 4.26 Displacement of the centre point of the offset beam as a function of the offset value h

6.7.4 is recommended to read in this context, since it explains the speciality of thermoelastic compared to other structural loading.

Attachment or physical connections between components are implemented in a wide variation of methods, such as bolted joints, adhesive joints and use of inserts in sandwich panels. All these attachments cause local disturbances of the continuous structural components (for instance, a panel with attachments) and require local modelling to simulate a physical introduction of the loads from one component to the other.

For the modelling of these attachments, two main challenges can be identified.

The first one is to capture with sufficient accuracy the stiffness of the local reinforcements. Which level of detail is needed for representation of the machined parts that are bolted or glued to the panel? How is the embedding of the reinforcement in the panel represented? There are many ways to do this with different levels of modelling effort. It is in most cases not possible to obtain stiffness test data for each attachment configuration. It is therefore recommended to determine for different ways of modelling of the brackets and the reinforcements the impact of the thermoelastic responses. This will allow to find out how sensitive the response is for the change in stiffness that comes with different ways of modelling. This information can also be of use for determining the predicted range of variation of the performance.

The second challenge is to attach these local features to the modelling, i.e. the mesh, of the undisturbed panel. A single or a set of discrete connections cause singularity and may cause the stiffness of the joint to depend on the mesh density (see the example in Sect. 6.4.1). Typically, quite some local mesh modifications are needed to align the meshes of the two parts, e.g. the panel and the bracket, with often some compromises in geometry. As a remedy for both the singular discrete attachments and the struggle to get well aligned meshes, the contact feature of MSC Nastran can be used as a possibility of joining parts with non-congruent meshes through linear contact simulation. With this method of joining parts, the problem of changing stiffness with changing mesh density does not appear and therefore provides a reliable way of modelling of joints. The method allows to "glue" together parts of the model with physical dimensions without the risk to spoil the representation of stiffness. At the same time, the effective way of joining non-congruent meshes brings a significant reduction in modelling effort.

4.8.4 Mesh Resolution and Level of Detail

When is determined per structural element which modes of deformation may be occur, this can be used as the basis for the discretisation of the structure in thermal nodes and finite elements.

For the thermal model is suggested to keep the following in mind:

- The deformation ending up at the interface of the instrument or optical component is a cumulation of small deformations of often many structural elements. It is therefore important that each part of the structure is meshed with sufficient number of thermal nodes to allow for capturing even relatively small temperature variations. Note that this may go against the intuition of an experienced thermal control engineer (see also Sect. 5.7).
- It is rare that an item has a perfectly uniform temperature. Even a few degrees temperature variation over an item can provide a relevant contribution to the cumulation of deformation ending up at the instruments. So make sure that this temperature variation is captured with sufficient thermal nodes. Especially near interfaces, where higher-temperature gradients often are present, a higher resolution of the thermal mesh may be needed.
- When a temperature gradient can cause deformation (in some cases bending) of a structural element, make sure that this can be captured. Examples are:
  - Temperature gradients occurring in the plane of a panel, require sufficient resolution of thermal nodes especially along the directions of the gradients.
  - Gradients through the thickness of a sandwich (see example in Sect. 4.2) or any three-dimensional shaped component can only be represented by more than one layer of thermal nodes through the thickness.
  - Features like electronics boxes (see Sect. 4.6) with high stiffness require in general several thermal nodes along each dimension of the box.
- Make sure that the level of detail in the thermal model in terms of geometrical modelling and corresponding mesh resolution is compatible with the structural finite element model.

- Make sure that the model is set up parametrically, especially to parametrise the mesh resolution. This allows to run a mesh convergence exercise to determine what has to be considered as "sufficient" number of thermal nodes.

In a similar way, a non-exhaustive list with recommendations for the structural model is:

- When is identified that temperature gradients through the thickness of panels may cause out-of-plane deformations, then make sure that the model is able to capture the temperature variation through the thickness. For sandwich panels, it is in most cases recommended to represent these with solid elements for the core material and shell elements for the face sheets (see for instance the model used for the PANELSAT example in Chap. 10).
- Sandwich components often have their attachment at one side of the panel. This introduces interface forces that have an offset relative to the neutral plane causing bending moments and rotational deformations in the panel and interfacing components. Make sure that this effect can be well captured by the modelling of the panel and the geometrical implementation of the interfaces. Using a 3-D representation (see previous point) may be convenient here.
- Make sure that the stiffness of joints between parts is well simulated. Increasing the level of detail in modelling will improve the approximation of the real stiffness values (See also Sect. 6.7.4). Stiffness representation is critical for thermoelastic responses.
- For compatibility with the thermal model, align the level of detail of the structural model with the thermal model.
- Assure that features, especially such as large electronic boxes or large brackets and interface, are well represented in terms of stiffness to simulate the high stiffness ratio with the interfacing components and in terms of mesh resolution to capture temperature gradients.

For both the thermal model and the structural model, it is recommended to verify by mesh convergence checking whether the mesh resolution is high enough. This exercise may also indicate that in some parts of the model the mesh resolution may be reduced without impact on the results.

Also for the way details are represented in the model, such as electronic boxes and joints, it is recommended to verify the impact of simplifying and enhancing the modelling.

A final recommendation on the modelling and level of detail is related to the fact that spacecraft projects run for many years. During the course of the project, many subtle changes are implemented to the environment and the design of the structure. For budget reasons, not all changes are implemented in the models. It is true that a modification of the model requires to verify the impact of this change. As may become clear from various examples and discussion in this book, it is often not obvious to justify without analysis that a change has no effect. A decision for not addressing a change in environment or design through an update of the model may therefore often be not based on technical grounds.

4.8.5 Temperature Mapping

To limit the effort to generate the correspondence between the thermal model and the structural finite element model, it is for many temperature mapping methods useful that the meshes of both models are as good as possible aligned. The best way to achieve this is to develop both models in parallel with good interaction between the engineers responsible for these two models.

Especially the quality of conduction-based temperature mapping methods benefit from good alignment of the two models.

With increase of thermal mesh resolution, the differences between the results of the different temperature mapping methods reduce. The prescribed average temperature (PAT) method is however more tolerant for a lower resolution of the thermal mesh. As a consequence, thermal mesh convergence is obtained in most cases earlier with the PAT method.

A benefit of using the PAT method is that the same correspondence data or overlap data can be used for generation for linear conductors (see Sect. 9.3). This is then an additional benefit of aligning thermal and structural models.

4.8.6 Selection of Worst Cases

"Thermal cases" are mostly considered as specific orbits with corresponding orbital environments. These orbits are then combined with possible operational scenarios in which heat dissipations of the electronics of the instruments interact with the variations in the orbital thermal environment. An other variation comes from the distance of the earth to the sun that changes with the time in the year with corresponding variation in level of solar radiation.

In the verification of the performance of an instrument or the strength of the supporting structure, it is important to keep in mind that:

- A criticality may occur in different transient cases, at different moments in time and at different locations in the instrument or spacecraft. So there will be in general not just one or two cases that can be declared critical, but a range of cases.
- It is in general not possible to determine the criticality from the bare thermal analysis results. It is always required to run the end-to-end thermoelastic analysis to identify the criticality of a case or to discard a case from further evaluation.

A frequent occurring question in projects is about the number of thermal cases that has to be considered. This kind of questions can only be answered with confidence after a decent screening of all possible cases in which the full thermoelastic analysis chain is run. In the early phase of a project, when the driving thermal cases have to be identified, the resolution or the number of points around the orbit can be kept low and also the variation of the inclination of the orbit can also be done with a lower resolution.

An important task at the start of a project is to define the criteria for identifying the worst-case conditions and implement these in an automated process. During the course of the project, these criteria may change or become more sophisticated. This is an important first step in the project and forms a solid basis for the thermoelastic assessments throughout the whole project. When this functionality is implemented, significant limitations for the scope of the screening are removed and the efforts are mainly reduced to the preparation of the different cases.

4.8.7 Uncertainties

Especially in cases where high thermoelastic stability is required, it is essential to obtain confidence in the obtained results. Sensitivity analysis is then recommended. The $2k + 1$ PEM method (see Sect. 10.6) is a rather light stochastic method that could be applied to determine the ranges of variation in the responses due to variation of the design variables and enhance the confidence in the produced results. One of the important parameters that are mentioned several times are the stiffness values of the joints, which are without mechanical test data, important uncertainties. Also different modelling approaches may introduce variation of the responses and need to be taken into account for determining the uncertainty ranges of the responses.

4.8.8 Concluding Recommendations

Invest time in automating the analysis chain from the beginning to the end. Each project may have some specificities that would require some ad hoc automation. This investment will pay off during the execution of the project.

An other important investment at the beginning of the analysis campaign or even at the start of the project is to parametrise as much property values or geometric dimensions as possible. This will ease the organisation of mesh convergence check runs and uncertainty analysis runs. A well-implemented parametrisation will make the difference between a feasible or an impossible mesh convergence check or uncertainty analysis exercise.

Problems

4.1 Verify all equations provided in the example "Off-setted beam", pp. 79.

Chapter 5
Thermal Modelling for Thermoelastic Analysis

Abstract A spacecraft is going through different phases of a mission. Each of these phases has its own characteristics in terms of thermal environment. The objectives of these different phases drive the difference in focus of thermoelastic analyses from one phase to the other. The lumped parameter thermal analysis method typically used for spacecraft thermal analysis is summarised. Transient thermal analysis is a common analysis approach for simulating time-varying thermal environments along the orbit around a planet. Special attention is paid to the difference between thermal analysis for thermal control and thermal analysis for thermoelastic.

5.1 Introduction

The main objective for thermal analyses in the thermoelastic analysis chain is to provide high-quality temperature fields needed for the prediction of reliable thermoelastic mechanical responses. These mechanical responses are then the basis of the verification of the performance of instruments on spacecraft and the strength of the spacecraft and its components.

This chapter aims to summarise the main aspects of space thermal analysis with emphasis on its role in the thermoelastic analysis process.

This chapter starts with an overview of the different thermal environments to which a spacecraft is subjected during the different phases of its life and what the focus is of the thermoelastic analysis for these phases. Supported by a summary of heat transfer mechanisms and space thermal environments, the lumped parameter method is introduced. This method is the most used analysis method for space thermal analysis.

The harsh space environment may lead to extremely high or low temperatures on the spacecraft. It is therefore required to verify whether components and material on the spacecraft can survive these circumstances. The discipline of thermal control is essential here to make sure that nowhere in the spacecraft allowable temperature limits are exceeded. The thermal control system is also responsible for limiting the temperature variation with time to provide a stable environment for the instruments.

S. Appel and J. Wijker, *Simulation of Thermoelastic Behaviour of Spacecraft Structures*, Springer Aerospace Technology, https://doi.org/10.1007/978-3-030-78999-2_5

For the verification of the thermal control systems obviously dedicated thermal analyses are required. In general, the objective of this type of thermal analysis is different from the objective of thermal analyses for thermoelastic. In principle, it can be stated that thermal analysis for thermal control verifies how well the thermal control system is able to limit the temperature variation in the spacecraft. Thermal analysis for thermoelastic then provides input to the process in which the spacecraft deformations and stresses due these temperature variations are determined to allow for the verification of the impact on the performance of instruments.

The importance of the difference in objective for these two types of thermal analyses is not always obvious. For that reason, the last section of this chapter is devoted to this topic in which the differences for several aspects of the thermal analysis process are discussed in detail.

5.2 Space Thermal Environment

In every phase of its life, a spacecraft is subjected to different thermal environments. In phases which the instruments are not operating, non-operating phases, the thermoelastic analysis is mainly used to determine the thermal stresses with the aim of verifying the strength of the structural components and to ensure that no damage can occur. For the phases in which the payloads are running, operational phases, it is also important to verify whether thermal deformations are affecting the alignment of optical elements or antennas. The focus of thermoelastic in the operational phases is therefore also on the prediction of the distortion of the structure and corresponding impact on the performance of the instruments.

Three main phases in the life of a spacecraft can be identified as follows:

- On ground phase
- Launch and ascent environment
- Orbital phase.

In the orbital phase, there are long periods during which the instruments are operating, operational phases.

5.2.1 *On Ground Phase*

On ground, the spacecraft is assembled under strictly controlled environment (temperature and humidity) in the clean rooms. This ensures that temperatures are not changing during the time of the assembly. Since the dimensions of the components may change under fluctuation of temperature, it is important that the assembly of the spacecraft takes place under controlled temperature and humidity. If temperatures are not well controlled, then close tolerance components could fit and be aligned nicely in the morning, but would be out of tolerance in the afternoon. For most CFRP components, a similar problem occurs due to the uptake of humidity. Humidity changes may cause as well deformation of these structural components. Also the transport of spacecraft and instruments is done with custom made transport containers, in which climate is controlled and effects of shocks minimised.

In some cases, the relative shape differences between assembly at room temperature and the operational temperature, often a low temperature, are taken into account during the manufacturing of the parts. Especially for optical elements, like lenses, this can be important.

Important to note is that the temperature on ground, at which the structure is assembled and parts are manufactured, becomes the reference temperature for subsequent in orbit analyses (see Sect. 3.2).

5.2.2 *Launch and Ascent Phase*

During this phase, three stages can be identified:

- At the launch pad
- During launch under the fairing
- During launch without fairing.

When the spacecraft is mounted on the launcher and waiting for the launch on the launch pad, some of the equipment may already be switched on. Depending on the season and location of the launch site, the external environment on ground may be harsh. The launcher authority can control the thermal environment under the fairing to satisfy temperature requirements prescribed for selected equipment of the spacecraft.

During the first portion of the flight, when the launcher crosses the denser layers of the atmosphere, the fairing protects the spacecraft from the aerodynamic heating that occurs at high Mach numbers. The heating of the fairing produces infrared (IR) fluxes on the spacecraft of the order of 1000 W/m^2, but its duration is very short, only a few minutes.

Once the fairing is jettisoned, the external surfaces of the spacecraft are heated by the environmental heat fluxes incident directly on them. These fluxes might include the heat generated by the impingement of the free molecular flow of the rarefied atmosphere, the sun and earth albedo illumination (radiant heat fluxes mainly in the visible spectrum) and the heat radiated by the earth (radiant heat flux in the IR spectrum). In some cases, it is needed to include the heat flow through the interface of the spacecraft and launcher in the thermal assessment. Depending on the launcher, also heat radiation of the plume of the engines needs to be taken into account.

Specifications concerning the thermal environment during launch and ascent can be found in the user's manual of the selected launch vehicle (e.g. Ariane 6, VEGA-C, Falcon, etc.).

After separation from the launcher, the "launch early orbit phase" (LEOP) and commissioning of the spacecraft take place according to the established timelines. It may also occur that for sometime the whole spacecraft is turned off. The spacecraft is then put into survival mode. This means that the thermal control system has to ensure, for instance with the help of heaters, that sensitive equipment is not getting too cold such that damage is prevented. For this phase, the thermoelastic focus is on the strength of the spacecraft and payloads.

5.2.3 Orbital Phase

As soon as the spacecraft is placed in its right orbit, or its designed escape trajectory if the spacecraft is an interplanetary probe, and all equipment and payloads are brought to their operational temperature ranges and are started, the operational phase is entered. Once the, often low, operational temperatures are reached, temperature variations have to be kept as small as possible in order to limit the variation of deformation of the structure that is supporting pointing sensitive items. The operational phase is therefore, in general, the focus of the thermoelastic analysis on the prediction of the deformations.

The environment experienced in orbit is the result of the impinging orbital environmental fluxes (solar, albedo, IR), the heat dissipated by internal equipment and the heat that the spacecraft rejects by radiation to the cold space background, as summarised in Fig. 5.1.

The orbital thermal environment is dominated by the solar radiation (mostly in the visible light frequency bandwidth) and the deep space radiation sink temperature. Solar fluxes are reflected by the planet, which is for most spacecraft our planet Earth. This reflected solar radiation is called albedo radiation, which is therefore also mostly in the visible light frequency range. The warm planet is radiating infrared radiation

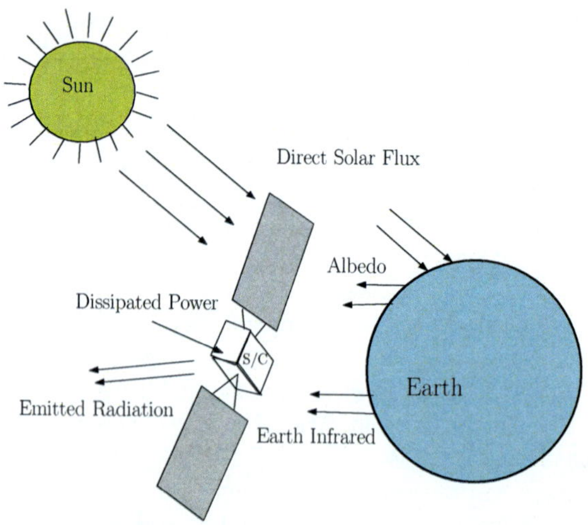

Fig. 5.1 Space thermal environment

that is called planet or Earth radiation. The following sections will touch upon each of these radiation sources.

In NASA's preferred practices guideline "Earth Orbit Environmental Heating" [1] currently accepted values for solar flux, Earth reflected solar flux (albedo) and Earth radiation are provided.

5.2.4 Direct Solar Flux

For spacecraft orbiting around the Earth, the nominal direct solar flux is equal to the solar constant value of about 1366 W/m^2. The solar constant is defined as irradiance of the sun at one astronomical unit (AU) from the sun, which is roughly the distance from the sun to the earth.

The variation of the Earth–Sun distance causes a $\pm3.5\%$ seasonal variation from nominal [8].

The intensity of the solar radiation is inversely proportional to the square of the distance to the sun.

5.2.5 Planet Reflected Solar Flux (Albedo)

The albedo is the solar flux, which is reflected by the surface of the planet. The annual Earth's average albedo factor, being the portion of the solar flux that is reflected by the planet, over the global surface is 0.3 (nominal [1]) of the solar constant. Depending on the position above the Earth, the albedo radiation may be different. For instance, the ice on the north and south pole reflects the solar radiation stronger than for instance the rain forest of South America. Depending on the required accuracy of the thermal analysis, these variations in albedo have to be taken into account.

5.2.6 Planet Flux, Infrared Radiation

The infrared radiation is the thermal radiation that the planet emits.

Earth's radiation levels at the planet surface vary approximately between 150 and 300 W/m^2 depending on the latitude, season and weather condition, etc., at the point directly below the spacecraft around the earth. The annual average of the infrared radiation over the global surface is 237 W/m^2. Since there is a large variation of temperature over the earth surface, it may also be important to take the corresponding variation in planet infrared radiation into account in the thermal analysis.

5.2.7 *Internal Dissipation*

Although this internal heat source does not fit well in the list of above external heat sources for the spacecraft, it is also essential to be taken into account. This heat source is driven by the equipment specification and by the operational timelines of the instruments. The instruments, equipment and electronics inside and outside the spacecraft dissipate heat. Their energy source is in most cases electrical power provided by the solar arrays. This internal heat dissipation is exchanged with the structure and components through radiation and conduction. The thermal control design often has radiators included to reject excess heat to deep space. These radiators form heat sinks and are connected by conduction, in many cases dedicated conductive elements, and heat pipes to the most dissipating items in the spacecraft. Quite often extra electrical heaters are introduced, which are controlled to prevent temperatures of equipment to drop below their qualified temperature levels.

5.3 Heat Transfer Mechanisms

This section aims to summarise in a compact way the main heat transfer mechanisms. It is by far not complete and tries to provide some context. The various references are recommended for further reading on the topic.

A simple and compact definition of heat transfer is: "Heat transfer (or heat) is thermal energy in transit due to a spatial temperature difference" [46].

Heat is a thermal energy and is measured in Joule. The exchanged heat rate or the dissipated energy of equipment is measured in J/s or W, which is also called power. For brevity, in many instances the term "heat" is used where it is actually meant "heat rate" being Joule per second. Also the term "heat loads" has the same meaning as heat rate. Finally, the term "heat flux" is also referred to, but this has the meaning "heat (J) per second per unit surface", thus power per unit surface.

There are different modes of heat transfer: conduction, convection and thermal radiation depending on the state of the spacecraft system [56].

5.3.1 *Conduction*

Conduction is the mode of the heat transfer when temperature gradients exist in a stationary solid or fluid medium. Energy is transferred from warmer to colder parts of the solids by interaction between them. Actually, the heat is transferred by the vibrations of adjacent atoms in the solid. Conduction in fluids takes place when fluids are at rest, without convection.

The heat flow vector (q) (W/m^2) by conduction per unit area is expressed by Fourier's law and is proportional to the temperature gradient

$$(q) = -k\nabla T, \tag{5.1}$$

where ∇T (K/m, °C/m) is the temperature gradient, i.e. the derivative of temperature with respect to the spatial coordinates. The parameter k (W/mK, W/m °C) is the isotropic thermal conductivity and is a characteristic property of the material [23, 45]. The material conductivity may be temperature dependent. Note the minus sign in Eq. (5.1) to express that heat is flowing from a location with a high temperature to location with a low temperature.

In case of a one-dimensional rod, the heat flow $\dot{Q}$ (W) by conduction through the cross section of the rod from end 1 to end 2, which becomes a scalar, is given by

$$\dot{Q} = kA\frac{T_1 - T_2}{L} = \frac{kA}{L}(T_1 - T_2) \tag{5.2}$$

where the conductivity k and the area A (m^2) are constant over the length L (m) of the rod. The temperatures T_1 and T_2 are the temperatures at the two ends of the rod. An important condition for this one-dimensional assumption is that no heat is flowing through the outer surface of the rod.

With the use of the symbol GL, as used by the thermal analysis tool ESATAN-TMS Eq. (5.2) is written as

$$\dot{Q} = GL(T_1 - T_2) \quad with \quad GL = \frac{kA}{L} \tag{5.3}$$

The symbol GL is the value the user has to provide as conductor value (W/°C) (thermal conductance coefficient) in ESATAN-TMS.

5.3.2 Contact Conductance

Contact conductance is the amount of heat per unit temperature difference that can flow across an interface between two surfaces in contact. The contact conductance is dependent on how intimate two surfaces are in contact with each other and has therefore a relation with the surface roughness of the two surfaces in contact. For larger surfaces, also the planarity of the two surfaces helps to increase the contact conductance. The contact conductance can also be raised by increasing the contact pressure. This has the limitation that the effect is mostly noticeable in the vicinity of the fasteners responsible for the contact pressure. Another method to improve the contact conductance is through the application of thermal fillers. Thermal fillers are usually soft materials and may also have the form of past. It can fill the roughness of the surfaces in contact and compensate to some extent lack of planarity.

The heat transfer $\dot{Q}$ (W) over a thermal contact can be described with

$$\dot{Q} = h_c A(T_1 - T_2) = \frac{(T_1 - T_2)}{R_c} \tag{5.4}$$

where A is the area of contact, h_c is thermal contact heat transfer coefficient, and R_c is the thermal contact resistance [23]. T_1 and T_2 are the temperatures of the two surfaces in contact.

It is important to realise that contact conductance can have a rather high level of uncertainty, which needs to be kept in mind. For instance for a box, the value of h_c is an average over the contact surface of the base plate. As the contact conductance increases with pressure, it will be high in the area around bolts often located at the edge of base plate and reduces away from the fasteners. For units with bolts along its perimeter, the minimum contact conductance is in the central area of the box.

It is therefore recommended to gain knowledge of the impact of the predicted thermoelastic responses due to the variation of the contact conductance values.

5.3.3 *Convection*

Convection is a mode of heat transfer that occurs between a surface and a moving fluid when they are at different temperatures. Heat is transferred from one place to another by movement of fluids. Whatever the nature of the flow is, the process is described by Newton's law of heat transfer. The heat flow q per unit area (W/m^2) through the surface of the structure can be expressed as

$$q = h\Delta T = hA(T_f - T_s) \tag{5.5}$$

where ΔT (K, °C) is the difference between the surface temperature T_s and fluid temperature T_f. The parameter h (W/mK, W/m°C) is the convective heat transfer coefficient. The value of the heat transfer coefficient is mostly based on empirical correlation incorporating surface geometry, temperature of the wall, temperature of the fluid and flow speed. The dependence of value h on these parameters has been measured and tabulated for typical fluids and regimes [23, 45].

This heat transfer mechanism may only be relevant on the launch pad or during thermal tests in ambient conditions.

5.3.4 *Thermal Radiation Heat Transfer*

Thermal radiation is a mode of heat transfer between two surfaces at different temperatures by means of electromagnetic waves that do not need medium to propagate. Even better, these waves are most efficiently propagated in vacuum. The theoretical upper limit of thermal radiation power per unit area q_b that a surface can emit is formulated by the Stefan–Boltzmann law.

$$q_b = \sigma T_s^4, \tag{5.6}$$

where T_s (K) is the absolute temperature of the surface and $\sigma = 5.67 \times 10^{-8}$ (W/m^2 /K^4) is the Stefan–Boltzmann constant. Such a surface is referred to as an ideal emitter or blackbody [46].

The emitted power by a real surface or grey surface per unit area, q_e, is a portion of the power emitted by a blackbody at the same temperature

$$q_e = \varepsilon \sigma T_s^4, \tag{5.7}$$

where ε (−) is a radiative property of the surface referred to as emissivity with $0 \leq \varepsilon \leq 1$.

A surface may also be subjected to radiation from its surroundings. This incident radiation power per unit surface is referred to as irradiation G. A blackbody surface absorbs the complete irradiation power. However, like with the emission, a grey body only absorbs a portion which makes the absorbed irradiation G_{abs}

$$G_{abs} = \alpha G, \tag{5.8}$$

with α being also a radiative property of the surface, which is referred to as absorptivity with $0 \leq \alpha \leq 1$.

It must be noted that both α and ε are dependent on the wavelength of the radiation waves. For space thermal analysis, typically two spectral bands are considered:

- Radiation in the infrared band
- Radiation in the solar band.

For both spectral bands, the emissivity and absorptivity are in principle different surface properties. Kirchhoff's law however states that for a specific wavelength [46]

$$\varepsilon_\lambda = \alpha_\lambda, \tag{5.9}$$

with λ indicating the wavelength. From this, it follows that when these two values are integrated over the same band and divided by the width of the band, the average over a certain band is obtained. This then implies that the average of ε and α is the same for that band.

In space thermal analysis, the absorptivity (=emissivity) in the infrared band is often indicated with the symbol ε, while for the solar band the absorptivity (=emissivity) with the symbol α. However, α is still referred to as absorptivity, but it is meant to refer to the solar band only.

The above has been implicitly considered that the reflected radiation is evenly distributed over all directions. This type of surface is referred to as a diffuse surface. Specular surfaces reflect the portion of the irradiation energy, that is not absorbed, under an angle with respect to the normal equal to the incidence angle: angle of incidence equals angle of reflection.

Finally, transparent surfaces (typically used for lenses) allow part of the irradiation to be transmitted through the surface.

By comparing all modes of heat transfer, two essential differences can be noticed. Both conduction and convection require the presence of a medium to transfer energy. However, radiation does not require a medium for the transfer of energy. Conduction and convection heat transfer are roughly linearly proportional to the temperature difference in case there is no strong dependence to the temperature. Radiative heat transfer is proportional to temperature to the fourth power. Thus, radiative heat transfer becomes more important at higher-temperature differences. An other difference is the effect of distance: conduction is very effective over short distances, while radiation is still effective over a large distance.

5.4 Spacecraft Thermal Modelling with the Lumped Parameter Method

Traditionally, the lumped parameter method is used in spacecraft thermal analysis. The tools in Europe are mainly ESATAN-TMS and SYSTEMA THERMICA. These tools are based on the electrical network principles that allows to simulate the individual items to be represented by single thermal nodes to which all the thermal properties are lumped.

In NASA Preferred Reliability Practices "Guidelines for Thermal Analysis of Spacecraft Hardware" [74], the general-purpose heat transfer computer program "Systems Improved Numerical Differencing Analyser" (SINDA/FLUINT) is mentioned. This tool is based on the same principles, but is more often used in the USA.

The thermal model is normally divided into two models with a different function in the thermal analysis:

- The geometric mathematical model (GMM)
- The thermal mathematical model (TMM)

The geometric mathematical model (GMM) represents the geometry of the spacecraft or the item that is subject of the thermal analysis and is mainly used to generate the radiative analysis data for solving the thermal problem. With this model, the radiative couplings between the thermal nodes (see Sect. 5.4.2 below) or also called radiative conductors are calculated. The GMM is also used for the computation of the radiative heat fluxes on the spacecraft in orbit. These orbital fluxes change from one position in the orbit to the other due to the change of orientation of the spacecraft relative to Sun beam and normal to the planet surface. In the last years, the GMM is also used for computing linear conductors.

The thermal mathematical model (TMM) is the thermal network model consisting of nodes connected by conductors representing conductance, radiation and convection couplings. Especially, the radiative conductors are computed with the GMM. The linear conductors are often computed by the user and more and more with the help of the GMM modeller. Chapter 9 provides the methods for systematic generation of the conductors for a complete conductive network with the use of the finite element method.

The TMM is used to solve the thermal system and to produce the temperature fields and the heat flows in the structure.

5.4.1 Thermal Network Modelling with the Lumped Parameter Method

A thermal network model or thermal lumped parameter model of a system represents a network of interconnected nodes analogous to an electrical circuit. Electrical quantities are replaced by their thermal counterparts which are compared in Table 5.1.

This electrical analogy is often discussed in heat transfer textbooks, mainly relating to the development of thermal resistance networks. These networks are normally constructed to determine steady-state temperatures and heat fluxes at points of interest in a system. Some RC element configurations are illustrated in Fig. 5.2.

The conductor value G is the reciprocity of the resistor value R.

Table 5.1 Thermal–electrical systems: analogy [38, 66]

Quantity	Thermal system	Units	Electrical system	Units
Potential	T	°C, K	E	V
Flow	$\dot{Q}$	W	I	A
Resistance	R	°C/W, K/W	R	Ω
Conductance	G	W/°C, W/K	$\frac{1}{R}$	$\frac{1}{\Omega}$
Capacitance	C	J/°C, J/K	C	F
Ohm's law	$Q = T/R = GT$	W	$\frac{E}{R}$	A

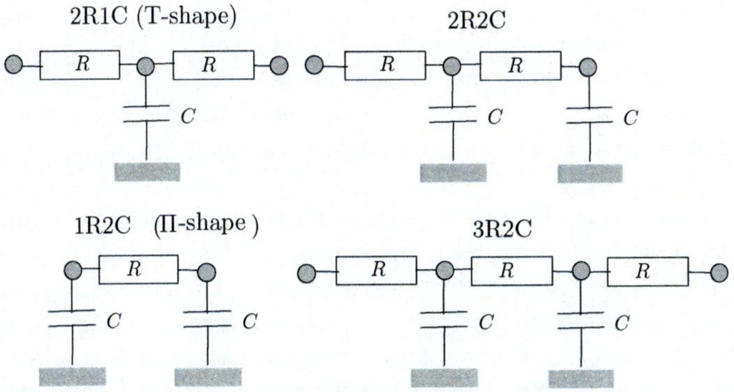

Fig. 5.2 RC elements configurations used in building thermal models, [38]

5.4.2 Thermal Node in a Thermal Lumped Parameter Model

In the previous section, the analogy with electrical circuit modelling is explained. Both network types are based on the interconnection of nodes. It is important to understand what a thermal node in thermal lumped parameter model represents. This section aims at clarifying this.

In an electrical network, the nodes are typically a representation of the electrical components. Depending on the level of detail, a node in the thermal lumped parameter model can be a complete electrical unit that is exchanging heat with its surrounding systems, or a small part of a plate to capture, with neighbouring thermal nodes, the thermal gradients in such a panel.

In the philosophy of the thermal lumped parameter method, the nodes represent the effective properties of an amount of material. Thermal properties, such as heat capacity, of a part of a sandwich plate containing the face sheets and core material are all lumped into a single node representing this part. Of course, when the level of detail in the thermal model is decided to be increased, it may be chosen to have one thermal node for the top face sheet, one for the core material and one for the bottom face sheet. This would then imply three nodes representing one part of the sandwich. Especially in the case of thermal analysis for thermoelastic, it may be needed to increase further the number of nodes either in the through thickness direction as well in the in-plane direction. Since thermal analysis for thermal control has a different objective than thermal analysis for thermoelastic (see Sect. 5.7), it could be sufficient for thermal control analysis to represent an electronics box with a limited number of thermal nodes, or even down to a single thermal node, lumping all the thermal properties of the different items of this box into this single node.

In the thermal network, the temperatures of the thermal nodes are the unknowns that need to be solved with a numerical procedure. As the thermal properties of a thermal node are the effective properties of all material represented by this thermal node, the thermal node temperature represents the average temperature of material that is represented by the thermal node. It is good to realise that this is different from a finite element node that represents the solution at a discrete location (see also Sect. 7.2).

For the radiative part of thermal analysis, a surface area may be associated with a thermal node with corresponding thermo-optical properties. The calculation of the radiative couplings is assumed that the temperature at any location of the surface is equal to the thermal node temperature. This implies that the whole thermal node surface is considered to have a uniform temperature equal to the average temperature of the thermal node.

From the above, it must be deducted that the thermal solution provides the average temperatures in the thermal node volume or material it represents. It does not provide information on the temperature distribution internal to the thermal node. When for thermoelastic analysis, the thermal node temperatures have to be converted to temperatures at the finite element nodes, and various assumptions can be made about this internal temperature field. These assumptions form the basis of thermal model temperature mapping approaches and are the topic of Chaps. 7 and 8.

5.4.3 Geometric Mathematical Model

Originally, the only objective of the geometric mathematical model (GMM) was to determine the radiative couplings between the thermal nodes and radiative heat loads (solar, albedo and planet radiation) at different locations around the orbit. It has to be noted that also conductive couplings are more and more often computed with the geometrical model.

The GMM represents the geometrical aspects (the shape) of the spacecraft. It also includes the orbit information and the attitude of the spacecraft relative to the orbit. The primitive surface shapes are meshed, and surface sub-segments are associated with thermal nodes. The nodal surfaces have thermo-optical properties assigned as input for the computation of the radiative couplings and heat load computation.

Most modellers use collections of primitive shapes (cylinders, cone, rectangles, ...) to represent the spacecraft geometry and its components relevant for mainly radiative modelling. The reason that most of the tools for space radiative modelling have chosen to base the geometry on a collection of primitive shapes is to optimise the calculation time and also to improve the accuracy of analysis with the often reduced number of thermal nodes. In most cases, the main shapes of the spacecraft and its components are segments of primitive shapes, which are more accurately approximated with primitives than through, for instance, a tessellation with triangles. However, with increasing meshing density, this criterion loses its relevance.

The computational results of the GMM are the radiative couplings and environmental heat loads. These results are input into the TMM, which will be applied to compute temperatures and heat flows.

5.4.4 Thermal Mathematical Model

The thermal mathematical model (TMM) refers to the mathematical representation of the thermal characteristics of the spacecraft, instruments or other components, which are used for temperature prediction within the thermal environment conditions [96].

1. The spacecraft system is discretised in a network of
    - *Nodes*: Isothermal volumes where heat can be stored; they are characterised by their thermal capacitance and optionally by a heat source.
    - *Links*: A link is a path between two nodes i and j that allows heat to flow from one to the other. It can be conductive (and it is characterised by its thermal conductance) or radiative (characterised by its radiative exchange factor). Convection is in most cases not considered.
2. The heat conduction (Fourier's law) and radiation (Stefan–Boltzmann's law) equations are applied to this thermal network. This yields a nonlinear system of differential equations, one for each node, of the form

$$\begin{aligned} C_i \frac{dT_i}{dt} =& Q_i + \sum_{i \neq j} GL_{ij}(T_i - T_j) \\ &+ \sum_{i \neq j} GR_{ij}\sigma(T_i^4 - T_j^4) - GR_i\sigma(T_i^4 - T_o^4), \\ &i, j = 1, \cdots, n, \end{aligned} \tag{5.10}$$

where n is the number of nodes in the lumped parameter network, T_i, C_i and Q_i the temperatures (°C or K), capacitance (J/°C) and the heat source (W) of node i, respectively, GL_{ij} is the conductors (W/°C) of the conductive link between nodes i and j, and GR_{ij} the radiative heat exchange factor (m^2) between nodes i and j. Besides the internal dissipative power sources Q_{di}, the heat source Q_i shall include the radiative solar flux Q_{si}, the albedo flux Q_{ai} and Earth infrared flux Q_{ei} absorbed by the external surfaces. This is illustrated in Fig. 5.1. The ith node coefficient of radiation to the environment is given by $GR_i = A_i\varepsilon_i\sigma$, where A_i denotes the outward facing area, ε_i its (IR) emissivity and σ the Stefan–Boltzmann constant. The temperature of the space environment is $T_o = 2.7$ K, the cosmic microwave background radiation. The values C_i, GL_{ij}, GR_{ij} and GR_i depend on the physical properties and the geometry of the system and can be variable with the time. In the space thermal domain, only conductive and radiative coupling matrices are used.

3. The system of Eq. (5.10) is integrated in order to provide values for the temperature vector $(T(t)) = (T_1(t), T_2(t), \ldots, T_n(t))^T$ at different times. Several methods exist for that purpose, usually based on the finite difference schemes, e.g. the explicit forward time integration method of Crank–Nicolson [95]. In case of steady state, the left-hand side of Eq. (5.10) is zero, and stationary values of thermal node temperatures (T_i), $i = 1, 2 \ldots, n$ have to be solved.

5.5 Thermal Transient Analysis

5.5.1 *Transient Phenomena in Space Thermal Analysis*

The thermal lumped parameter network model can also be used for transient thermal analyses. Transient analysis allows to simulate changes in time of the responses due to changes in the environment. A typical example of changing thermal environments with time is a spacecraft orbiting around the Earth. For instance, an observation satellite keeps its optical instruments oriented towards the Earth surface during each position in the orbit. As a consequence, the orientation of the spacecraft with respect to the Sun has to change, which leads to changes in heating of the external surfaces. Also the albedo radiation changes with the position of the spacecraft relative to the so-called sub-solar point of which the position is also changing over the day.

This kind of changes is relatively gradual, however more abrupt changes occur when the spacecraft is entering or leaving eclipse, the shadow of the Earth. The direct solar flux drops to zero in a matter of seconds when the spacecraft enters eclipse and direct solar heating ramps equally fast when the spacecraft leaves the shadow of the planet.

Due to the changing environment, obviously the temperature field changes with time along with the position in the orbit (see as an example Fig. 4.11 in Sect. 4.5). As a consequence, also the thermoelastic deformation is changing with time. Especially for instruments that need to be kept pointed on specific points in space or on a planet surface, these variations in deformation need to be kept within limits to maintain the quality of the observations and therefore maintain the needed performance.

The thermal control system aims at limiting the variation of the temperature field over time. For this purpose, passive thermal control elements are used, such as insulation materials. Also controlled heaters are used as part of the active thermal control system.

The thermoelastic analyses and subsequent performance impact assessment can then predict whether the thermal design succeeded in limiting sufficiently the temperature field variation. These results are then fed back to the thermal design for a possible design iteration.

5.5.2 *Solution Approach for Thermal Transient Problems*

In this section, the principles and equations to be solved for a thermal transient problem are explained on the basis of a simple example.

Using the similarity between the thermal and electrical diffusion equations, the thermal network model can be sketched in Fig. 5.3.

Heat capacitance is also referred to as thermal inertia, because high thermal capacitance reduces the rate of temperature change of a system, when it is subjected to a changing thermal environment. With reference to rate of temperature change, or speed of temperature change, the time effect is introduced into the system. For node 1 in Fig. 5.3, representing a volume V_1 with a material with density ρ and a specific heat c_p, the capacitance C_1 (J/°C, J/K) is written as

$$C_1 = V_1 \rho c_p. \tag{5.11}$$

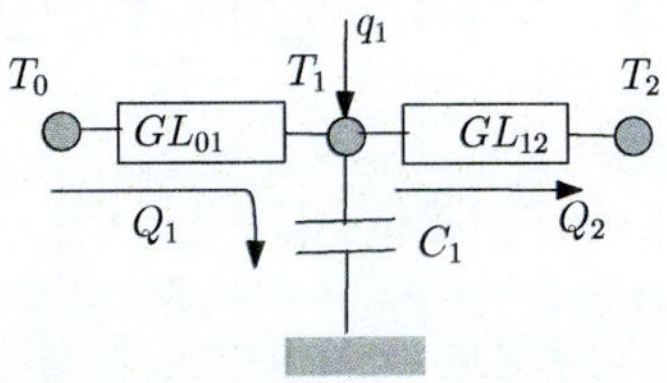

Fig. 5.3 Heat flow at the section [48]

Node 1 also receives heat flow Q_1 (W) from node 0 and loses heat Q_2 (W) to node 2. Also external heat flow q_1 (W) is applied to node 1. With this, the thermal balance equation for node 1 can be written as

$$C_1 \frac{dT_1}{dt} = Q_1 - Q_2 + q_1. \tag{5.12}$$

The left-hand side of Eq. (5.12) contains the thermal inertia term that shows the rate with which the thermal energy is stored or released from capacitance C_1 (J/°C, J/K). This term introduces the time effect into the thermal balance equations. The inertia term at the left-hand side must be in equilibrium with the right-hand side terms, being the internal conductive heat flows Q_1 and Q_2 and the external heat load q_1. With GL_{01} (W/°C, K) being the conductance between node 0 and 1 and GL_{12} (W/°C, K) being the conductance between node 1 and 2, the expressions for Q_1 and Q_2 become

$$\begin{aligned} GL_{01}(T_0 - T_1) &= Q_1 \\ GL_{12}(T_1 - T_2) &= Q_2. \end{aligned} \tag{5.13}$$

To solve Eq. (5.12) at different moments in time, incremental numerical solution procedures are typically used. Starting at t_0 with the temperatures of node 0, 1, and 2 being T_0^0, T_1^0 and T_2^0, the equilibrium equation for time step $p+1$ can be written as

$$C_1 \frac{T_1^{p+1} - T_1^p}{\Delta t} = q_1^p + GL_{01}(T_0^p - T_1^p) + GL_{12}(T_1^p - T_2^p) \tag{5.14}$$

The time step Δt has to be kept within limits to maintain a stable solution [45]. It is recommended to keep the the time step lower than the smallest value of CSG in the model (no specific acronym, most likely the C represents capacitance, the S represents the sum, and the G represents the conductors, [35]). The value CSG is defined as the ratio of capacitance to the sum of connected conductances for a thermal node.

For this example, the time step has to be

$$\Delta t \le \frac{C_1}{GL_{01} + GL_{12}} = \text{CSG}. \tag{5.15}$$

5.6 Thermoelastic Analysis for Transient Problems

Most of the thermal analysis "cases", which is the term often used to refer to a specific orbit around a planet, involve changes in the thermal environment with time. These problems then require a thermal transient analysis to determine the temperature fields as a function of time. The incremental numerical solution methods can produce temperature fields in the form of thermal node temperatures at each time step. In most cases, the thermal numerical solution method requires a higher resolution in time than is needed for the thermoelastic verification. This implies that not each computed temperature field needs to be subjected to the thermoelastic analysis process. The number of temperature fields to be verified through the thermoelastic analysis chain may typically vary between 50 and 100 for a single orbit.

Around moments of strong changes in the thermal environment, for instance entry and departure of eclipse, it is recommended to increase the resolution in time of the thermoelastic verification. During these moments, temperature gradients are potentially the strongest.

For the thermoelastic verification, the temperature fields produced by the thermal model at each selected time step for thermoelastic assessment must go through the process of mapping to the structural finite element model and subsequent structural thermoelastic response analyses, from which the instrument performance impact can be assessed. This is the only reliable way to scan the many temperature fields and identify criticality in terms of meeting the requirements. It is essential that the selection of the critical time steps is left to the end of analysis chain at which the performance impact is assessed. An instrument is too complicated to apply the famous "engineering" judgement for this purpose.

An important condition for applying this approach is that the automation of the analysis process is in place. Without proper automation and data handling, it is not possible for a human being to handle this kind of tasks.

An advantage of the automated scanning of the temperature fields for the each selected time step is that all requirements can be evaluated at the same time. This may have as outcome that for one requirement the temperature field at one specific time step is most critical, while another requirement shows up to be mostly addressed at another time step, corresponding to another position in the orbit.

This approach allows for a full coverage of all requirements for the complete orbit.

5.7 Thermal Analysis for Thermoelastic Versus Thermal Control

5.7.1 Objectives of Thermal Analysis for Thermal Control

For every spacecraft, thermal analyses are performed at system, sub-system and at unit levels. The objective of these thermal analyses is to predict the expected extreme temperature levels that units, components and structural items will experience, either low or high, to verify whether these levels remain within the applicable allowable temperature ranges in the worst cases occurring during the mission. This is obviously a very relevant verification exercise. Based on the outcomes, the thermal design of the spacecraft will incorporate different measures to control the temperatures, such as

- Radiators to reject heat through radiation to space
- Insulation material to limit heat exchange with space
- Selection of coating materials based on thermo-optical properties
- Heaters to compensate for loss of heat to space
- Heat pipes to enhance heat transport from dissipaters to radiators.

The sizing and selection of these measures are an iterative process based on thermal analysis with the objective of controlling the temperatures inside the spacecraft within an allocated budget of resources (mass, required heating power, etc.). The verification of the thermal control objectives is based on a well-established logic for assessing temperature uncertainty values that need to be added to the temperatures computed with thermal model (see Sect. 5.7.6).

In this section, this type of thermal analysis is referred to as "thermal analysis for thermal control" in order to distinguish from "thermal analysis for thermoelastic", which is the topic of this book. Generally, analysis with objective to support the thermal control does not require any knowledge or concern about the deformation or stresses in the structure. However, one could say that thermal control up to a certain level prevents thermal stresses in a component to exceed a level that could cause damage to the component. This is also at more macroscopic scale the objective of the thermoelastic analysis chain.

5.7.2 *Objectives of Thermal Analysis for Thermoelastic*

As may have become clear from Chap. 4, thermal analysis for thermoelastic has a different objective and therefore a different focus. Where the thermal analysis discipline can support the design for thermal control in principle without the involvement of other disciplines, thermal analysis for thermoelastic is the starting point of the full thermoelastic analysis chain, possibly extending to instrument performance impact evaluation. In the context of thermoelastic, the temperature fields produced by the thermal analysis drive the consequences for the structure or the instrument performance.

As will be discussed in more detail in the subsequent sections and has been demonstrated in Chap. 4, the thermal analysis is part of a multiphysics problem, in which the quality and level of detail of the temperature fields produced by the thermal analysis have a direct impact on the quality of the prediction of the structural responses. This consequence is not present for the temperature fields produced with models for the thermal control objective.

The multiphysics nature of thermoelastic implies that the major objective of the thermal model for thermoelastic analysis is to provide temperature fields that have the potential to excite the structure and provide temperature loading information for the structure represented by the finite element model. The structural finite element model shall also have the objective to simulate well the thermally induced responses based on the temperature field provided by the thermal model. The thermal mesh resolution and level of detail of representation of features in the thermal model must therefore be able to induce accurately the structural responses simulated with a structural model. This implies as well that the structural model must have a corresponding level of detail in order to capture this detailed temperature field.

Also the selection of design driving thermal environments must be focussed on thermally induced responses of the structure and not on the temperature levels, which is only relevant for thermal control. In line with the difference in objectives between thermal analysis for thermal control and thermoelastic, the logic for establishing the uncertainties affecting the temperature field and subsequent structural responses need to be different as well.

5.7.3 Selection of Worst Case Temperature Fields

From thermal control perspective, it is needed to identify the so-called hot cases and cold cases. Under the thermal environments for these cases, many of the components of the spacecraft or instrument are experiencing their extreme temperatures. From thermal control perspective, these cases are obviously important with a subset of them representing design driving cases. Unfortunately, from thermoelastic perspective, it is a bit more complicated. A common mistake is to assume that these worst hot and cold cases are also the worst cases to be considered for the thermoelastic analyses. As is stressed several times in Chap. 4, it is in general far from trivial to judge, based only on the temperature levels and even complete temperature fields, whether one temperature field is more severe than another.

Most structures have varying stiffness and CTE values combined with temperatures varying with the position on the structure. The complex interaction between stiffness and thermally imposed strains requires adequate finite element models to evaluate the severity of each temperature field in terms of structural responses and consequences for the performance. The case presented in Sect. 4.6 and detailed in Appendix A illustrates this complex interaction. In most cases, the local temperature levels are not driving the thermoelastic distortion or stresses. This applies very often also for the temperature differences in a structure. However, a certain temperature difference over a shorter distance is likely to have a higher impact than the same temperature difference over a larger distance. It is therefore important to recall the definition of "gradient" from Sect. 4.2. In the thermal control domain, there are frequently requirements specifying limits to temperature differences between two locations. In the context of thermal control, there is therefore a tendency to also use the term "gradient" for this temperature difference. The example discussed in Sect. 4.6 showed that the observation of temperature difference did not point us to the worst case configuration. Basically, the higher the temperature change per unit length, the higher the chance on finding high levels of distortion or stresses in the structure.

Transient effects, responsible for changes with time in the thermal state of the spacecraft or instrument, are of equal importance for both thermal control as thermoelastic, but for different reasons. For a spacecraft, common changes in the environment are related to the mission (changes of attitude, entry and exit of eclipse) and changes of internal configuration (typically switch on/off of units, heaters, etc.). From thermal control perspective, the results from the transient analysis are used to verify whether the thermal design is able to limit the effects of changes in the environment for the internal of the spacecraft or instrument. For instance, it is verified whether the rate of heat loss to the environment is adequately compensated, e.g. with sufficient heater power, or whether the selected insulation material is limiting sufficiently the heat flow from and to the environment.

From the thermoelastic perspective, transient effects are characterised by the difference in rate of temperature change at different locations in the structure. This leads to thermal gradients (see Sect. 4.2) in the structural components that are depending on the thermal inertia, the conductivity and of course the rate of change of thermal environment. Typical moments with strong transient effects are during entry and exit of eclipse in an orbit around a planet. The challenge for thermoelastic is therefore to capture these gradients in the thermal model, which puts requirements on the thermal mesh, which will be discussed in the next section.

The more one becomes conscious of the complexity of the interaction between temperature, thermal gradients, thermally induced strains and stiffness, the more one realises that only for very trivial cases the worst temperature condition can be identified without detailed analyses. Although it is not a simple job to set up a sophisticated scanning process for the many temperature fields of the many transient cases that includes for each temperature field the full end-to-end simulation of the thermoelastic responses (see Sect. 4.7), it is the only reliable way to find the worst case temperature fields.

5.7.4 *Thermal Mesh Convergence for Thermoelastic*

The mesh for thermal control has to ensure that it can adequately simulate the heat flows inside the spacecraft, the exchange with the external environment and the temperatures of especially the sensitive components, the so-called temperature reference points (TRP). The ECSS thermal analysis handbook [35] writes therefore as guideline in the section on spatial discretisation and mesh independence: “Make sure that the spatial discretisation used for thermal models is fine enough that key model outputs are no longer dependent upon it within an acceptable range”. From thermal control perspective, this means mainly that thermal mesh refinement has to lead to converged temperature values at the important locations (key model outputs) and possibly also the heat flows through import elements of the thermal design.

Thermal mesh convergence for thermoelastic has a similar approach. An important difference is that instead of using a result of the thermal model, one of the structural model is used to confirm the thermal mesh convergence. This means that after each thermal mesh refinement, increment is verified whether the change of a displacement, rotation or stress produced by the structural finite element model is becoming less than a beforehand agreed value. So for each thermal mesh convergence iteration, the full thermoelastic analysis chain has to be run.

Since the final objective of thermoelastic analysis in many cases is to verify that the performance of the spacecraft with its instrument is not impacted too much, the thermal mesh convergence criterion can also be based on an instrument performance quantity, such as optical image distortion level or RF signal quality level. Summarising, thermal mesh convergence for thermoelastic must be based on responses that are a consequence of the thermal analysis results (such as deformations, optical image quality) and not the responses of the thermal model itself.

5.7.5 Level of Detail in Models for Thermoelastic

Depending on the tightness of the requirements for thermoelastic stability, both the system-level structural model and thermal model for thermoelastic need to include details that are definitely not relevant for thermal control and equally not relevant for dynamic structural models. In Sect. 4.6, it is explained what a single dissipating box under mild conditions can do with a rather stiff sandwich panel. This example shows that it can be important to represent the structure at a higher level of detail in the thermal model for thermoelastic than is needed for thermal analysis for thermal control. Also in the structural model, the classical simplified structural representation of such a unit for dynamic analysis with a rigid body element connecting a lumped mass to a spacecraft panel has to be replaced by a more physical representation with all panels included and meshed with rather detail. Mesh resolution and level of detail in both models have to be compatible to ensure the best use of the modelling efforts in both disciplines.

5.7.6 Thermal Analysis Uncertainties for Thermoelastic

The uncertainty approach that is generally applied to results of the thermal analysis for thermal control recalls well the objective of thermal analysis for thermal control, being the prediction of the temperature limits or temperature range that components (or in general temperature reference points) will experience during their life on the spacecraft. The ECSS thermal control general requirements [33] includes the scheme that is presented in Fig. 5.4 to extend the computed temperature range for a specific temperature reference point with uncertainty margins. A similar scheme is provided by the JAXA guidelines [8].

The left-hand side of Fig. 5.4 shows well how the temperature range for a temperature reference point is extended to include the uncertainty values. Both the "ECCSS thermal control general requirements" [33] and the "ECSS thermal analysis handbook" [35] provide guidelines for determining these uncertainty values based on established methods.

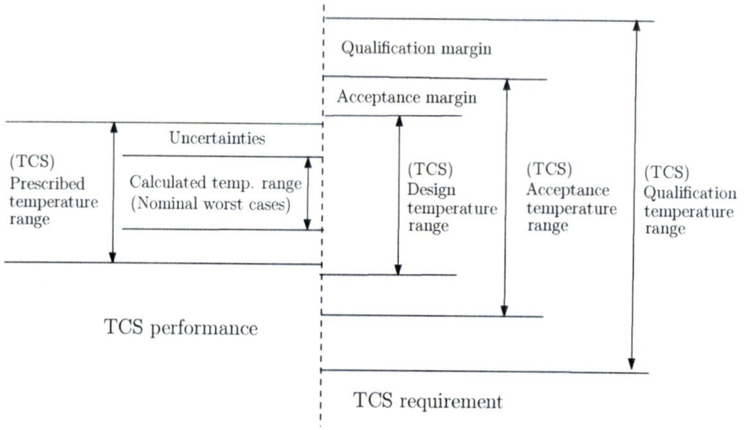

Fig. 5.4 Uncertainty approach applied to results from thermal analysis thermal control: Uncertainty margins added to the computed temperature ranges (taken from [33])

This implies that the uncertainty approach does not operate on the temperature for a specific point that is computed for a certain moment in time during the transient, but on the two extreme values (minimum and maximum) of the temperature for this point. This uncertainty margin does not help to generate more severe temperature fields in the sense that these would lead to higher thermoelastic responses. Adding a temperature margin to a complete temperature field will lead to a meaningless temperature field which is not obeying thermal energy balance requirements. In other words where the margin approach for thermal control helps to increase the confidence that components will not experience temperatures outside their qualification range, this approach does not work for thermoelastic.

In Chap. 10, two approaches are described. The first approach assumes that the uncertainties and modelling errors could cause the thermoelastic responses to be higher than the predicted responses. This effect is simulated with amplification factors ≥ 1 for different contributors to be applied to the thermoelastic responses. The values of these factors have an important extent and an arbitrary character. It is however also the approach that is followed with the structural factors of safety K_p and K_m as defined by the ECSS standard "structural factors of safety for spaceflight hardware" [34]. In Chap. 10, two factors dedicated to the error and uncertainties in the thermal results are mentioned: K_{mt} and K_{pt}.

As mentioned, this approach with the factors of safety has some arbitrary elements. It also assumes that the shape of the temperature field with its relative temperature values is correct, and implementation of uncertainties only requires scaling of the temperature field. For that reason, Chap. 10 explains also stochastic methods, including the "Rosenblueth $2k + 1$ Point Estimates Probability Moment Method" (see Sect. 10.6). Chapter 10 discusses an example of a panel orbiting around the Earth (PANELSAT) to which this method is applied and shows that it can be a very efficient alternative for the well-known computational heavy Monte Carlo method.

5.7.7 *Concluding Thermal Analysis for Thermal Control Versus Thermoelastic*

In the above, it was explained that the thermal analysis for thermal control is focussed mainly on demonstrating the compliance of the temperature of the spacecraft items with respective allowable limits. Thermal analysis for thermoelastic, however, is the starting point of an analysis chain with the structures discipline and performance-related disciplines, such as optics and RF. Thermal analysis for thermoelastic tries to capture as much as possible all relevant temperature information for structural analysis in which as accurate as possible details of the thermal gradients are included.

For the selection of worst cases in the context of thermal control, only the results of the thermal model for thermal control need to be analysed. The worst cases for thermoelastic have to be based on the responses of the structure and possibly on how these structural responses impact the instrument performance.

For thermoelastic, the thermal mesh density has to be driven by convergence of the structural responses. For thermal control, the thermal mesh convergence is based on mainly convergence of the temperature of the important model outputs (key model outputs). The mesh resolution requirements for the thermal model for thermoelastic will, in general, be more demanding with respect to the requirements on the model for thermal control verification. This may lead to two different thermal models or the use of a single model that is adequate for thermoelastic and therefore expected to be also suitable for thermal control.

Similarly, the structural function of components, and stiffness of units on the spacecraft and their potential to induce thermal deformation, is not of much relevance for the development of the thermal model for verification of the thermal control design. For thermoelastic, it is important that the thermal model captures the temperature field well enough, especially on the stiff structural components, for the simulation of the deformations.

In line with the above, uncertainties in the thermal control domain only affect the temperature range of the TRPs, while for thermoelastic the uncertainties in the temperature field have consequences for the structural responses and the performance of the instruments.

With Table 5.2, it has tried to summarise the comparison of thermal analysis for thermal control with thermal analysis for thermoelastic.

Problems

5.1 In which category of phases of the life of a spacecraft shall the thermoelastic analysis focus on strength of the spacecraft and in which category of phases shall the focus be also on deformation? Explain the reason for the different focus.

Table 5.2 Comparison of thermal analysis for thermal control with thermal analysis for thermoelastic

Analysis for thermal control versus thermoealstic	
Thermal control	Thermoelastic
Thermal phenomena in isolation	Starting point of a multiphysics analysis chain
Calculation of temperature extremes at TRPs	Providing temperature fields for computation of structural responses (distortion and stresses)
Temperature levels are driving	Spatial temperature gradients are driving
Hot and cold cases are worst cases (based on temperature levels)	Worst cases selection based on scanning over structural and instrument performance responses for all temperature fields
Thermal mesh convergence based on key (thermal) model outputs	Thermal mesh convergence based on structural response and instrument performance impact
No need for system-level model to capture temperature field of units[a]	Detailed temperature field needed to simulate well the deformation of stiff units
Uncertainties affect the individual TRP temperature range	Uncertainties affect complete temperature field and responses in subsequent models in the analysis chain

[a] Large units with non-uniform distribution of highly dissipating items, may be violating the isothermal assumptions and require also the appropriate attention in terms of mesh resolution

5.2 For the heat transfer through radiation, two spectral bands can be identified. How are are these spectral bands referred to? Through which spectral band is a radiator emitting heat to space?

5.3 Spacecraft thermal analysis can be performed with the objective to support and verify the thermal control design or to generate input data for the thermoelastic analysis chain. What are the main differences? Explain with this comparison why a thermal analysis for thermal control would not be adequate for thermoelastic.

Chapter 6
Structural Modelling for Thermoelastic Analysis

Abstract The thermal stresses and thermal deformations are the responses of a structure due to temperature fields as one of the applied loads. These responses of a structure are computed through thermoelastic analyses. As for many structural problems, also for simulating the thermoelastic response of a structure, the finite element (FE) method is the method mostly used. Using the fundamentals of the FE method and implementations in the finite elements, this chapter discusses the different finite element types and their adequacy to simulate thermoelastic responses. Within a space project, several types of structural analysis, in general all based on the FE method, are run. For practical reasons, there is a preference to use the same model for all these analyses. The single model that is commonly used is the dynamic model, developed for simulating the structural dynamics, which is as well an important design driving aspect for a spacecraft. However, a model, adequate to simulate the dynamic responses of a structure, may not always be able to represent properly the thermoelastic behaviour. Some limitations of typical dynamic models are explained, and suggestions are provided to adapt or refurbish the dynamic model to enhance the quality of the computed thermoelastic responses.

6.1 Introduction

The structural analysis model, for computing the structural responses due to applied changes in the temperature field in the simulated structure, is often referred to as the "thermoelastic model". Also in this book this term will be mostly used.

As for many structural analyses, the finite element method is also the method mostly used for computing the responses of a structure to a change of temperature field. Like any finite element model, it is created or generated from known geometry with corresponding dimensions, uses the material properties of the applied materials and has the relevant constraints and boundary conditions implemented.

The objective, that is specific for a thermoelastic model and specifically for a spacecraft structure, is to compute the mechanical responses due to the applied temperature fields. Typical responses are as follows:

S. Appel and J. Wijker, *Simulation of Thermoelastic Behaviour of Spacecraft Structures*, Springer Aerospace Technology, https://doi.org/10.1007/978-3-030-78999-2_6

- Stresses in the structure for strength verification.
- Variation of the deformations of the structure, especially during operational phases, to verify for instance:
  - the alignment or pointing angle of antennas or optical items in instruments
  - the shape distortion of reflectors, antennas or lenses and mirrors.

The thermal loading is represented by a temperature distribution over the structure that is discretised by assigning temperatures to the individual FE nodes and, if needed, complemented with temperatures assigned to the finite elements. In this context, all the responses of the structure originate from the thermally induced strains. The resulting deformation and stress fields become in general non-trivial due to [15]

- Non-uniform temperature distributions
- External constraints that prevent free deformation of the structure
- Variation of CTE values due the use of different materials that appear in heterogeneous structures
- Non-uniform distribution and locally abrupt change of stiffness due to different Young's modulus values and joints between components.

Besides the temperature levels in the temperature field, the induced strains also depend on the reference temperature field (see Sect. 3.2), and this is the temperature field at which the structure is considered free from deformation and stresses. Typically, this is the temperature at which a structure is manufactured or assembled.

Detailed temperature distributions are necessary as input for the thermoelastic structural analysis to predict thermoelastic stress distributions and deformations.

In Chap. 3, several aspects involved in the physics of thermoelastic were discussed:

- The coefficient of thermal expansion (CTE)
- The Young's modulus
- Constitutive laws (stress–strain relationship) of thermoelasticity.

These aspects are essential components of the material models that are implemented in the finite element formulations in finite element software tools.

In this chapter, the limitations and adequacy of common finite element types to simulate thermoelastic responses are discussed with supporting modelling suggestions.

Within a space project, several types of static and dynamic structural analyses are run. Traditionally, the model used for dynamic response analysis forms the basis for structural analyses and thus also for thermoelastic analyses. A dynamic model may not always be fully suitable for thermoelastic analyses and may need to be adapted or refurbished to become more adequate for the thermoelastic analyses. Aspects of this model refurbishing are discussed.

6.2 The Finite Element Method for Thermoelastic Simulations

The finite element method (FE method) has been developed at the end of 1950s early 1960s [25] and became more and more popular due to the increase of computer power. Instead of solving systems of ordinary and partial differential equations, the finite element method transforms these systems into a set of linear algebraic equations that could be solved numerically with a computer. Within the FE method, structures are idealised with a mesh of nodes connected with elements. Depending on the type of structural item to be represented, a combination of different element types can be used, covering one-dimensional elements, such as scalar, rod and bar elements, 2-D elements, like membrane, bending plate and shell elements, and finally, the three-dimensional solid elements. The system matrices and the loading vectors are assembled from element matrices, and the system of simultaneous equations is finally solved by numerical algorithms.

Over the years, many textbooks and journal papers on the FE method are published, e.g. [14, 36, 60, 83]. The accuracy of a finite element model and applied finite elements is often checked by comparing the finite element analysis (FEA) results with theoretical solutions of simple structural elements (bar, plate). For that purpose, the National Agency for Finite Element Methods and Standards (NAFEMS), founded in 1983, established various benchmarks to validate the quality of software codes based on the FE method. NAFEMS also introduced various guidelines for good use of the FE method.

Today, the FE method is by far the most popular and most used numerical method to solve all kinds of linear and nonlinear structural problems in mechanical engineering. Well-known FEA software packages used in spacecraft design applications are several version of the original NASA developed program NAsa STRuctural ANalysis (Nastran) and SIMULA/ABAQUS.

The FE method has demonstrated to be a versatile structural analysis method covering a wide range of structural analyses problems, such as linear and nonlinear static analyses, stability problems and a whole range of dynamic analyses. Also, implementations based on the FE method are available for solving thermoelastic problems in most FEA software tools.

In this chapter, the focus is on the application of the finite element method for simulation thermoelastic mechanical responses due to applied temperature fields.

6.3 Characteristics of Finite Elements for Thermoelastic Analysis

In general, the structure of the spacecraft consists of rods (struts), bending beams, membranes, shells, sandwich structures, machined parts (brackets, housings), etc. Scientific instruments and electronic boxes are often contained in machined housings.

Many types of structural items can be modelled with associated types of finite elements, 1-D, 2-D and 3-D finite elements, to perform the thermoelastic analysis. Besides these elements that have a physical geometrical representation, also scalar element (0-D) and constraint equations are applied. Since spacecraft structures and instruments are 3-D structures, 2-D axi-symmetric elements are not used in the models of these structures.

The choice of element type and the resolution of the discretization should be driven by the required accuracy of the model and gradients in the temperature field that need to be captured. Thorough preparation, consisting of close interaction with the thermal colleague and the systems engineer of the instrument, is essential for the design of the finite element model.

This book is not intended to provide elementary background on the finite element analysis method. There are other authors who produced excellent textbooks covering the philosophy and theory of this method, e.g. [14, 61, 83]. However, because many examples in this book illustrate various topics with the help of one-dimensional finite element models, Appendix C is dedicated to the description of a one-dimensional rod element and the derivation of the equations describing the thermoelastic characteristics of this element.

6.4 Elastic Finite Elements

6.4.1 0-D, Scalar Element

The scalar element is a stiffness connectivity between two FE nodes. The most basic variant of the scalar element is a spring element connecting specific degrees of freedom of these two FE nodes. In this basic configuration, the FE nodes are coincident resulting in a zero-length element, that is therefore lacking geometrical characteristics and as a consequence does not contain enough information to compute thermoelastic strains and thus deformation of the element.

There are implementation of zero-length scalar elements that offer the possibility to provide an artificial length for the computation of thermoelastic strains.

FE models of structural parts can be connected with the aid of scalar elements. For thermoelastic applications, special attention must be paid to the fact that connections between structural parts made through scalar elements are injecting loads at discrete location into these parts. Mathematically, this kind of load introduction forms a singularity. The simulated overall stiffness of the assembly will therefore depend on the mesh density of the models of the connected parts. As a consequence, the use of springs may not only introduce local inaccuracies, but also global effects may be observed. This also applies to the use of 1-D elements, such as bar and rod elements, that are connected to a shell or 3-D solid mesh.

Irrespective of the thermoelastic context, it is in general recommended not to attach spring elements at discrete locations of 2-D or 3-D continuous parts like panels or solid pieces. Although it is quite convenient to represent the stiffness of a connecting part with a single spring element, it is better to represent the physical connecting part with shell or solid elements, because this allows for establishing the connection via the "glued" contact option that is available for instance in MSC Nastran. This way of modelling does make the representation of the stiffness of the connection insensitive to changes of the mesh density.

A way to limit the effect of the singularities that are introduced with the discrete attachments through spring elements is to what may be called “load spreaders”. In practice, these are the local reinforcements at the physical interface points, such as brackets or doublers. It is recommended to at least reproduce the physical dimensions of “load spreaders” and their corresponding stiffness to minimise the effect of mesh resolution dependency when spring elements are used. It must be noted that the modelling of load spreaders reduces the effect of the singularities, but it is not a real solution for the problem.

The above-mentioned mesh dependence phenomenon of the attachment to 2-D or 3-D continuous structures is illustrated in the example “Mesh dependence for spring to plate connection”. It is a basic example that nicely shows the potentially important influence of the mesh resolution on the stiffness of the connection, especially for thermoelastic problems.

Spring type elements are quite popular for use as “load sensor” in a model. Highly, non-physical, stiffness values are assigned with the intention to provide an almost rigid joint that should not affect the surrounding deformation field. The recovered spring forces provide the interface forces. In many cases, the dynamic properties of the structure (eigenfrequencies and mode shapes) are not sensitive to the stiffness level of the joint, and the stiffness values of the springs used to model those. However, if it is decided to keep this way of modelling for thermoelastic simulation, then the sensitivity of thermoelastic responses to the spring stiffness is recommended to be verified. Also, it is advised to tune the stiffness of the springs such that these approximate the physical stiffness as good as possible and refrain from using numerically infinite stiff values.

In this context, it is good to take note of the effects that joint stiffness values may have on the thermoelastic response as is explained in Sect. 6.7.4.

There are implementations of spring elements that have the limitation that it does not contain corrections for misaligned displacement coordinate systems of the connected degrees of freedom. In addition, not all spring element implementations support well the connections of non-coincident finite element nodes. Problems in the model, due to these limitations, may be detected by model health checks such as the ones described in Sect. 6.8.

Mesh dependence for spring to plate connection

Two configurations of a constrained plate connected to a spring, as are shown in Fig. 6.1, are assessed in this example. The assembly of plate and spring can represent any plate connection that is often found in the finite element model of spacecraft.

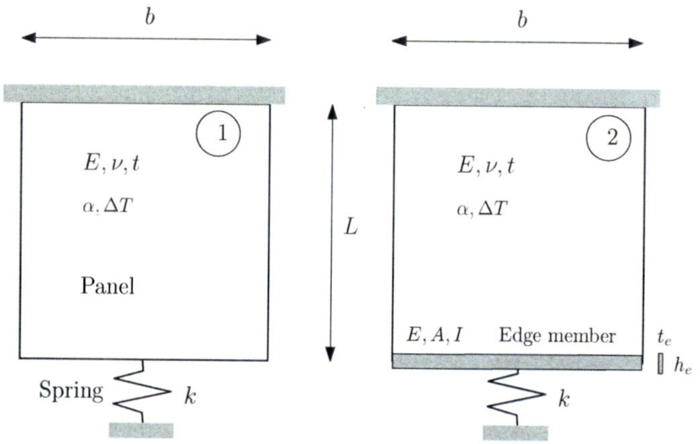

Fig. 6.1 Constrained plates connected to spring

One end of the spring is connected to the middle of the lower edge of the plate as indicated in Fig. 6.1. The other end of the spring is constrained. Of the two configurations shown, configuration "1" (picture at the left in Fig. 6.1) has no reinforcement at the edge of the plate to which the spring is attached. In configuration "2" (picture at the right in Fig. 6.1), an edge member (load spreader) is added to introduce the spring force more smoothly into the plate.

The load is in this example introduced by reducing the temperature in the plate, causing the plate to shrink. Due to the thermoelastic displacement of the edge, a force is built up in the spring that is reacted by the plate.

The square plate has a length and width of $L = b = 0.5$ m, and the thickness is $t = 2$ mm. The material properties are as follows: Young's modulus is $E = 90$ GPa, Poisson's ratio is $\nu = 0.33$, and CTE is $\alpha = 25 \times 10^{-6}$ m/m/°C. The spring stiffness $k = 87{,}964{,}594$ N/m (derived from Al-alloy bolt, diameter 4 mm and elastic working length of 10 mm).

The temperature change of the plate relative to the reference temperature is $\Delta T = -20$ °C. The cross section of the edge member has a thickness $t_e = 2$ mm and the height $h_e = 50$ mm.

For each of the configurations, three different FE models with different mesh densities are shown in Fig. 6.2. Both linear CQUAD4 and quadratic CQUAD8 FE elements are used for the plate, leading in total to 12 FE models. The edge member is idealised with CBAR elements. The thermoelastic FE analyses are done with MSC Nastran.

Table 6.1 shows the spring force for the two configurations, for the three different mesh densities and for two element types CQUAD4 and CQUAD8. Physically, the problem for all models is the same. However, for both configurations "1" and "2", the spring force changes with the number of finite elements in the FE model. With increasing number of elements, the thermoelastic spring force is decreasing. Applying CQUAD8 finite elements, with quadratic shape functions, the reduction of the spring forces with increasing mesh resolution is stronger.

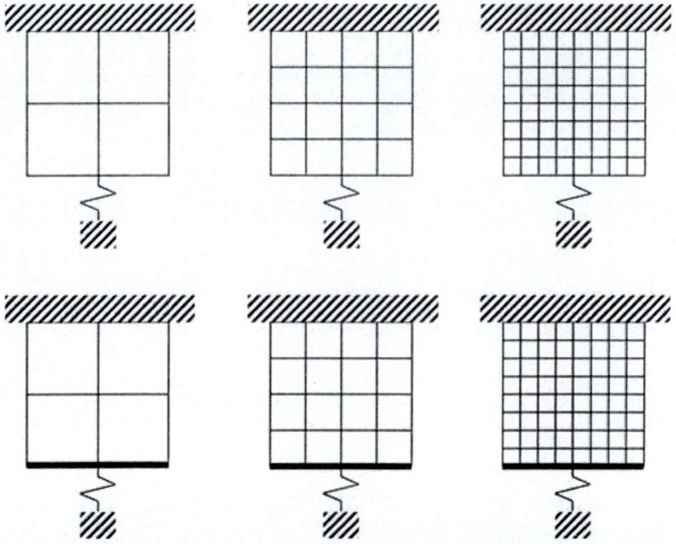

Fig. 6.2 FE models of constrained plates connected to a spring with three different mesh densities

Table 6.1 Spring forces for two configurations of plate and spring for three different mesh densities

Configuration	# CQUAD4 elements	Spring force (N)	# CQUAD8 elements	Spring force (N)
1	2 × 2	12,126	2 × 2	11,726
	4 × 4	11,316	4 × 4	10,485
	8 × 8	10,241	8 × 8	9537
2	2 × 2	12,206	2 × 2	11,907
	4 × 4	11,611	4 × 4	11,125
	8 × 8	11,092	8 × 8	10,866

The reduction of the spring force with increasing mesh resolution indicates that the stiffness of the attachment is influenced by the mesh resolution. In this example, only three mesh configurations are analysed. The results do not show a sign of mesh convergence, which typically indicates that the different models try to simulate a singularity. Although less strong, the effect is also present for the model with the edge member included.

With this, simple exercise is shown how the stiffness of the assembly changes with the mesh resolution when discrete attachments are made to 2-D models.

Another illustration of mesh dependency is the stress singularity in the neighbourhood of a point support.

A plate with length $L = 1.0$ m, width $B = 0.5$ m and thickness $t = 2$ mm is shown in Fig. 6.3.

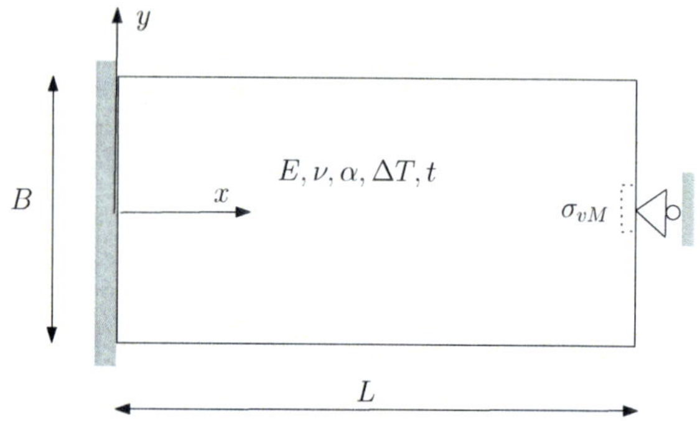

Fig. 6.3 Plate with point constraint at $x = 1.0$, $y = 0$ m

Table 6.2 Von Misses stresses

# CQUAD4 elements	σ_{vM} (MPa)	σ_{vM} (0.5, 0.0) (MPa)
20 × 10	2.5824×10^2	2.2150×10^1
40 × 20	4.6133×10^2	1.9812×10^1
80 × 40	8.3310×10^2	1.7948×10^1

Young's modulus is $E = 70$ GPa, Poisson's ratio is $\nu = 0.33$, and the plate thickness is $t = 2$ mm.

The temperature decrease is $\Delta T = -20\,°\text{C}$, and the CTE is $\alpha = 25 \times 10^{-6}$ m/m/°C.

The von Mises stress at the constrained node in X-direction ($x = 1.0$, $y = 0.0$) is dependent on the mesh density of the FE model of the plate. The mesh dependency of the von Mises stress σ_{vM} is illustrated in Table 6.2.

As a reference, the von Mises stresses at the middle of the plate ($x = 0.5$, $y = 0.0$) are given too.

Considering that the von Mises stress contains contributions of all stress components, it can be used to characterise a singular point in the stress field with $\sigma(r) \rightarrow \infty$ for $r \rightarrow 0$, with r being the distance from the singular point [82].

In this example, the singular point in the stress field is located at the constrained FE node. With the increase of the mesh density near the constraint point, the distance r between centre of the FE element and the constrained FE node is becoming smaller. With further refinement, this distance will go in the limit to zero, resulting into infinite stress.

The contour plot of the von Mises stress in the FE model of the plate with 40 × 20 CQUAD4 elements is shown in Fig. 6.4.

The high value of the von Misses stress is clearly noticeable in the neighbourhood of the constrained FE node.

The stress level away from the singular point is, however, dropping with increasing mesh density. For externally applied loads, i.e. load types different from thermoelastic loads, one would expect that the effects of the boundary condition would vanish with increasing distance from the stress concentration. This is still true in the sense that

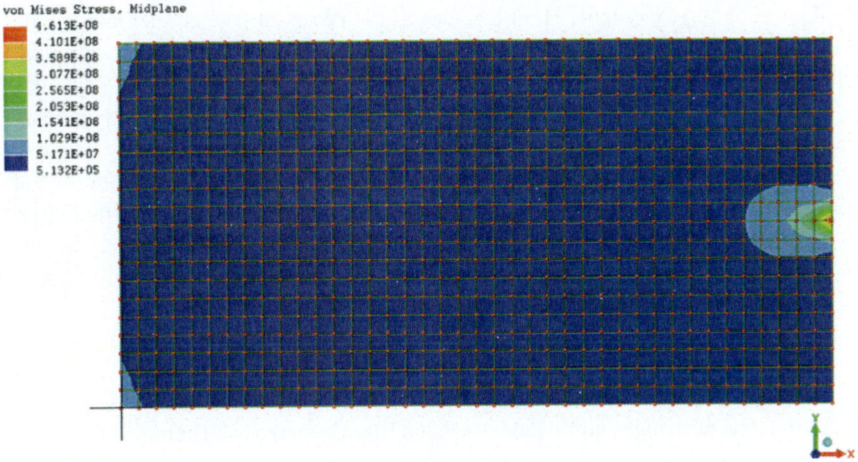

Fig. 6.4 Von Mises σ_{vM} stress σ_{vM} contour plot

the strong gradients in the stress field near the singular boundary condition are not any more noticeable in the centre of the plate.

The behaviour that is observed in this example illustrates the difference between external loads and imposed deformation (see also Sect. 6.7.4). With the experience of external loads, one tend to expect, based on the Saint Venant principle [42], that the simulated stress field away from the singularity approaches the real physical stress level. However, since the free thermoelastic deformation of the plate is not affected by the mesh density, the only property that can modify the stress field in the undisturbed part of the plate is the simulated stiffness of the plate near the constrained FE node.

This clarifies the observed reduction in the reaction force in the spring in the first model in this example. The mathematical singularity is not only disturbing the simulated stress field in the plate, but also the simulated stiffness of the plate near the attachment.

6.4.2 1-D, Rod Element

A rod element connects two FE nodes and has in most cases only translational and rotational stiffness along the axis of the element, that is defined by the line connecting the two FE nodes. The axial stiffness is derived from the length L of the rod element, i.e. the distance between the two connecting FE nodes, the cross-sectional area A and Young's modulus E. The torsional stiffness is also defined by the length L of element and, in addition, by the torsional constant J and the shear modulus G. The rod element can simulate axial force and torsional moments. Bending loads cannot be taken by the rod element. The FE nodes at both ends of the rod element should be considered as hinges with free rotation around the two directions perpendicular to the axis of the element.

The temperature is assumed to be constant within a cross section of a rod element. Rod elements with two FE nodes simulate a linear variation of temperature between the two ends of the element, which makes that the average of the temperatures at the two ends of the element, T_{av}, is adequate for the calculation of the thermoelastic elongation (see also Eq. (C.21) and Eq. (C.47) in Appendix C). In addition, the reference temperature T_{ref}, the length L and CTE α are required. With this information, the axial thermoelastic deformation of the rod becomes

$$\Delta L = \alpha L(T_{av} - T_{ref}) \tag{6.1}$$

A temperature change cannot induce torsional strain in a rod and hence no torsional moment.

When connecting a rod to 2-D or 3-D elements, the same consideration as for the scalar spring element shall be taken into account (see Sect. 6.4.1).

The rod finite element can be used to model rods in a truss frame.

Truss frame

A truss frame is idealised with rod finite elements and shown in Fig. 6.5. The connection between the trusses form hinges and cannot transfer bending moments.The spring on the right-hand side represents the stiffness of an adjacent structure. No thermal properties are assigned to that scalar element. The temperature field is applied to the FE nodes 1 to 8. The material properties of the rod elements E, α and A are to be specified.

This model shows the limitations of the rod element that is not able to take any loads normal to the axis of the element. As a result, the structure in this example requires that the displacements normal to the plane of the structure are suppressed and the problem is reduced to a two-dimensional problem. Alternatively, the rod elements

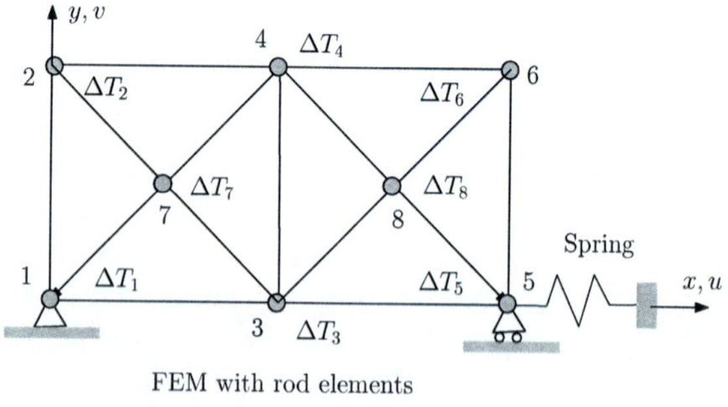

Fig. 6.5 FE model of truss frame build from rod elements

are replaced by beam elements (see Sect. 6.4.3) to transfer bending moments to handle out-of-plane loading of the structure.

6.4.3 1-D, Bar and Beam Element

Bar or beam elements are, like rod elements, used for structural components of which one dimension is significantly larger than the other two dimensions. Rods are not to be able to carry bending moments and therefore are elements with ony axial stiffness (and often torsional stiffness). Bars and beams, however, do have axial, bending and transverse shear stiffness properties. As a result, these elements can be considered to be more versatile.

Implementations of the bar element consist in most cases of a combination of a rod formulation and a classical (Euler–Bernoulli) bending beam formulation. The geometry of most bar or beam elements is defined by only two FE nodes at both ends of the bar. Some finite element codes have implementation of bar elements with more than two FE nodes, which allows to define curvature in the shape of the bar.

The bar is characterised by its length L, cross-sectional area A, the second moments of area I_{yy}, I_{zz}, I_{yz}, Young's modulus E and CTE α. In general, all six degrees of freedom at the FE nodes, i.e. three translations and three rotations, have stiffness associated with corresponding ability to respond with forces and bending moments at the FE nodes. With the definition of the bar in a finite element model, it is important to take care of the orientation of the bar, since the bending stiffness is often not the same in the two main axes of the cross section.

The simplest use of bar elements in a thermoelastic simulation is by assigning temperatures to both end nodes. In this way, the bar is behaving like the rod with the elongation based on Eq. 6.1. In such a case, the temperature is uniform in each cross section of the bar.

Different from a rod element, can a bar element bend due to a temperature variation through the cross section. Basically, when the temperature in a bar section is not uniform, a thermal gradient is present through the cross section of the bar that induces different elongation at different sides of the cross section and thus introduces a bending moment as reaction. For the calculation of such thermoelastic responses, the non-uniform temperature field needs to be defined in each cross section of each beam element that is subjected to these cross-sectional temperature gradients. The definition of such temperature field can be quite tedious, since the orientation of the bar needs to be related to the orientation of temperature field that is intended to be applied. The specification of temperature gradients as a loading condition to bar elements is not supported by most FE model preprocessors. This means it has to be done manually, and considering its complexity, it is likely to be error prone.

When a bar type structural component is subjected to cross-sectional thermal gradients, it is recommended to model the bar with 3-D solid elements. Without any doubt, the computational cost will increase significantly, but the risk on human input errors is drastically reduced. It also allows for visual verification of the applied temperature field within a preprocessor. An other advantage is that the 3-D modelling can also be applied in the thermal model. It also simplifies the set-up of the correspondence between the thermal model and structural finite element model. As a consequence, it will improve the quality of the temperature mapping process.

6.4.4 2-D, Membrane Element

The membrane elements are applied in case of plane stress situations. These situations are not so common in a spacecraft structures. A possible application of these elements could be for modelling of a thin foil. Since these elements do not have out-of-plane stiffness by themselves, this out-of-plane stiffness can be created through pretension. Often nonlinear iterative solvers are required to deal with these elements in a 3-D structure.

The element geometry is defined by the thickness and the positions of the corner FE nodes and possible mid-side FE nodes. The simplest version of the membrane element is the triangular linear version with three corner FE nodes. A slightly more sophisticated linear version connects four FE nodes. Versions with six or eight FE nodes are higher order formulations of the element, using quadratic shape functions. The relevant material properties for this type of elements are Young's modulus E, Poisson's ratio ν and CTE α. Also, orthotropic material models can be handled by most implementations of the membrane element.

Since this element is a 2-D element, also the applied temperature field can vary in two directions. The temperature is considered constant through the thickness, which implies that only in-plane temperature gradients can be simulated. With T_{ref} being the reference temperature of the structure, the thermoelastic strains in every position in the element are derived from the principle of $\varepsilon_t = \alpha(T - T_{ref})$. The actual simulated temperature variation is depending on the applied formulation of the shape functions that strongly depend on the number of connected finite element nodes.

6.4.5 2-D, Plate, Shell, Sandwich Element

The plate and shell elements are a collection of two-dimensional element types that have membrane properties combined with the ability to simulate bending and deformations due to transverse shear load. Depending on the implementation of the element, these three different properties can be coupled or uncoupled. This means practically that a plate element does not need to use the same thickness for membrane stiffness calculation as for the bending stiffness or the transverse shear stiffness formulation. This feature can typically be used for representing the properties of sandwich panels. For the membrane stiffness, only the thickness of the thin face-sheets is relevant, while for bending not only the face sheet thickness, but also the distance between the face sheets is important. This face sheet distance is the height of the core. The transverse shear stiffness is in general provided by the core material that has significant other dimension than the face sheet thickness.

Like the 2-D membrane elements, the plate elements may be triangular and quadrilateral. Also, the variants in connectivity are similar to the membrane elements. Again depending on the implementation, the properties of a plate element may vary over the element surface. In case of coupled properties, the plate thickness and material properties are sufficient. As explained above, in case of uncoupled properties also, other characteristics of the plate or sandwich are needed.

For spacecraft structures, a plate is often manufactured from a stack of thin unidirectional fibre-reinforced polymer plies. The stack can be complemented with a layer of core material. The finite element code assembles the individual ply properties with consideration of the fibre orientation into a system representing the combined membrane, bending and transverse shear properties. The stiffness of such a system is often represented by the ABD matrix introduced by the classical laminate theory.

Since in general stiffness values are associated to all modes of deformation of the element, corresponding forces and bending moments are resulting at the connected FE nodes. Some implementations of the plate element may not have any stiffness associated with the rotation around the normal to the element surface, the so-called drilling degree of freedom.

The simplest temperature field that can represent a load for the element is the one with a uniform temperature through the thickness of the element. In those cases, the temperature field is represented by the temperatures at the FE nodes of the element. This will only induce thermoelastic strains in the plane of the plate element, i.e. membrane thermoelastic strains.

The implementation of cross-sectional thermal gradients may be different for each finite element code, for instance:

- With MSC Nastran, it is possible to specify the average element temperature and a single gradient value through the thickness. This implies only a linear temperature variation through the thickness.
- ABAQUS allows to specify the temperature at each FE node location at multiple positions through the thickness of the plate. This allows for approximating any temperature profile through the thickness of the plate.

The ability to represent cross-sectional gradients is very useful. However, it also comes with some practical limitations. These limitations become mostly apparent at junctions of panels. Imagine panel A (see Fig. 6.6) with different temperature fields at both sides. Panel B, also with different temperatures at both sides, is interfacing with panel A. Figure 6.6 shows that first of all the two plate elements have some overlapping material, but most of all the definition of the temperature field in the joint is ambiguous. Away from the junction, this problem disappears. Therefore, in cases where the modelling of the temperature field near the junctions is of importance, it is recommended to represent the plate elements with solid elements. One could decide to limit this refinement in representation to the junction area. This allows for

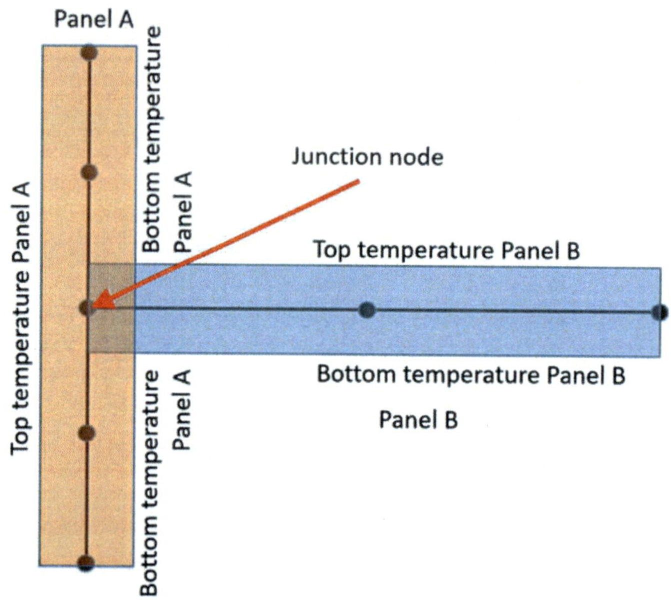

Fig. 6.6 Ambiguity of the specification of cross-sectional thermal gradients at a junction of two plates

a more physical representation of the temperature field and removes the problem of artefacts in the responses of which the contribution to the global effect is hard to estimate.

Although the formulation of plate elements is very powerful and versatile, they lack the possibility to connect structural elements to only one side of the plate that it is representing. It is also not possible to let mechanical equilibrium sort out how the load is distributed through the cross section of a plate element. Examples where this may be relevant are indicated in Fig. 6.7. In the top picture in Fig. 6.7, it is obvious that with any loading to the sandwich applied by the bracket, the lower face sheet will experience far less load than the top face sheet. Maybe less obvious, but also for the configuration of the lower picture in Fig. 6.7, the loading of the two face sheets is not the same. For this kind of situations, it is strongly recommended to switch to three-dimenaional modelling of the panel with solid elements for the core and shell elements for the face sheets. In Sect. 4.6 and Appendix A, it is shown how important this local modelling can be for the global deformation of the panel.

6.4.6 3-D, Volume (Solid) Element

Where 1-D and 2-D elements had to sacrifice on geometrical representation at the benefit of computational efficiency, volume or solid finite elements basically allow

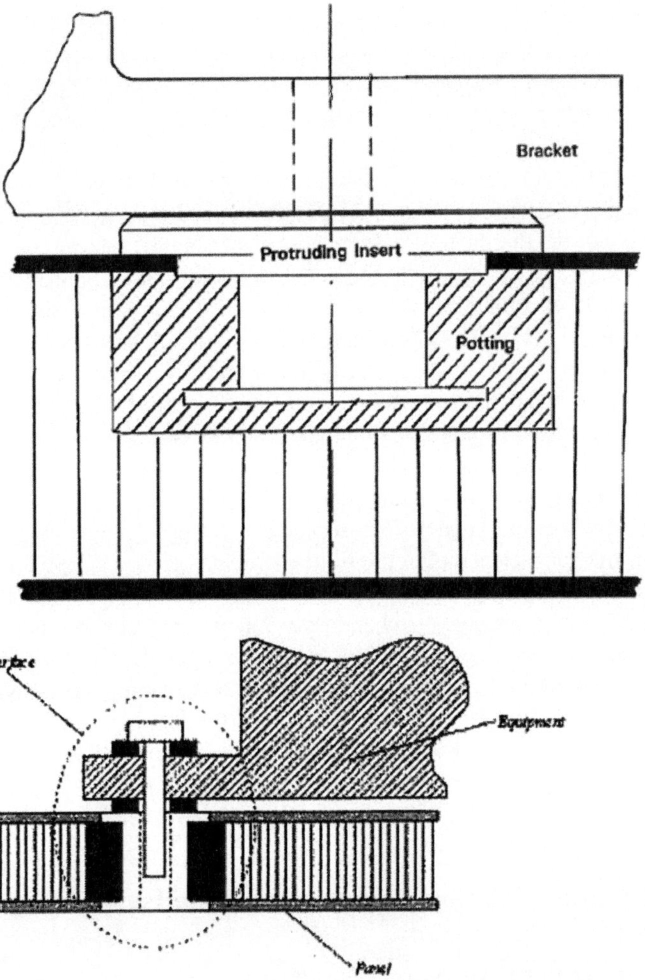

Fig. 6.7 Examples of typical attachment of equipment to a sandwich panel with a non-through and through insert [2]

for removing all geometrical simplifications in the model at the cost of computational effort. As was discussed with the previous element types, limitations of the 1-D and 2-D elements become especially evident when it comes to details around joints and assemblies for which it is recommended to switch to solid elements, at least locally. These local details are often not only important for capturing local thermoelastic effects, but due to the fact that the locally improved capturing of the stiffness, it often also has an importance for the global structural response that can make a difference in the thermoelastic context.

Most preprocessors can generate meshes of solid elements on 3-D CAD geometry. For complex shapes like the ones for machined parts, this is very convenient. In general, modern meshing algorithms are able to fill the solid geometry nicely with a regular mesh of tetrahedra elements. The more advanced mesh generators are also able to generate hexagonal- or brick-shaped elements for complex solid geometries. For the tetrahedra element, it is in general recommended to use the version with parabolic shape functions, and this is a 10-node tetrahedra element.

Irrespective of the thermoelastic application, one should take care that sufficient solid elements are used to represent the occurring deformation fields. For instance, in a cross section of a plate or bar that is subjected to transverse shear, the shear flow function has a parabolic shape as a function of the thickness coordinate with a zero shear at the free edges of the sections. This indicates a third-order deformation field. At least four linear elements are recommended for this kind of sections. When elements with parabolic shape functions are used, less elements are expected to be sufficient. When one wants to be sure to have the right mesh resolution, it is recommended to do mesh convergence checks.

The supported material models (iso-tropic, orthotropic, ...) depend on the implementation of the solid elements in the finite element software. The FE nodes connected solid finite elements only receive translational stiffness from these elements. This implies that only forces and no bending moments are produced at the FE nodes by these elements.

The application of the temperature field is done by assigning the temperatures to the FE nodes of the solid elements. This allows then for defining a three-dimensional temperature field and basically capturing any thermoelastic phenomena with the potential to represent thermal gradients in all three directions.

6.5 Constraint Equations and Rigid Elements

6.5.1 *Principle of Constraint Equations*

A constraint equation or multipoint constraint (MPC) defines a linear relation between degrees of freedom (displacements or rotations) of one or more FE nodes of the form

$$\sum_{k=1}^{N} a_k u_k = \delta_j, \tag{6.2}$$

where a_k is the weighting factor associated to the displacement u_k of FE node k. δ_j may be a prescribed relative displacement or thermally induced extension or reduction of a leg of a rigid element (see below) for constraint relation with sequence number j. In most applications of multipoint constraints, $\delta_j = 0$.

Traditionally, the concept of multipoint constraints is based on the principle that one degree of freedom, the slave or dependent degree of freedom, can be expressed as a linear combination of other degrees of freedom, the independent or master degrees of freedom.

Constraint equations are frequently introduced through the application of rigid body elements (RBE). In the case of rigid body elements, the finite element code generates, based on the relative position of the connected FE nodes, the weighting factors for the constraint equation Eq. (6.2). MSC Nastran allows also for the definition of explicit MPC equations through which the user can manually enter the weighting factors.

The commonly used RBE elements of MSC Nastran are the RBE2 element, which is a real rigid element , and RBE3 element, which is an interpolation element. Both are based on the principle of linear constraint equations of the form of Eq. (6.2).

Appendix D provides background information on mathematical methods for implementing these constraint relations into the system equations of the finite element model.

6.5.2 The Interpolation Element

The RBE3 element generates a set of constraint equations that makes the displacements and rotations of one FE node dependent on the displacements and rotations of the other connected FE nodes. The relation defines the motion at the dependent FE node as the weighted average of the motions of the independent FE nodes. This multipoint constraint is not simulating a rigid body behaviour. Its characteristic is therefore often referred to as interpolation element, rather then rigid element.

The RBE3 is popular in cases where the mass of an equipment is known, but one would like to ignore the (sometimes unknown) stiffness contribution of such an item. In dynamic analysis, this will produce a lower eigenfrequency of the structure. In such a case, the use of the RBE3 is considered to provide conservative predictions for the stiffness of the structure.

Since the RBE3 is not introducing any stiffness, it is missing an important property to contribute to the thermoelastic response of a structure. However, because of its interpolation property it is frequently used in thermoelastic problems. A typical example is the use of RBE3 for computation of the effective response of the interfaces of an item represented by sometimes many FE nodes. Optical elements of an instrument, such as lenses or mirrors, are not always explicitly modelled, but the FE nodes at the interfaces of the optical item are included in the model. The change of position or orientation of this kind of items is important for the functioning of the instrument. The dependent FE node or slave FE node of the RBE3 is for these kind of circumstances used to compute a sort of average or effective translation and rotation of the interface on the basis of the movement of independent FE nodes at this interface.

This use of RBE3 may look convenient; however, the user documentation of MSC Nastran points to some limitations that may be relevant to be aware of for this purpose. It is therefore recommended to use the RBE3 with care. The way the weighting factors are determined is by far not obvious. For this reason, it is suggested to compute the effective translations and rotations of the interface plane through geometrical operations such as constructing the interface plane and determining the tilted and translated versions of this plane from the translations of the FE nodes at the interface. In this way, it is exactly known how the effective or average translation and rotation of the interface plane is determined and, not less important, also the level of approximation is known. The latter information is difficult to obtain from the MSC Nastran documentation.

6.5.3 The Rigid Body Element

In the RBE2 element implementation of MSC Nastran, there is one FE node of which all six degrees of freedom can have the role of independent degrees of freedom. The RBE2 causes all connected other FE nodes to translate and rotate with the independent FE node as a rigid body. This formulation prevents any movement of the connected FE nodes relative to the independent FE node.

In dynamic analysis models, especially the system-level models for spacecraft, stiff units are represented by rigid body elements. Each rigid element connects the mass of the unit, represented by a lumped mass at the centre of gravity of the unit, to the interface points at the panel to which it attached. A typical way of representing such stiff units is shown in Fig. 6.8. In some cases, the connection of the rigid element with the panel is extended with some springs, joining the FE nodes of the rigid element to the FE nodes at the panel, allowing to extract interface loads. For dynamic analysis, this method has proven to be adequate.

As mentioned before, when the information on the stiffness of the equipment is not yet available, an RBE3 element provides the ultimate lower bound of the stiffness of the item. In early design phases, units are frequently modelled with RBE3.

MSC Nastran supports the possibility to assign a CTE to the rigid elements. At first sight, this may be quite convenient and could allow the use of the simplified representation of an electronics box also for thermoelastic analysis. However, one should realise how the RBE2 elements simulate the thermal expansion.

Consider the temperature field that can typically occur for a electronics box mounted on a sandwich panel as is visualised in Fig. 6.9. This temperature field at the FE model is obtained through temperature mapping of the thermal analysis results (see Chaps. 7 and 8). What can be observed is that, thanks to the highly conductive baseplate of the equipment, the temperature at all six interface points is more or less the same. The FE nodes at the feet of each leg of the RBE2 of Fig. 6.8 get therefore the same temperature assigned.

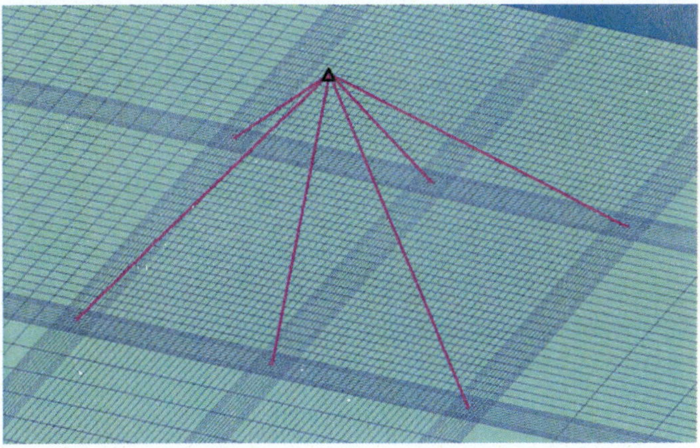

Fig. 6.8 RBE2 or RBE3 on a panel

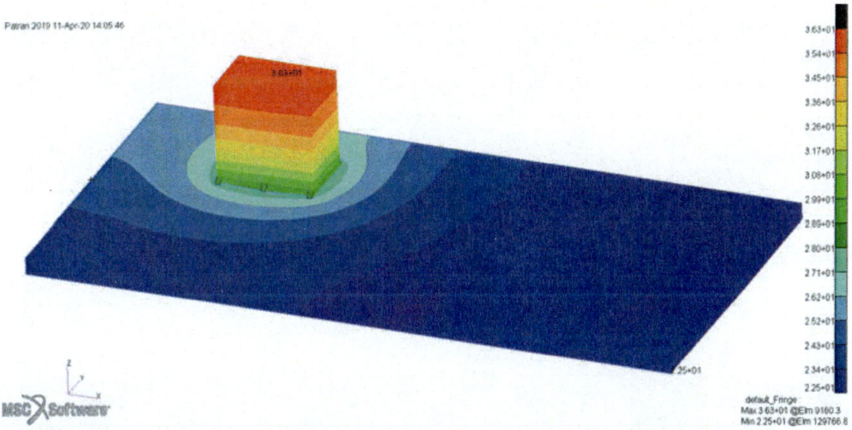

Fig. 6.9 Temperature field for an electronics box mounted on a sandwich panel

The temperature of the FE node at the centre of gravity is for each leg the same, being the average temperature of the unit. MSC Nastran computes the thermal expansion of a leg using the average temperature of that leg, being the average of the temperatures of the foot and centre point. Since the average temperature of each of the legs in the RBE2 in Fig. 6.8 is more or less the same, the shape of the RBE2, and especially the relative out-of-plane position of the interface points, is hardly changed. As a consequence, the interface plane between the box and the panel is forced to stay flat. Is this correct? Would not the thermal gradient in the box as is shown in Fig. 6.9 suggest a bending of the box? Would the locally higher temperature of top skin not also indicate a potential to induce bending of the panel?

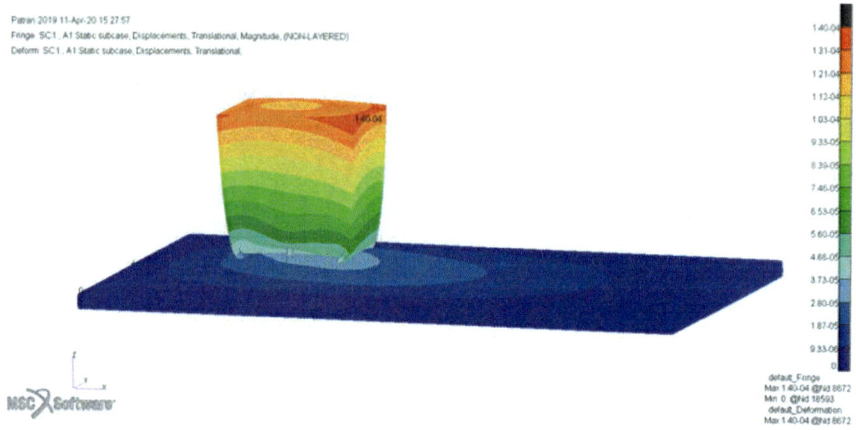

Fig. 6.10 Thermally induced deformation on an electronics box mounted on a sandwich panel

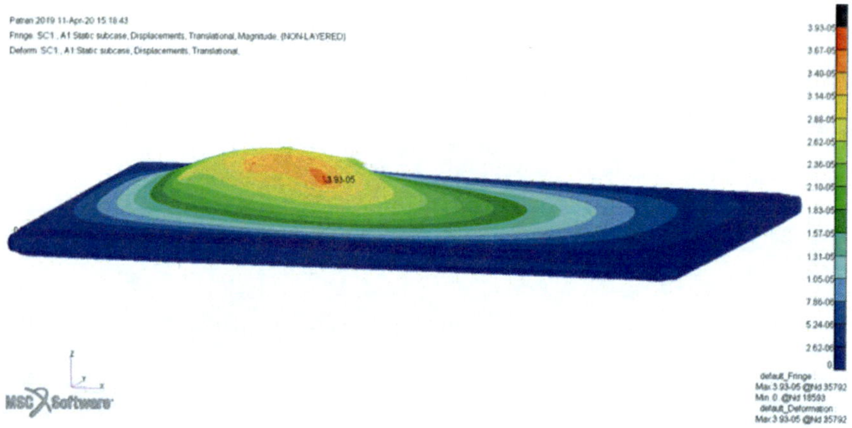

Fig. 6.11 Thermally induced deformation on a sandwich panel induced by a dissipating electronics box

Figures 6.10 and 6.11 show that indeed the box is changing shape due to the temperature field in the box and that this effect in combination with an especially locally important temperature difference between top and bottom skin induces bending of the sandwich panel.

As is discussed in great detail in Sect. 4.6, rigid body elements are not suitable for simulating the thermoelastic response of equipment. The predicted influence of the stiff units on the deformation of the structure on which these are mounted is in general simply wrong. The curvature that these boxes due to their internal temperature field induce on the panels is not captured. Essential effects on pointing sensitive items, especially on the same panel, are then not simulated and therefore ignored.

The simplified modelling of units with rigid body elements is a way to limit the computational effort. Section 4.6.3 is providing recommendations to model these units in more detail to capture the physical effects, but still with a limited impact on the size of the models.

A special application of the RBE2 element is its use to connect two parts in the model by tying two coincident FE nodes, each being part of the different parts of the model. Depending on the type of joint, that is intended to be represented, all or just a limited set of degrees of freedom of the coincident FE nodes are connected by the RBE2. In the context of thermoelastic simulation, this application of the RBE2 elements is not introducing any non-physical behaviour. Due to the fact that the two connected FE nodes are coincident, the length is zero. Thermal expansion of a zero length item is zero, and the thermoelastic response is also zero. As a result, this use of RBE2 is not introducing any non-physical constraints on the surrounding parts in the FE model. However, one should not ignore the singularity problem that is introduced by connecting to a two- or three-dimensional mesh at a discrete location, as was discussed in Sect. 6.4.1.

Application of MPC

A fixed-free rod is modelled with six FE nodes and five rod elements as shown in Fig. 6.12. All degrees of freedom of FE node 1 are fixed (constrained). The active degrees of freedom of the FE nodes 2 to 6 are the axial displacements. The other degrees of freedom of the FE nodes are fixed.

The nodes are placed equidistant with a distance ΔL. The CTE is α, and Young's modulus is E. The temperature raise ΔT is constant along the length of the truss.

A linear constraint is introduced with the objective to keep the axial displacement of FE node 3 and 5 the same. The linear relation for this constraint is

$$u_3 - u_5 = 0 \tag{6.3}$$

with u_3 and u_5 being the axial displacements of the corresponding FE nodes as indicated in Fig. 6.12.

The displacement along the length of the rod is visualised in Fig. 6.13

The finite elements between FE nodes 1 and 3 and between FE node 5 and 6 are free to expand without any stresses. In these elements, no stresses are built up by the temperature increase. The total displacement of node 6 is $u_6 = 3\alpha\Delta T\Delta L$.

Due to the constraint of Eq. (6.3), there is no relative displacements between FE node 3 and FE node 5. As a consequence, also FE node 4 is not moving relative to FE

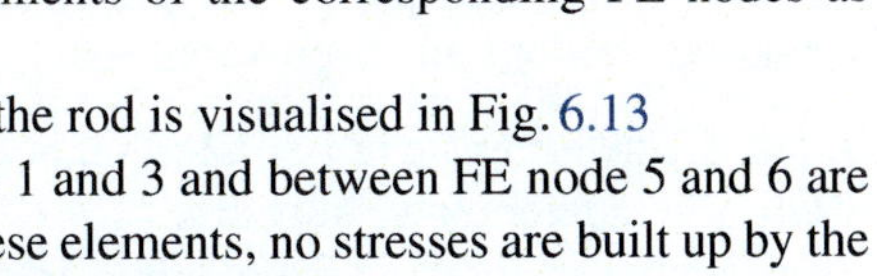

Fig. 6.12 Fixed-free rod

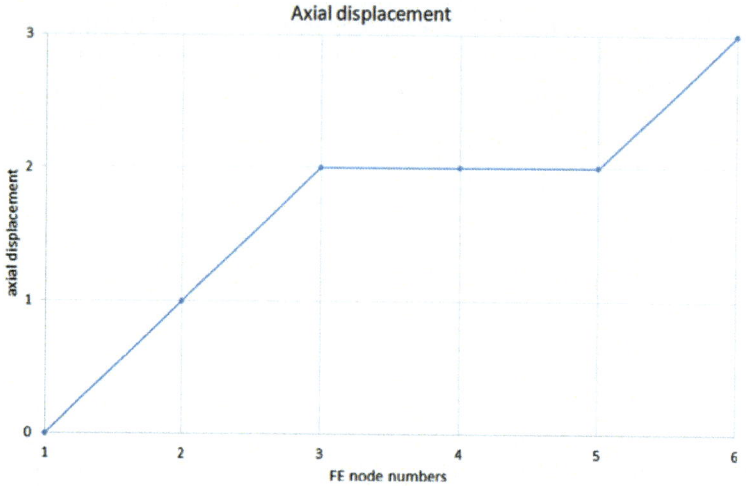

Fig. 6.13 Axial displacement along the rod. Units of displacements are $\alpha \Delta T \Delta L$

nodes 3 and 5. This makes that the linear constraint suppresses the strain ε between node 3 and node 5, but the thermoelastic stress is $\sigma = -E\alpha\Delta T$.

This example shows how linear constraint equations can affect the simulated thermoelastic displacement field.

6.6 Boundary Conditions

In general, boundary conditions have a significant influence on results of structural (thermoelastic) response analyses. In this section, the following types structural boundary conditions are discussed:

- Statically determinate or iso-static supports
- Statically indeterminate supports
- Inertia relief method in case of a free-free structure.

6.6.1 Iso-static Supports

Iso-static boundary conditions also referred to as as statically determined boundary conditions are boundary conditions that allow the reaction forces at the constraint locations to be determined with only the equilibrium equations [42]. This kind of

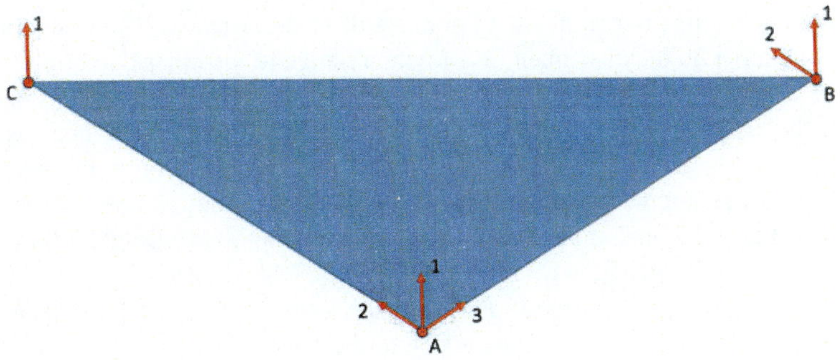

Fig. 6.14 Principle of constraints to simulate an iso-static support with three points

boundary conditions is often implemented in the support system of optical instruments, because an ideal or perfect iso-static support system of a unit prevents that deformation of the underlying structure is able to inject loads into the unit.

In theory, a structure is already supported in a statically determined way through just a single attachment point. This single point should then take all three reaction forces and three reaction moments. In principle, this single point can be designed to handle well the translational forces. However, the bending moments will lead to high stresses at this single interface point which would in general result in an infeasible design. Therefore, for an ideal iso-static support, it is needed to support the structure at exactly three points.

The three support points of an iso-static boundary condition span a plane (see Fig. 6.14) that could be imagined as a schematic representation of a structure of an instrument. The most basic definition of an iso-static support is presented in Fig. 6.14, in which arrows are used to indicate which constraints need to be applied at each of the individual support points to simulate an iso-static support. Three sets of constraints are applied, each with its own function:

- Suppression of translation of the surface in the direction normal to its plane. This is achieved with the translational constraints marked with “1”.
- Suppression of rotation of the surface in its own plane, i.e. around an axis normal to the plane of the surface. This is achieved with the translational constraints marked with “2”.
- Suppression of last available free translation of the surface in its own plane. This is achieved with the single translational constraint marked with “3”.

Although the constraints presented in Fig. 6.14 only constrain translations, it is important to note that iso-static constraints do constrain rotations in an indirect way. This can be observed from Fig. 6.14 in which at the attachment points only translations are constrained and no rotations. The attachment points act as hinges with rotational freedom around all three axes. These kind of supports are also called “ball-bearings”.

In case one of the three attachment points, due to deformation of the supporting structure underneath the instrument, is moving relative to the other two points, then no forces as reaction to this relative movement of the points will occur. Consider for instance the case that point A in Fig. 6.14 moves normal to the plane and point B and C do not move. This will cause a rotation of the structure around the line B-C without introducing any loads on the structure. Other rotations can be imagined as well; for instance, a translation of point B in the direction of constraint 2 will give a rotation around point A with a rotation axis normal to the surface.

From above observation follows that supports with more than three attachment points will not be able to respond stress free to the relative movement of one of the attachment points.

One important drawback of the basic iso-static support system is that the movements as a rigid body of the supported unit have to be reacted by only one or two support points. In the configuration as presented in Fig. 6.14, the support at point A will therefore experience the highest amount of loading when the supported unit is subjected to for instance accelerations in direction "3".

Because the iso-static support has the characteristic that the supporting structure cannot introduce deformation in the supported unit, also the opposite is true: Thermal expansion of the unit is not constraint by the supporting structure.

In Fig. 6.14, all movement of the surface is relative to point A. This would cause a de-centred expansion of the structure of the supported unit. For that reason, a concept of iso-static mounts is often used in which the structure is free to expand relative to an (imaginary) centre point. Figure 6.15 shows an example of the application of iso-static mounts in the form of bip-pods which are used in sets of three, each oriented under 120° relative to the other two. This keeps the optical unit and its optical elements centred.

Quite often designers have to accept the compromise to introduce more attachment points in order to reduce internal responses under dynamic loading through more support points.

6.6.2 Statically Indeterminate Supports

When the number of equilibrium equations is not sufficient to solve the reaction forces, the structure is denoted as statically indeterminate. Such a system is also called "overdetermined". In other words, there are more than six degrees of freedom constrained. This requires that in addition to the equilibrium equations, also compatibility relations, involving the stiffness of the structure between the interface points, need to be taken into account to determine the reactions forces at the constrained degrees of freedom.

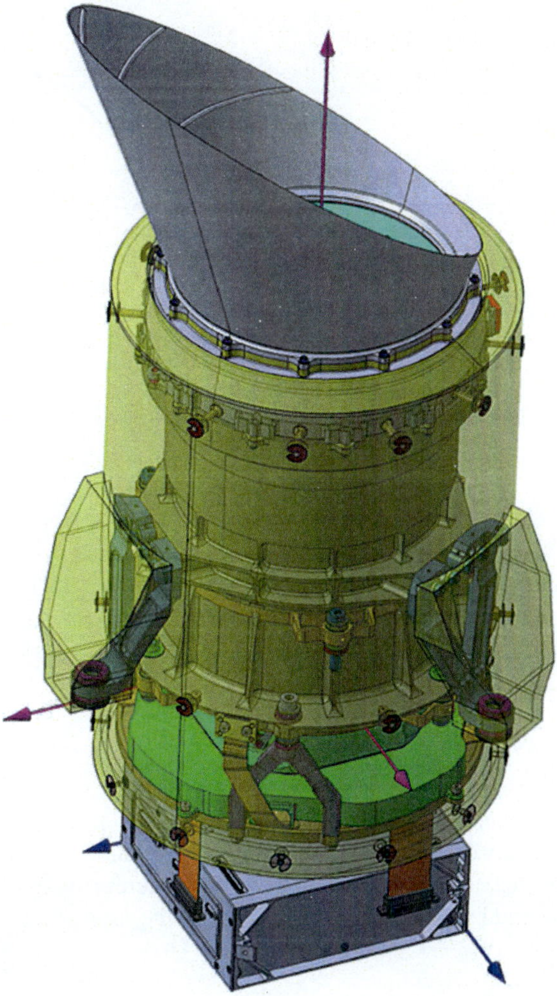

Fig. 6.15 Application of iso-static mounts in the form of bi-pods in the PLATO-TOU [10]

Iso-static and statically determinate truss frame

In Fig. 6.16, two 2-dimensional examples of an iso-static and a statically indeterminate truss frame are given. For a two-dimensional case, a free-free structure has three degrees of freedom and three corresponding equilibrium equations. The iso-static structure (top picture in Fig. 6.16) has three degrees of freedom constrained, i.e. two at node 1 and one at node 5. So, the reaction forces of this system can be solved completely with only the three equilibrium equations. The additional spring in node 5 for the statically indeterminate case (bottom picture in Fig. 6.16) causes that the number of constraints for the structure becomes four. This makes that the number of equilibrium equations is not sufficient any more to compute the reaction forces.

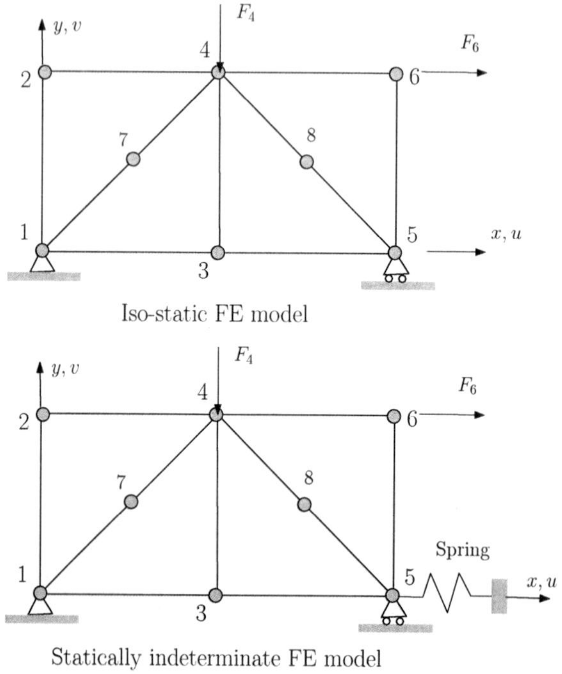

Fig. 6.16 Iso-static and statically indeterminate truss frame

Provided that the CTE of all materials used in both structures is the same, the statically determined structure is able to expand stress free under a uniform temperature increase. However, the statically indeterminate structure will experience stresses induced by the additional constraint of the spring.

6.6.3 Intertia Relief Method

For the calculation of thermoelastic responses, a static finite element analysis needs to be run. This requires normally that the model does not contain mechanism and therefore the model requires the supports to remove these mechanisms that mathematically represent singularities in the system to be solved. There are, however, circumstances, where there are simply no constraints. The most clear cases are a free-flying spacecraft, launch vehicle or aircraft. There may also be cases where any choice of boundary conditions, only for the sake of getting a solution, is not convenient. For this purpose, the inertia relief method is introduced.

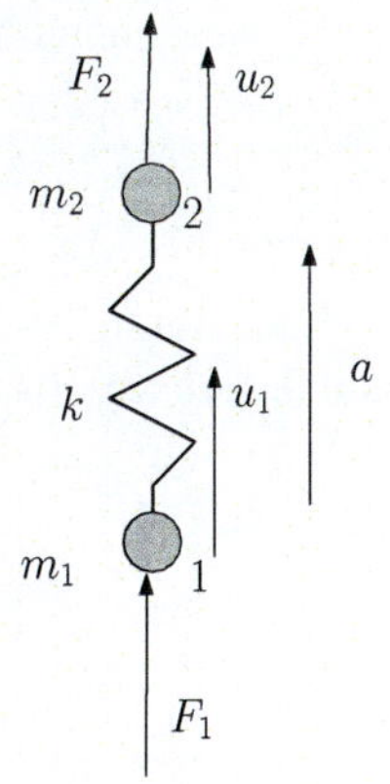

Fig. 6.17 Mass spring structure

The inertia relief method can be applied to unconstrained structures. The theory about the inertia relief method is explained in [62, 93]. Also, the MSC Nastran Linear Statics Analysis User's Guide [69] provides a short introduction to the method that is also implemented in MSC Nastran. The principle of the method is explained below with the help of a one-dimensional system of two masses connected with a spring.

Consider a mass spring structure as shown in Fig. 6.17. The spring has a stiffness k (N/m), and the discrete masses are attached to nodes 1 and 2, m_1, m_2 (kg). The displacements in the nodes are u_1 and u_2. A constant acceleration of a (m/s^2) is assumed. The d'Alembert forces or inertia forces in the FE nodes are $m_1 a$, and $m_2 a$, respectively [67].

To show the principle of inertia relief, the approach applied in [62] will be used. During a steady-state acceleration, the sum of the inertial forces of the spring mass system is in equilibrium with the external forces

$$F_1 + F_2 = m_1 a + m_2 a \tag{6.4}$$

which can rewritten into

$$a = \frac{F_1 + F_2}{m_1 + m_2} \tag{6.5}$$

The internal forces in the system are those in the spring:

$$F_s = k(u_1 - u_2) \tag{6.6}$$

At steady state of the mass spring system with acceleration a, the external, the internal and the d'Alembert's forces are in equilibrium at every FE node (the sum of applied force and inertia force is called the effective force):

$$\begin{aligned} F_1 - F_s - m_1 a &= 0 \\ F_2 + F_s - m_2 a &= 0 \end{aligned} \tag{6.7}$$

In matrix notation, Eq. (6.7) can be written as

$$\begin{bmatrix} k & -k \\ -k & k \end{bmatrix} \begin{pmatrix} u_1 \\ u_2 \end{pmatrix} = \begin{pmatrix} F_1 \\ F_2 \end{pmatrix} - \begin{pmatrix} m_1 a \\ m_2 a \end{pmatrix} \tag{6.8}$$

Equation (6.5) is now substituted in Eq. (6.8), but Eq. (6.8) is a singular system, and the relative displacements can be solved by restraining either u_1 or u_2

$$\begin{aligned} u_1 = 0,\ u_2 &= \frac{F_2 m_1}{k(m_1+m_2)} - \frac{F_1 m_2}{k(m_1+m_2)} \\ u_2 = 0,\ u_1 &= -\frac{F_2 m_1}{k(m_1+m_2)} + \frac{F_1 m_2}{k(m_1+m_2)} \end{aligned} \tag{6.9}$$

It can be seen that the displacement u_1 relative to u_2 is the same for either constraint. As a consequence, the spring force $F_s = k(u_1 - u_2)$ is for both constraints the same, showing that for the inertia relief method not the magnitude of the displacements of the individual FE nodes are relevant, but the relative values of the displacements.

Since the inertial load distribution relies on the assumption of constant rigid body acceleration a, it is important to have in the model the correct mass distribution. This is important even for a static thermoelastic analysis when the inertia relief method is used.

The following basic example illustrates how the application of inertia relief works for thermoelastic analysis. The results are compared with results obtained through the application of iso-static boundary conditions.

Application of Inertia relief to a small one-dimensional thermoelastic problem

Considered is a FE model consisting of two rod elements and three FE nodes. This model is shown in Fig. 6.18.

The rod elements are assigned the material properties of an Al-alloy with Young's modulus $E = 70$ GPa, CTE $\alpha = 24 \times 10^{-6}$m/m/°C and density $\rho = 2700$ kg/m^3. The cross section of the rod elements is $A = 0.0004$ m^2, and the total length of the rod is $2L = 0.5$ m. The temperature elevation at all three nodes is $\Delta T = 50$ °C.

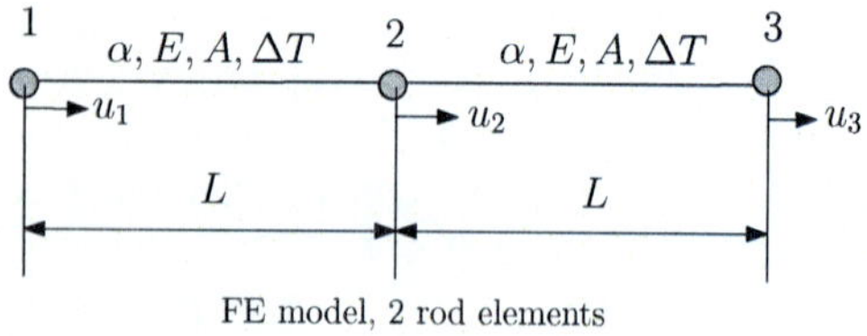

Fig. 6.18 Simple rod model

Table 6.3 Thermoelastic deformations

	Displacements (m)		
Method	Node 1	Node 2	Node 3
Iso-static	0.0	0.3×10^{-3}	0.6×10^{-3}
Inertia relief	-0.3×10^{-3}	0.0	0.3×10^{-3}

Two cases are analysed:

- Case 1: Using an iso-static boundary condition by constraining the displacement in node 1 ($u_1 = 0$)
- Case 2: Without any constraints, thus free-free, with the application of the inertia relief method.

Both the free-free case (Case 2) and the case with the iso-static boundary condition (Case 1) allow the rod to expand freely. The displacements of the three FE nodes are calculated and presented in Table 6.3.

The results show nicely that for both cases the relative displacements between node 1 and 3 are 0.6×10^{-3}. In these two cases, the rod is free to expand, and hence, no thermoelastic stresses occur.

6.7 Refurbishing a Dynamic Finite Element Model for Thermoelastic

6.7.1 Introduction

In most spacecraft projects, the first structural models that are made in the project aim at demonstrating the compliance to the stiffness requirements in terms of dynamic modal properties of which an important one is the first eigenfrequency of the structure. These models, here referred to as dynamic models, evolve over the time of the project and are mostly used for dynamic modal and dynamic response analyses in the time and frequency domain [21, 27, 93]. As is explained at various locations in this book, simplifications in the representation of details of the structure that are fully acceptable for the dynamic response prediction may not always be adequate for the simulation of the thermoelastic responses. It is therefore recommended to review the dynamic model on adequacy for thermoelastic analysis.

In this section, it is explained how the dynamic model for different aspects is falling short to predict properly thermoelastic responses. This is complemented with suggestions on how to modify the dynamic model. Applying the modifications is in this section referred to as "refurbishing" and could be considered as the conversion of a dynamic finite element model into a thermoelastic finite element model.

This section does not claim to be complete, and there are definitely other important aspects that are not mentioned that cause the dynamic model produce less reliable thermoelastic results. The objective of this section is to make the reader aware of these limitations of the dynamic model and inspire him or her to have a better look at the dynamic model before applying it for thermoelastic analyses. This section is recommended to be read together with Sect. 4.8.

6.7.2 *Required Mesh Resolution for Dynamic and Thermoelastic Models*

In an early phase of a spacecraft-associated project, the stiffness characteristics of the spacecraft, scientific instrument, solar arrays (fixed and deployed), etc., have to be verified against stiffness requirements. In general, the stiffness requirements for spacecraft or its sub-systems are specified for fixed or hard mounted interface (I/F) configurations expressed in undamped minimum allowed natural frequencies and associated modal effective masses. For spacecraft and large instruments, lowest natural frequency requirements are specified in lateral and longitudinal (launch) direction, where for smaller instruments, minimum natural frequency is specified in all directions. Calculation of natural frequencies is in fact solving the the undamped eigenvalue problem, eigenvalues and associated eigenvectors or vibration modes taking into account the boundary conditions (most limes fixed-free). The eigenvalue problem is given by

$$(-\omega^2[M] + [K])\,(\phi_i) = (0)\,, \tag{6.10}$$

where ω^2 is the eigenvalue, $[M]$ is the mass matrix, $[K]$ is the stiffness matrix, and $(\phi_i)\,,\ i = 1, 2, \ldots$ are the eigenvectors. The natural frequency (Hz) can be obtained by $f = \omega/2\pi$. The mass and stiffness distributions need to be carefully estimated or calculated. Assembly of the mass and stiffness matrices is performed using FEA by creating a dynamic FE model. In general, the calculation of the first number of eigenvalues and mode shapes can be still performed accurately with a rather coarse finite element mesh. The question is whether these meshes are also adequate for the thermoelastic analyses. The following examples aim to illustrate the limitations for thermoelastic predictions of a mesh sufficiently fine for dynamic analysis.

Cantilevered beam

This example considers a cantilevered beam for which the theoretical expressions for the natural frequencies and mode shapes are provided by Blevins [17]. The bar has a length $L = 1$ m and a cross section $A = 0.02 \times 0.02$ m^2, leading to a second moment of area is $I = 1.33333 \times 10^{-8}$ m^4. The Young's modulus is $E = 70$ GPa, and the density of the material is $\rho = 2700$ kg/m^3. With this data, the mass per unit length is $m = A\rho$ (kg/m), and the bending stiffness is EI (Nm^2).

The expression for the natural frequency f_i (Hz) is

$$f_i = \frac{\lambda_i^2}{2\pi L^2}\sqrt{\frac{EI}{m}}, \quad i = 1, 2, \ldots \tag{6.11}$$

and for associated mode shape:

$$\phi_i(x) = \cosh\left(\frac{\lambda_i x}{L}\right) - \cos\left(\frac{\lambda_i x}{L}\right) - \sigma_i\left(\sinh\left(\frac{\lambda_i x}{L}\right) + \sin\left(\frac{\lambda_i x}{L}\right)\right) \tag{6.12}$$

The constants λ_i, σ_i, the theoretical natural frequencies and associated modal effective mass are presented in Table 6.4.

Now, the same cantilevered beam is represented by a simple FE model with N beam elements and $N + 1$ nodes. The classical stiffness and consistent mass matrix of the beam element are used [61]. The beam is clamped at node 1. Varying the number of finite elements, the first three natural frequencies and associated modal effective masses are compared with the theoretical values presented in Table 6.4. The analysis results are presented in Table 6.5.

Even with only one beam element, the first eigenfrequency is engineering-wise accurate enough. However, the modal effective mass deviates significantly from the theoretical value. The results show that for the calculation of the first three eigenfrequencies and associated modal effective masses a coarse FE model with only five elements could already be considered adequate. In the following, it is investigated if this mesh of the FE model is also sufficient to capture a temperature field.

The temperature field for which this is verified originates from the TMM that is presented in Fig. 6.19. Assumed is that the thermal analysis has produced the temperatures for the thermal nodes as is indicated in the figure. In same Fig. 6.19, two FE models for the beam structure with different mesh resolutions (5 and 10 beam elements) are shown along the thermal model. The PAT temperature mapping method (Chap. 8) was used to map the temperature field of the thermal analysis on the two FE models.

Figure 6.20 presents the temperature mapped on the model with five beam elements that showed to be adequate to represent the modal properties of the beam.

As was explained, the same temperature field produced by the thermal model is also mapped on the FE model of the beam structure with ten elements. This temperature field is presented with the graph in Fig. 6.21.

Comparing the temperature fields in Figs. 6.20 and 6.21 reveals that, even with such a simple model, the lower resolution FE mesh is literally cutting corners in representing the temperature field. The FE model with ten beam elements is able to capture both the temperature gradients and the change of temperature gradients much better.

This example shows in a simple way that the mesh resolution requirements for dynamic modal analysis and thermoelastic for the same structure may differ quite strongly. Therefore, before a dynamic model is used for thermoelastic analysis, a

Table 6.4 Constants λ_i,σ_i, natural frequencies and modal effective masses, [17, 93]

Mode #	λ_i	σ_i	Frequency (Hz)	Modal effective mass M_{eff_i} (kg)
1	1.87510407	0.734095514	16.45043	0.66212
2	4.69409113	1.018467319	103.0931	0.20336
3	7.85475744	0.999224497	288.6637	0.06991

Table 6.5 Natural frequencies f_i (Hz) and modal effective masses M_{eff_i} (kg) [17, 93]

Mode	Theoretical		$N = 1$		$N = 2$		$N = 5$		$N = 10$	
#	f_i	M_{eff_i}	f_i	M_{eff_i}	f_i	M_{eff_i}	f_i	M_{eff_i}	f_i	M_{eff_i}
1	16.45	0.662	16.53	0.396	16.46	0.619	16.45	0.659	16.45	0.662
2	103.09	0.203			103.97	0.115	103.24	0.193	103.10	0.202
3	288.66	0.070					289.70	0.057	288.74	0.068

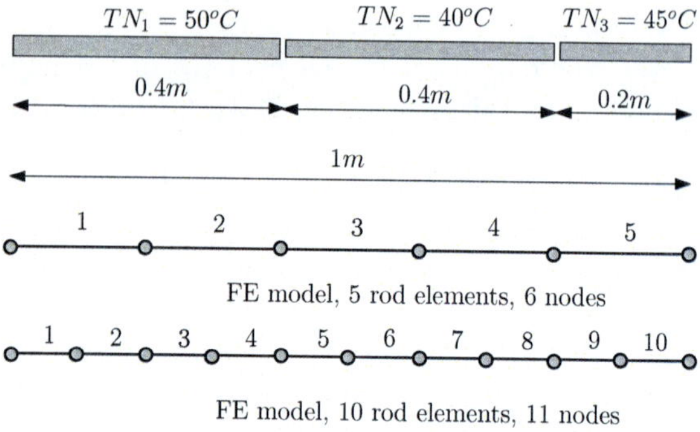

Fig. 6.19 TMM, thermal node temperatures and corresponding FE models

good understanding of the ability of the FE mesh to capture the temperature field is needed. In most real-life cases, it is not that obvious as in this example that the mesh is not adequate to capture a temperature field. For that reason, it is recommended, as explained in Sect. 4.5, to run a mesh convergence checking exercise.

Capturing gradients

Considered is now a square aluminium plate of 0.9×0.9 m with a thickness of 2 mm. In the centre of the plate, a square patch of 0.1×0.1 m and a thickness of 50 mm are representing a heat dissipating box.

For a modal dynamic analysis, in general, a finite element model with a resolution as presented in Fig. 6.22 would be adequate. The plate is considered to be simply supported.

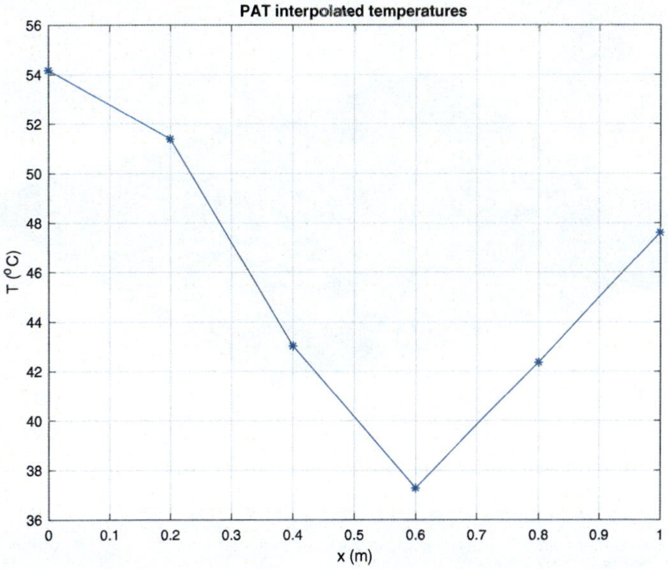

Fig. 6.20 Temperature distribution along the length the FE model with five rod/beam elements

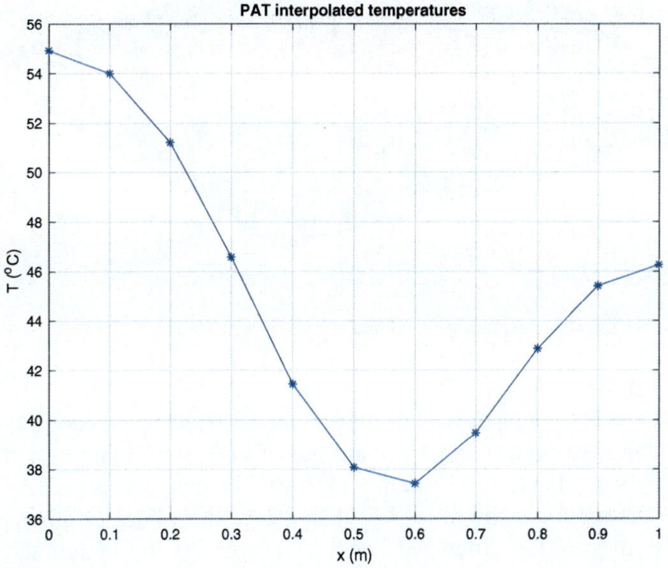

Fig. 6.21 Temperature distribution along the length the FE model with five rod/beam elements

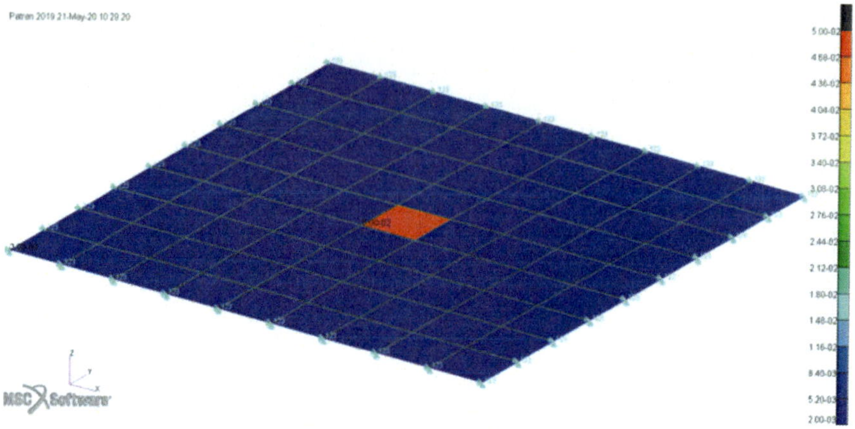

Fig. 6.22 FE model of square plate with typically adequate for modal analyses

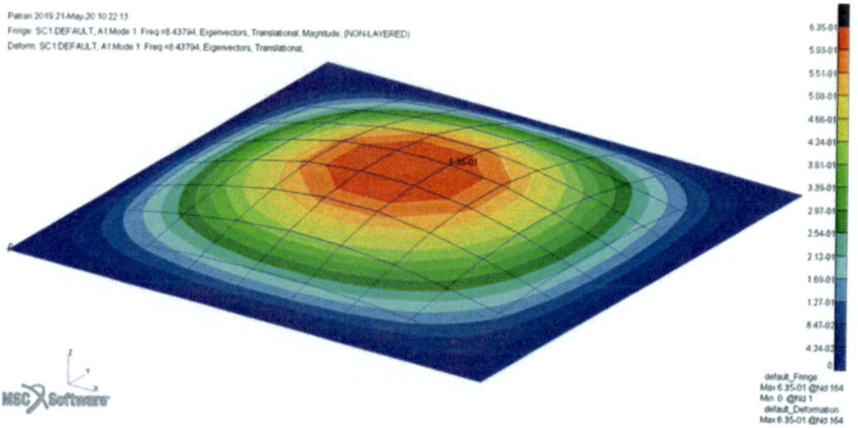

Fig. 6.23 Mode shape 1 of square plate with simplified box at 8.43 Hz

The first two mode shapes with significant modal effective masses [94] obtained with this model are mode 1 and 4 and are shown in Figs. 6.23 and 6.24. The mesh is reasonably well able to represent these modes. The eigenfrequencies of a model with twice the mesh density differ less than 1% of the values of this model. For the effective masses, the differences are a bit bigger, but are limited to 7%. This confirms that the mesh has converged for the purpose of modal dynamic analysis and is therefore considered adequate for this purpose.

The question now is, whether this mesh is also good enough to capture the temperature field produced by the thermal model.

The box is dissipating 60 W, and the edges of the plate are constrained to 20 °C. The only heat exchange mechanism considered is conduction.

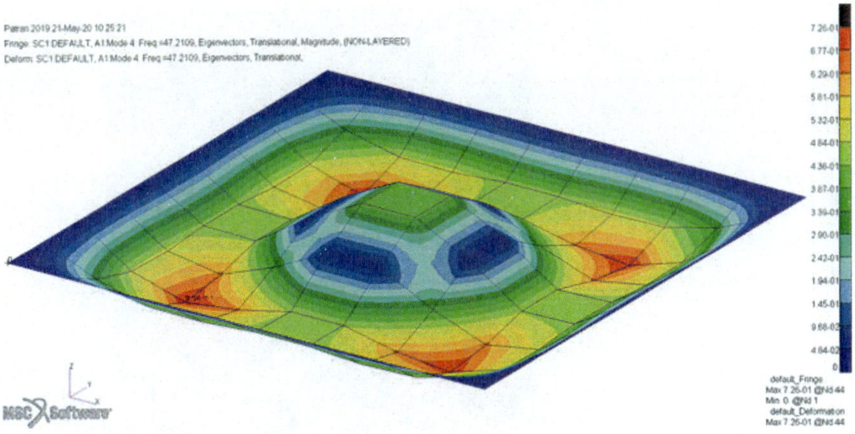

Fig. 6.24 Mode shape 4 of square plate with simplified box, 47.2 Hz

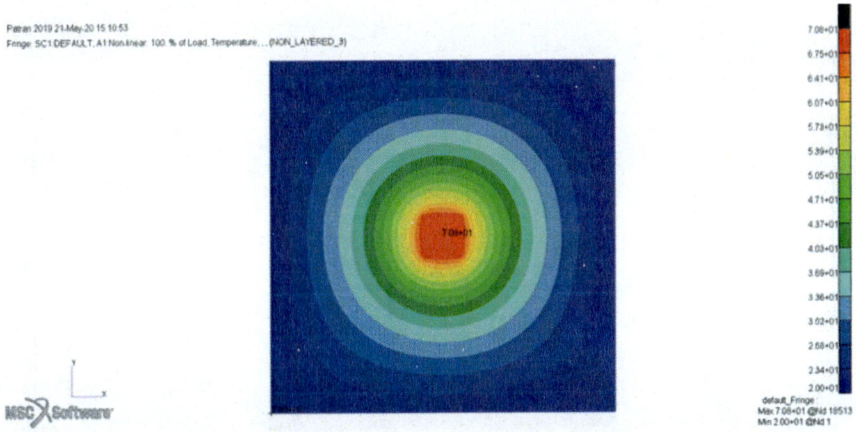

Fig. 6.25 Reference temperature field computed with a high-resolution FE model

As a first step, a reference temperature field is computed. For that purpose, a FE model with a (extremely) high mesh resolution of 900 × 900 CQUAD elements is prepared. The result is a nice symmetric temperature field as presented in Fig. 6.25. As can be expected, high thermal gradients are present at the edge of the simplified box.

Considered is now a rather high-resolution thermal lumped parameter model, which a thermal engineer, who is apparently aware of the thermoelastic needs, has made. The thermal mesh is presented in Fig. 6.26 in which every colour represents a different thermal node. The linear conductors (GLs), used by the thermal model, are computed with the method described in Sect. 9.3.2. For that purpose, the thermal node geometries are meshed with FE elements, which are also shown in Fig. 6.26.

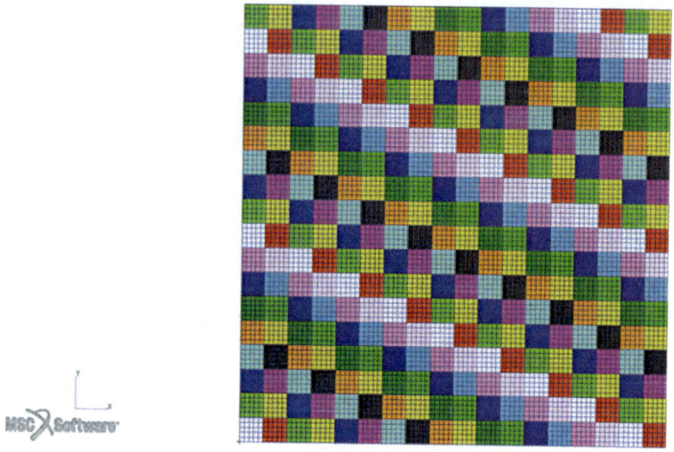

Fig. 6.26 Thermal lumped parameter mesh of the plate

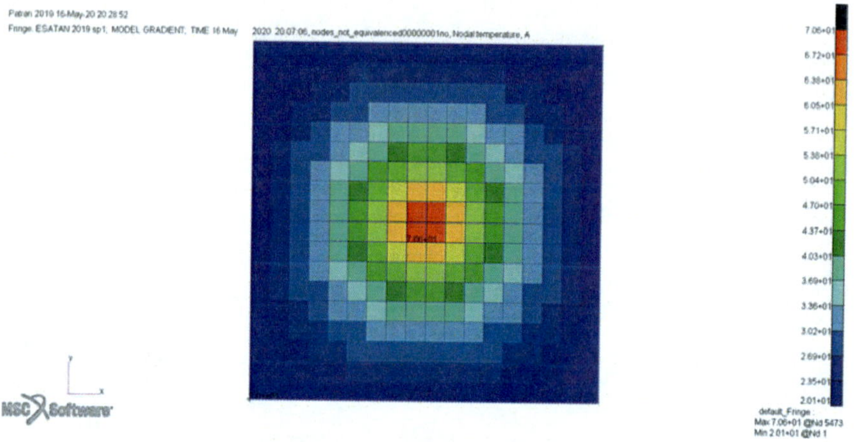

Fig. 6.27 Thermal node temperatures without interpolation

With this thermal model, the thermal node temperatures are computed and presented in Fig. 6.27, in which the thermal node temperatures are mapped without any interpolation on the same mesh that is used for the computation of the linear conductors (see Fig. 6.26). It can be noted that the maximum temperature computed with the lumped parameter model (Fig. 6.27), is not far off from the maximum temperature in the reference temperature field (Fig. 6.25). This could be an indication that the thermal lumped parameter mesh resolution is well able to represent the reference temperature. It is not an evidence that the thermal mesh does not need further refinement to get convergence of the thermoelastic responses.

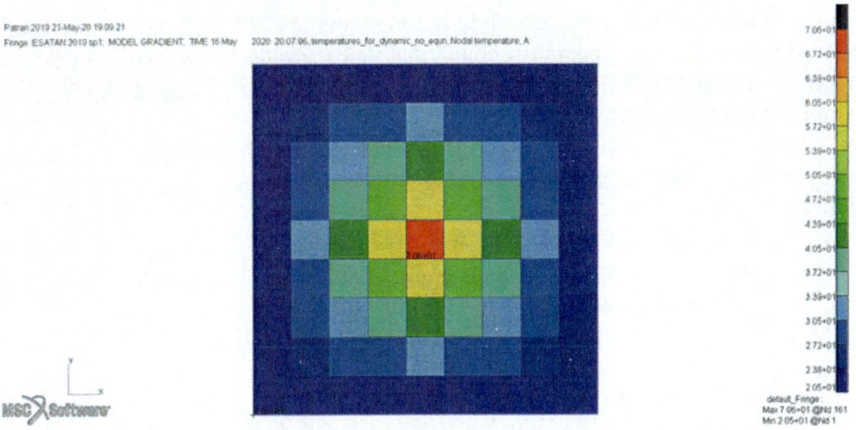

Fig. 6.28 Aggregated thermal node temperatures without interpolation

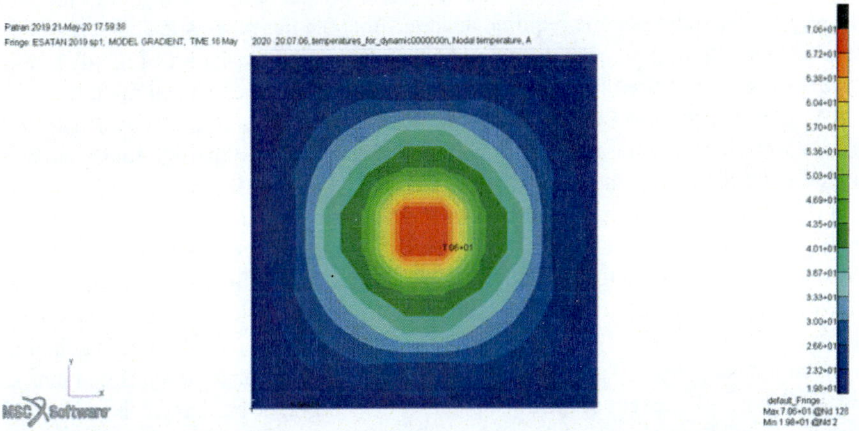

Fig. 6.29 Aggregated thermal node temperatures mapped on the mesh of the dynamic model

The question in this example to be answered is, whether the mesh of the dynamic model of the plate is good enough to capture the temperature field. For this purpose, the lumped parameter temperature field of Fig. 6.27 must be mapped to the mesh of Fig. 6.22. When both figures are compared, one could notice that the thermal mesh is denser than the structural finite element mesh: Four thermal nodes correspond to one finite element. This implies that none of the methods of section 7 can be applied directly, because all of these methods assume a higher mechanical FE mesh resolution than the thermal mesh resolution. In this example, it is the other way around.

As a workaround, to allow for the application of the PAT method described in Sect. 8.2, the results of the four thermal nodes corresponding to a single finite element are combined or aggregated. This means that the temperature of four thermal nodes are replaced by the average of the temperatures of the original four thermal nodes.

This then leads to the temperature field as shown in Fig. 6.28. With this aggregation operation detailed information of the thermal model results is lost. The fact that is operation is needed could already be considered as not such a good sign for the quality of the thermoelastic results.

With the use of the PAT method (Sect. 8.2), this temperature field is mapped on the mesh of the dynamic model. This mapping results in the temperature field presented with Fig. 6.29.

For a better comparison of the reference temperature field in Fig. 6.25 with that for the dynamic model in Fig. 6.29, the temperature profile along the diagonal of the plate (bottom left to top right) is presented in Fig. 6.30.

Near the locations with the highest gradients, the curves tend to deviate from each other. Figure 6.31 zooms in into this area, where the difference may be more clear.

As could be expected, the much coarser dynamic finite element model has difficulties to follow the reference temperature field, especially in the areas where gradients are high and rapidly changing. In Fig. 6.31, it can be seen how the reference temperature field shows how, already "inside" the box, the temperature field drops at the edges. Depending on the required accuracy that the thermoelastic analysis has to reach, it may be important or not to make the effort to re-mesh part of the dynamic model to better capture the temperature field produced by the thermal model.

What can also be observed is that the PAT method is able to recover a lot of information with so little resolution from both the aggregated thermal mesh and the dynamic finite element model.

Although differences in the temperature field are clearly there and shown by the deviation of the curves in Fig. 6.31, depending on the requirements, one could call these differences in the temperature field small. However, from these temperature curves, the consequences in terms of deformation and stresses cannot be estimated. In the end, it is the quality of the thermomechanical response that is relevant. In general, a coarser mesh is less good in predicting the stress levels than a more refined mesh. Especially, near the edges of the thick piece (representing the box), it can be expected that the refined mesh is producing higher stresses than the coarse dynamic model mesh.

Figures 6.32 and 6.33 show the stress fields computed with the very fine reference mesh and the mesh used for the dynamic model. Here, it becomes clear that, although the temperature field is not far off from the reference, the dynamic model mesh has severe difficulties to approximate the reference temperature field, both in terms of stress levels and shape of the stress field.

Without any doubt, a finer dynamic model mesh would do a better job. The question is what would be fine enough. To answer to this question is simple, but may require some effort: The required resolution needs to be derived from a mesh convergence checking exercise. In this case, the stresses have been considered as responses. So, for this case, the convergence checking process should stop when changes in the stress values are below an acceptable level. For stresses, this "acceptable" level of change can be driven by the strength margins that are produced. When the strength margins are getting smaller, the accuracy of the model has to be increased. It is, however, good to realise that larger finite elements smear out stresses peaks, and as

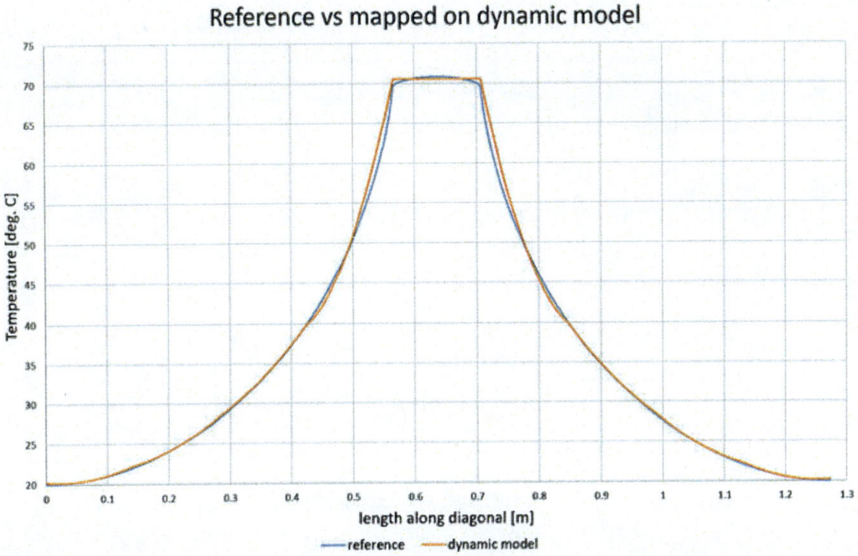

Fig. 6.30 Reference temperature field versus temperature field mapped on dynamic model

a result, these large elements tend to produce more optimistic margins. Therefore, it is recommended to do the mesh convergence exercise properly.

A similar discussion is applicable when the convergence checking exercise is based on deformation responses that are driving the performance of a pointing sensitive instrument. It has to be understood, whether an increase of mesh resolution in the areas of items that are causing the deformation leads to a degradation of the performance of the instrument. When this degradation is observed, it is worthwhile to investigate whether further refinement of the mesh can lead to predicting too low margins in the performance. When it is demonstrated that more accurate results are not showing problems in terms of negative margins, it could be justified to limit the mesh resolution to lower levels.

Concluding: The question that was raised at the beginning of the example can be answered by stating that the reference temperature field was reasonably well reproduced by the coarse dynamic model mesh, but it is clearly lacking the resolution to approximate the reference stress field. Discussed is that mesh refinement needs to be applied up to a level resulting from a mesh convergence exercise.

As is stated in several locations in this book, it is always important not to stop with the comparison of temperature fields, but to continue to the thermoelastic responses. Small changes in the temperature field may cause significant changes in the deformation or stresses! Only based on the comparison of the thermoelastic responses can the severity of a temperature fields be judged.

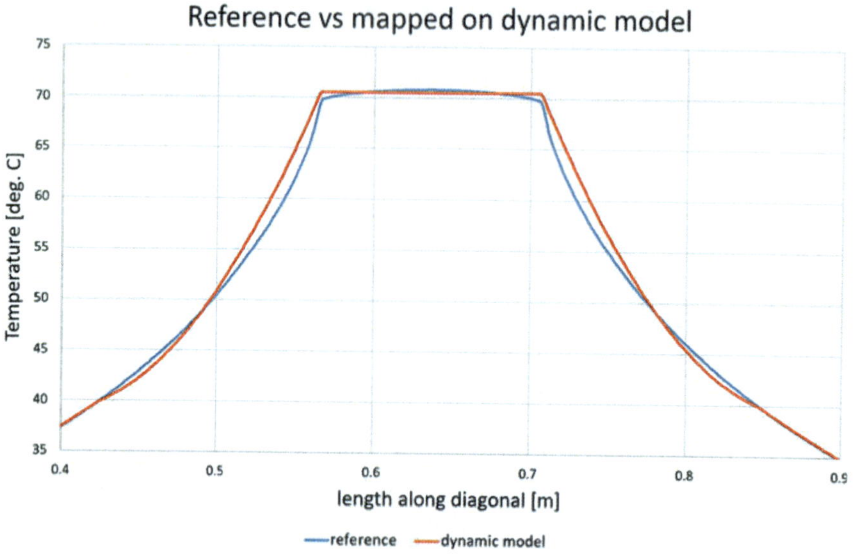

Fig. 6.31 Reference temperature field versus temperature field mapped on dynamic model, zoomed-in

Fig. 6.32 Reference von Mises produced with fine reference model

6.7.3 Finite Element Models for High-Frequency Response Analysis

In the above, the focus was mainly on the suitability of finite element model mesh resolutions for low-frequency dynamic analyses. For this type of analysis, in general, long wavelengths are considered, which can in general be well captured with a

Fig. 6.33 Von Mises stress field approximated with dynamic model mesh

limited mesh resolutions. This lower mesh resolution has a higher likelihood to have problems capturing (variation in) temperature gradients in the temperature field.

Dynamic analysis that aims to capture shorter wavelengths needs to have a correspondingly higher mesh resolution. High-frequency random, acoustic and shock analyses would typically require a higher mesh resolution, because higher frequency waves have shorter wavelengths. Similar to the above examples, adequacy of this finer mesh for thermoelastic applications still needs to be confirmed, and in general, it is therefore expected that high-frequency (and therefore short wavelength) dynamic models have mesh resolutions that are more suitable for thermoelastic analyses.

6.7.4 *Simulation of Joints*

Obviously, a spacecraft consists of many joints that mechanically connect the bigger and smaller components. Just to connect the panels of the spacecraft platform with each other easily, one hundred of those small brackets are needed. These brackets, often referred to as cleats, are in many cases metallic machined pieces.

For most system-level analyses, i.e. considering the whole spacecraft, the modelling of these many joints is simplified to limit the computational efforts. Quite often spring elements, combined with rigid body elements, are applied. In practice, there is not much tuning needed of the effective stiffness of these joints to reproduce the dynamic properties of the whole spacecraft measured with sine or modal survey tests. This tuning is often limited to setting the right order of magnitude, i.e. stiffness values of 1.E5, 1.E6 or 1.E7 (N/m or Nm/rad) to springs at the joints. Often differentiation is applied between the order magnitude for the rotational and translational stiffness values. In general, the rotational stiffness values are set to a lower order of magnitude than the translational values.

The springs are also frequently used as load sensors to extract from the finite element simulation the interface forces between the components for use in further detailed strength evaluations of the joints. Typically, spring forces are used as input to derive for a joint with multiple bolts the loads per bolt and assess the strength of the complete joint.

In the following explanation, it is important to realise that there are two ways to apply loads to a structure, that is, through:

- the application of external forces and moments
- the application of imposed deformations (displacements and rotations) or imposed strains.

External forces and moments are all mechanical loads originating from an external sources, such as inertial forces coming from a dynamic base excitation or gravity field and direct application of a force coming from an actuator. An important characteristic of external forces is that these do not depend on the deformation of the structure.

The other type is the one of imposed deformation or strain. Imposed deformation can be applied through prescribed displacements or rotations. The example of imposed strain, relevant for the context of this book, is the thermal strain coming from a temperature change combined with the CTE α of the material. The familiar expression for this thermal strain ε_t is given in Eq. (6.13)

$$\varepsilon_t = \alpha \Delta T \tag{6.13}$$

It is important to realise the fundamentals of this difference in loading to understand how these different loading types respond to stiffness.

In the case of external forces and moments, all interfaces respond to establish equilibrium with these external loads. The ratio of the stiffness values between the interfaces only influences the way the external forces and moments are distributed over the interfaces. The sum of the interface loads simply has to equate with the external forces and moments.

In case of imposed deformation and strains, the magnitude of the reaction forces are proportional to the combined stiffness of the interfaces and the two interfacing structural components. The essence comes down to the basic force-displacement relation in Eq. (6.14)

$$F = ku \tag{6.14}$$

Thermoelastic responses follow this principle. Resulting from the temperature field, the structure is subjected to the thermal strains based on Eq. (6.13). In an unconstrained situation (in a free-free situation of the component or structural part), these thermal strains cumulate to displacements and rotations at the interfaces. The combined stiffness of the interface and the adjacent structures limits this free deformation, leading to the forces at the interfaces based on the principle of Eq. (6.14).

The above tries to explain that for external loads the role of stiffness is limited to how this load is distributed over the structure. It is mainly force and moment equilibrium that has to be satisfied. For imposed deformation, such as thermal strains, stiffness is driving the response of the structure. As a consequence, a 50% change in the stiffness of a joint will in general not show up as an important change in the responses of an external structure, but it has the potential in case of imposed deformation to introduce a relative change of the same order as the relative change of stiffness.

In a finite element model, the stiffness of the structure is represented in roughly two ways that are often applied simultaneously in a single model. The first way is applicable to continuous parts represented with solid, shell or plate elements. The stiffness of such a part is in general quite accurately represented by specifying the (correct and reliable) geometrical and material data. In some cases, beam elements can also do a good job, except when these are used for modelling interfaces.

The other way of representing stiffness is used for discrete joints, which is the focal point of this section. As explained in the introduction of this section, discrete joints, e.g. the ones between spacecraft panels, are often modelled through combinations of rigid elements and spring elements representing the joint stiffness. In some cases of panel connections, simplified representations of the cleats are implemented, but then still in combinations of springs and rigid elements. The stiffness values that are specified for these spring elements should approximate as good as possible the real stiffness of the joint. Ideally, these values are derived from mechanical tests.

With the attachment of parts at discrete locations on panels or 3-D solid structures, incorrect representation of the stiffness of the attachment may be introduced without the model developer realising this. The two basic cases discussed in the example in Sect. 6.4.1 shows how the stiffness of the attachment of a spring element or discrete support is affected by the mesh resolution. This is due to the fact that a discrete load introduction represents a singular problem. In the second case of the mentioned example of Sect. 6.4.1, it is shown that, even away from the singular point, the stress field is dependent on the mesh density near the singularity.

Due to the singularity, the mesh refinement is reducing the simulated stiffness of the plate near the attachment. Using again the expression of Eq. (6.14) and keeping in mind the reduction of stiffness with an increase of the mesh density, the same free thermoelastic deformation leads to reduced reaction force. As a consequence, the stress in the plate away from the singularity also reduces with the mesh refinement, even if this refinement was done near the singularity.

The fact that dynamic responses during vibration tests are well reproduced by the analysis results may raise the impression that the stiffness properties of the joints are properly modelled. However, as explained above, the predicted dynamic responses are quite insensitive to the modelled stiffness of the joints [92]. This implies that a large range of stiffness values for the different components of a joint produce more or less the same predicted dynamic properties of the structure. This is quite convenient for the modelling of for instance a dynamic problem, but gives an invalid basis for simulating a thermoelastic effects. An other problem is the fact that during dynamic excitation not always all stiffness components are playing a role, which makes that the more or less unused stiffness components can have any value without affecting the dynamic responses. Also, for static cases with external loading, this insensitivity for the interface stiffness is present.

Due to the insensitivity of the responses to external loads for variation of interface stiffness values, it is difficult to discover, from for instance a dynamic test, whether or not the modelled interface stiffness values are accurately represented. Without any further verification of the stiffness properties of a joint, it is not unlikely that wrong joint stiffness values are used in the model. As a consequence, this will lead to wrong predicted thermoelastic responses.

In the example, "Joints: External loads versus thermoelastic loads" is illustrated with a simple problem that a wide range of interface stiffness values hardly affects the responses to external loads. It also shows what the implications are for the thermoelastic responses.

Joints: External loads versus thermoelastic loads

Considered is the simple model of a beam as depicted in Fig. 6.34. The beam is considered made of aluminium with a length of 2 m and a hollow box-shaped cross section with dimensions 2.5 × 5 cm and a wall thickness of 1 mm. In this example, the stiffness at both ends of the beam is varied in order to show what the effect is on the responses for two different external loading conditions and a simple thermoelastic environment, i.e. a uniform temperature increase.

The interface configurations that are considered are listed in Table 6.6. These range from fully clamped interfaces to suspended interfaces with different stiffness values. For the interface components ("IF component" in the tables), all the three rotational stiffness values (K_{rot}) have per interface configuration the same specified value. The same is the case for the three translational stiffness values (K_{trans}).

For each set of the different boundary conditions listed in Table 6.6, four eigenfrequencies are computed. The results are listed in Table 6.7. It can be seen that even with the strong changes in the stiffness values at the interfaces, the variation in eigenfrequencies are quite modest and would in many cases be considered as an acceptable deviation from the tests

The same beam models with variations in interface stiffness values were used for two static load cases with a gravitational load of 10 m/s^2 in two directions: one parallel to the axis of the beam and one perpendicular to the beam axis. The results are presented in Tables 6.8 and 6.9.

The resulting interface forces are a direct consequence of the applied inertial load. The interfaces need to make sure that their reaction forces and moments are in equilibrium with the external load. Only the bending moment is showing a light change with the change of the rotational stiffness at the interface for the lateral inertia load (see Table 6.9). In this example, the translational interface forces appear to be completely insensitive to changes of the interface stiffness values (see Table 6.8).

In the above cases, the responses appeared to have low or even no sensitivity for the stiffness of the interface for external loads. This basically means that any stiffness value within a wide range is producing more or less the same responses.

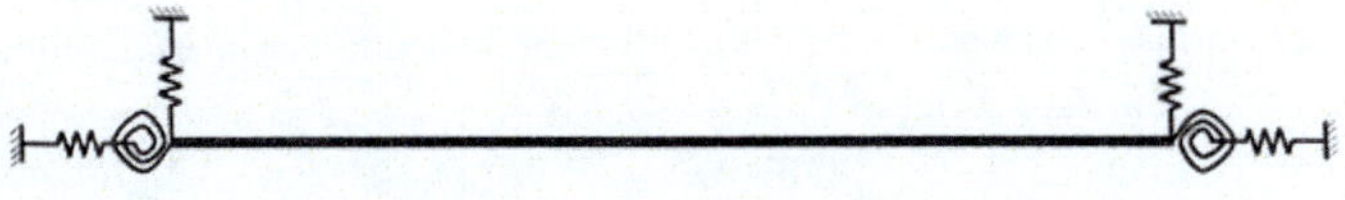

Fig. 6.34 Simple beam with schematic interface stiffness

Table 6.6 Interface configurations considered in the example

IF component	Interface stiffness values			
K_{trans} [N/m]	Clamped	1.0E+09	1.0E+07	2.0E+07
K_{rot} [Nm/rad]	Clamped	1.0E+09	1.0E+05	2.0E+05

Table 6.7 Natural frequencies for different I/F stiffness values

IF component	Interface stiffness			
K_{trans} [N/m]	Clamped	1.0E+09	1.0E+07	2.0E+07
K_{rot} [Nm/rad]	Clamped	1.0E+09	1.0E+05	2.0E+05
Mode #	Natural frequency (Hz)			
1	8.19E+01	8.19E+01	7.68E+01	7.92E+01
2	2.23E+02	2.23E+02	2.09E+02	2.15E+02
3	4.31E+02	4.31E+02	4.02E+02	4.15E+02
4	6.96E+02	6.95E+02	6.46E+02	6.68E+02

Now is verified how this works out for a case of a temperature increase of $\Delta T = 5\,°\mathrm{C}$. The results are shown in Table 6.10.

The results in Table 6.10 show that for the thermoelastic case only axial interface forces are produced and thus only the axial stiffness is used. The other stiffness components are not addressed. The axial force responses differ strongly between the interface configurations for the thermoelastic case. Although the axial gravitational load case was also addressing the axial stiffness of the interface, there was no change in reaction forces at the interfaces, and the level of the forces was completely driven by the external load.

Also, the other case with gravitational load appeared to be hardly sensitive for changes in the configuration of the support. The results of the normal modes analysis also appeared to have a very low sensitivity for changes in the stiffness values of at the interfaces.

This example illustrates that a model, with interface stiffness values correlated after a dynamic test, will not necessarily produce reliable thermoelastic results. The strong variation in interface forces for the thermoelastic case indicates that deformation of potentially connected components will also be different.

Table 6.8 Interface loads under gravity field in axial direction for different interface stiffness values

IF component	Interface stiffness			
K_{trans} [N/m]	Clamped	1.0E+09	1.0E+07	2.0E+07
K_{rot} [Nm/rad]	Clamped	1.0E+09	1.0E+05	2.0E+05
I/F force				
F_{ax} [N]	3.94	3.94	3.94	3.94
F_{lat} [N]	0	0	0	0
M [Nm]	0	0	0	0

Table 6.9 Interface loads under gravity field in lateral direction for different interface stiffness values

IF component	Interface stiffness			
K_{trans} [N/m]	Clamped	1.0E+09	1.0E+07	2.0E+07
K_{rot} [Nm/rad]	Clamped	1.0E+09	1.0E+05	2.0E+05
I/F force				
F_{ax} [N]	0	0	0	0
F_{lat} [N]	3.94	3.94	3.94	3.94
M [Nm]	1.3	1.3	1.26	1.28

Table 6.10 Interface loads under uniform temperature increase of $\Delta T = 50\,^{\circ}\mathrm{C}$ for different interface stiffness values

IF component	Interface stiffness			
K_{trans} [N/m]	Clamped	1.0E+09	1.0E+07	2.0E+07
K_{rot} [Nm/rad]	Clamped	1.0E+09	1.0E+05	2.0E+05
I/F force				
F_{ax} [N]	7.11E+03	7.04E+03	3.52E+03	4.71E+03

6.7.5 Check on Adequacy of Rigid Body Elements for Thermoelastic

Equipment housing, such as used for electronics boxes, and large brackets are relatively stiff items. In dynamic analysis, it is common practice that these kind of items are represented by a single lumped mass element connected by a rigid body element (RBE2) to the interface FE nodes at the model of the supporting structure. This way of modelling has proven to be sufficient for the representation of these items especially for system-level dynamic analyses.

In Sects. 4.6 and 6.5 and Appendix A, this type of simplification is discussed. In those sections, it is explained that these stiff items have the potential to play an important role in the thermoelastic response of the structure. The effect of these boxes is not just local, but may span the whole panel to which these are mounted and potentially will affect the tilting of pointing sensitive instruments and also the neighbouring structural elements. With the use of rigid body elements, these effects are ignored, and other non-physical effects are introduced.

From the modelling point of view, this kind of simplifications is convenient, because these require very little effort from the person building the structural FE model. In addition, the way of modelling hardly increases the computational effort for the in general already large system-level models. However, important physical properties are ignored, and results will be wrong and may have important consequences for highly pointing sensitive instruments. Section 4.6.3 provides suggestion on how to have still a computationally light representation of these items for both the structural and thermal model and have at the same time an adequate representation of the physics.

Rigid body elements are also used extensively for modelling the brackets of joints. Also, these brackets are relatively stiff compared to the surrounding structure. In general, the legs of these rigid body elements have a shorter length. From that perspective, it is expected that these RBE elements have more reduced negative effect on the quality of the thermoelastic predictions. This needs, however, to be verified. It must be checked whether there are strong thermal gradients that can cause the stiff bracket to deform for instance the attached surrounding structure.

Also, care must be applied to the overall stiffness representation of the joint. The use of the RBE plays an important role in the stiffness modelling. As is explained in Sect. 6.7.4, the thermoelastic responses are highly sensitive to the stiffness modelling. With the experience of modelling for dynamic simulation, this importance may be easily under estimated. In [92], it is illustrated how differences in modelling of a panel interface with different rigid element configurations, nicely tuned to obtained almost equal dynamic behaviour, had an important difference in thermoelastic responses as result.

Finally, the RBE3 elements are often used to prevent introduction of stiffness in the model and still introducing the mass of the item in the model. Also, the interpolation function of the RBE3 elements is often used. In both cases, it is needed to check whether this stiffness free modelling is correctly capturing the thermomechanical responses. It is also for these items recommended to consult the recommendations in Sects. 4.6 and 4.6.3 to introduce at least a light model version of these items in the model.

An other reason to introduce more "physical" modelling as replacement of the RBE elements is to capture the local temperature field that the thermal model provides. This information is available and should therefore be used to prevent overlooking effects that can easily be included in the model. It goes without saying that when the level of detail in the structural model is increased to enhance the thermoelastic predictions, this also requires the same level of detail in the thermal model.

6.8 Finite Element Model Health Checks Thermoelastic FE Models

6.8.1 Introduction

Most spacecraft projects require that, in conformance with the ECSS, the finite element model is subjected to a number of fundamental health checks to make sure that the model is not violating basic physical principles. Two of these finite element model health checks relevant for thermoelastic modelling are discussed in this section.

- Zero strain energy as rigid body
- Stress-free iso-thermal expansion.

A brief description of these checks is also found in ECSS "Structural finite element models" [32], Sect. 5.4.3 (Rigid body motion strain energy and residual forces check) and Sect. 5.6 (Stress-free thermoelastic deformation check for non-reduced models). An other ECSS document, "General Structural requirements", [31] (see Annex I of this ECSS), specifies which results these checks are to be reported in the description of the finite element model.

In this section, a bit more information is provided on the principles behind the two mentioned checks that are of special interest for thermoelastic simulations.

The check on zero strain energy as rigid body is also a useful and relevant check for other types of analyses, but is considered important enough to be explained here.

The ECSS [32] and for instance [20] recommend also other checks. These checks should be done anyhow and should be standard practice for every finite element method user. A few of these checks are as follows:

- Elements with offsets: Switch to an alternative way of modelling.
- Material properties without CTE specification: Add CTE values to capture correctly the thermomechanical responses.
- Nonzero length scalar elements (0-D, CELAS) may create unphysical strain (depends on the implementation).
- Bad element geometry (warping, aspect ratios, internal angles between element edges) causes non-physical strains.

The mentioned FE model health checks are recapitulated in subsequent sections.

6.8.2 Strain Energy as Rigid Body

The rigid body strain energy check has the objective to detect unintended and unconsciously introduced boundary conditions. These constraints are not always obvious and often do not show up directly when the results of a model are evaluated.

A rigid body movement of a structure, as the word "rigid" indicates, should not include any internal deformation in the structure and thus should not generate any strain energy. A well-defined finite element model should reproduce this fundamental physical behaviour. If the model, however, produces important values of strain energy, then there are one or more unintended constraints active in the model.

Different ways to generate rigid body displacement fields exist, for instance:

- Generation of three displacements and three rotations around a selected point to all the finite element nodes in the model
- Generation through a static analysis with enforced displacements and rotations of one single FE node
- Generation through a free-free eigenvalue analysis.

In MSC Nastran, a rigid body strain energy check is called "grounding check" and can be activated through the case control with the GROUNDCHECK entry. This check uses the first type of rigid body displacement fields in the list above. A unit translation in the three orthogonal directions is applied to all the FE nodes, and a unit rotation around the same three orthogonal axis through a selected point is applied to construct the six rigid body displacement fields.

The grounding check in MSC Nastran evaluates for each of the rigid body displacement fields the following expression

$$U_{Rk} = \frac{1}{2}\,(\phi_{Rk}\,)^T\,[K]\,(\phi_{Rk}\,) = \delta_k \leq \epsilon_k,\ \ k = 1, 2, \ldots, 6. \tag{6.15}$$

with $(\phi_{Rk}\,)$ being the k-th rigid body displacement vector and $[K]$ being the stiffness matrix of the finite element model.

ECSS "Structural finite element models" [32] states that for a 1 m translation or a 1 rad rigid rotation, typical acceptable values of U_{Rk} should be less than 1.E−03 J.

When the values obtained for U_{Rk} are significantly higher than the recommended maximum value, it is clear that a constraint is present in the model. A static analysis with enforced displacements, i.e. the second option listed above to generate the rigid body displacement field, can then be useful to identify the items in the model responsible for this unintended constraint.

Very often the value of U_{Rk} is just slightly higher than the recommended maximum acceptable value. In most cases, this is due to bad element shapes. It is then recommended to review the diagnostics on element shape quality. Most preprocessors also can help identifying badly shaped elements.

More details on this health check can be found in [94].

6.8.3 Free Iso-thermal Expansion

When the following situation is imagined,

- a structure is made of a single (dummy) material, and therefore, the CTE value is the same at every location in the structure.
- the structure is free to expand, for instance, through the use of iso-static boundary conditions (see Sect. 6.6.1).
- the structure has a uniform temperature increase from a uniform initial temperature.

then this structure should not be experiencing any thermal stresses. This is the physical principle of the FE model health check that is briefly discussed in this section.

A finite element model, with all boundary conditions, that are intended for the simulation of all physical supports, replaced by an iso-static support shall be able to simulate this physical behaviour. The iso-static support simply suppresses the rigid body movement of the model and allows for stress-free expansion of the simulated structure. Also, all deformations shall not contain any rotations, i.e. the structure should simply grow in all directions with the same relative amount.

The ECSS "Structural finite element models" [32] states in section 5.6 (Stress-free thermoelastic deformation check for non-reduced models) that if the model is "clean", there should be no rotations, reaction loads, element forces, or stresses. This standard declares that, when an aluminium alloy is used as dummy material with a uniform temperature increase of 100 K, 0.01 MPa is a maximum acceptable value for the iso-thermal von Mises stress. As acceptable maximum value for the rotation 10^{-7} rad is indicated.

With the use of stress and rotation contour plots, the locations in the model that are not compliant to these criteria can be identified. It is sometime not obvious what the problem is. In those cases, the suggestions given in the Sect. 6.8.2 may be of use to identify the sources of the problem.

The theoretical rationale of the zero stress-free expansion is given in [36, 44]. There is also explained that a stress-free state can also be obtained with a temperature distribution that is a linear function of the three spatial coordinates:

$$\Delta T = T - T_{ref} = a + bx + cy + dz \tag{6.16}$$

where $a,\ b,\ c\ and\ d$ are arbitrary constants.

Problems

6.1 A truss frame is shown in Fig. 6.35. This truss frame is 1 m long and 0.5 m height. The cross section of the trusses is $A = 0.05\ \text{m}^2$. The CTE $\alpha = 24 \times 10^{-6}$ m/m/°C and Young's modulus is $E = 70$ GPa. The spring stiffness $k = 1.0 \times 10^9$ N/m. The increase of the node temperatures is $\Delta T_1 = \Delta T_3 = \Delta T_5 = 40$, $\Delta T_2 = \Delta T_4 = \Delta T_6 = 50$ and $\Delta T_7 = \Delta T_8 = \Delta T_6 = 55$ °C. Calculate the thermoelastic displacements (u, v), stresses/forces in the rods and spring with aid of:

- Writing your own finite element programme
- Use your own favourite FEA software package.

Fig. 6.35 Truss frame

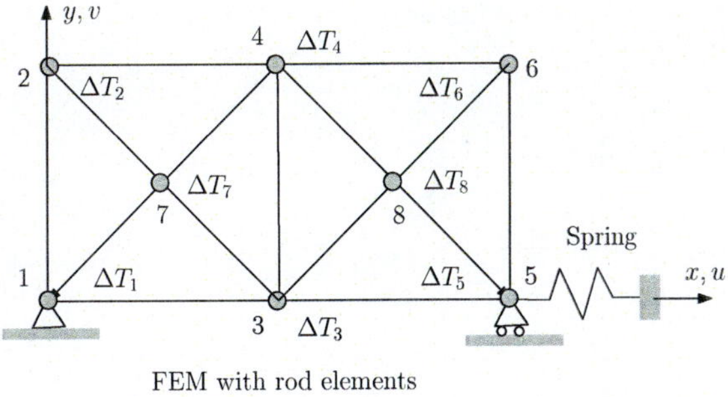

Fig. 6.36 Rigid bar supported by rods

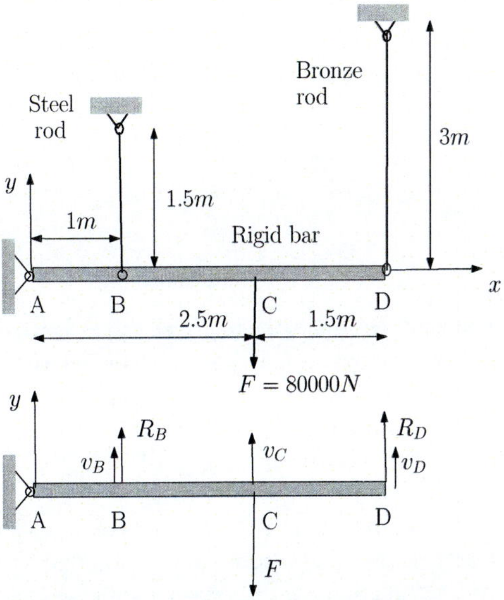

Table 6.11 Geometry and material properties of rods

Parameters	Dimension	Steel	Bronze
L	m	1.5	3
A	mm^2	320	1300
E	GPa	200	83

6.2 Hinged rigid bar AD is supported by a steel and bronze rod. The system is shown in Fig. 6.36. The geometry and material properties of the rods are provided in Table 6.11.

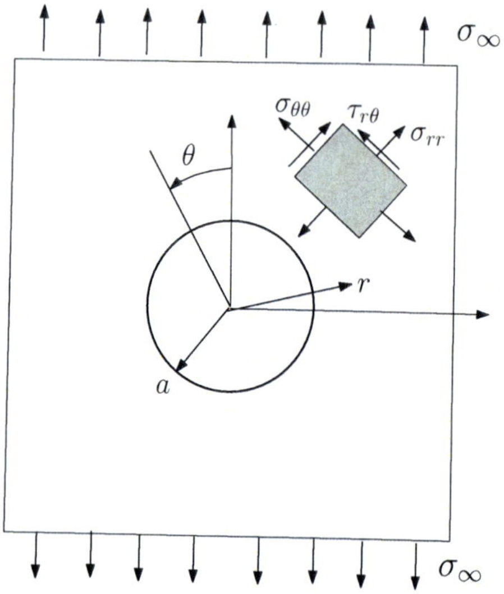

Fig. 6.37 Stress distribution around hole [65]

Perform the following assignments:

- Calculate (by hand) the interface forces R_B and R_D due to the external force F. The system is indeterminate! To solve the interface forces, the equation of equilibrium and the relation between the displacement v_B and v_D shall be used.
- Create a structural FE model of the given system, 11 nodes and 10 rod structural finite elements per supporting rod (total 20 elements and 22 nodes). Write down the MPC equations $(U_m) = (0)$ to incorporate the rigid bar and to create the possibility to apply the external load F. Can you use a RBE2 or RBE3 element?
- Carry out the FEA and obtain the internal forces in both the steel rod and bronze rod. Compare these FEA results with the hand calculations.

6.3 Gustave Kirsch's [16] solved in 1898 the problem of the stress distribution at a hole for the case of uniaxial tension σ_∞ in an infinite plate as illustrated in Fig. 6.37.

The solution for the stress state around a hole is

$$\begin{aligned}
\sigma_{rr} &= \frac{\sigma_\infty}{2}\left(1-\left(\frac{a}{r}\right)^2\right)+\frac{\sigma_\infty}{2}\left(1-4\left(\frac{a}{r}\right)^2+3\left(\frac{a}{r}\right)^4\right)\cos 2\theta,\\
\sigma_{\theta\theta} &= \frac{\sigma_\infty}{2}\left(1+\left(\frac{a}{r}\right)^2\right)-\frac{\sigma_\infty}{2}\left(1+3\left(\frac{a}{r}\right)^4\right)\cos 2\theta,\\
\tau_{r\theta} &= -\frac{\sigma_\infty}{2}\left(1+2\left(\frac{a}{r}\right)^2-3\left(\frac{a}{r}\right)^4\right)\sin 2\theta.
\end{aligned} \tag{6.17}$$

At $\theta \pm \frac{\pi}{2}$ and $r = a$, the stresses are

$$\sigma_{rr} = 0,\ \sigma_{\theta\theta} = 3\sigma_\infty,\ \tau_{r\theta} = 0.$$

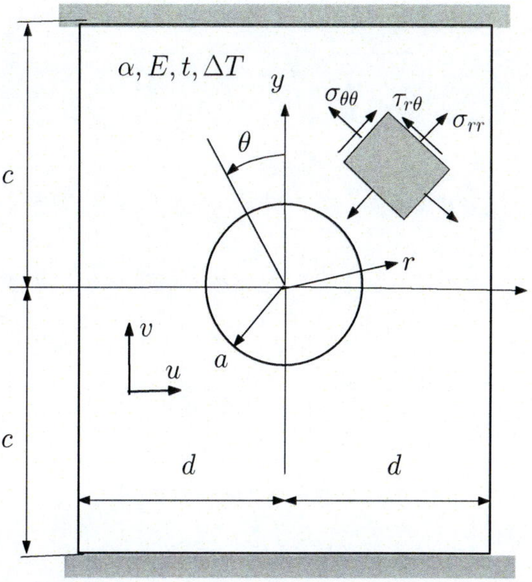

Fig. 6.38 Prestressed strip with hole

A strip with a hole in the centre is prestressed by fixing both sides (y-dir) at $y = \pm c$, and a constant temperature field ΔT is shown in Fig. 6.38. Calculate the stresses $\sigma_{rr}, \sigma_{\theta\theta}$ and $\tau_{r\theta}$ at the following locations ($x = a \cdots d$, $y = 0$) and ($r = a \cdots c, \theta = 0$) using previous Eq's (6.17).

The strip has the following dimensions, namely $a = 15$ mm, $d = 30$ mm and $c = 60$ mm and the thickness $t = 2$ mm. The Al-alloy strip has the following material properties: Young's modulus $E = 70$ GPa, Poisson's ratio $\nu = 0.33$ and CTE $\alpha = 24 \times 10^{-6}$ m/m/°C.

When the tensile stress is $\sigma_\infty = 1$ MPa, calculate ΔT (°C) to obtain the equivalent thermoelastic tensile stress σ_∞ in the strip at $x, y = 30, 60$ mm. Perform the following assignments:

- Create a FE model of the strip with central hole as shown in Fig. 6.38 making advantage of the symmetrical properties and with adequate detail to calculate the stresses at afore mentioned locations ($x = a \cdots d, y = 0$) and ($r = a \cdots c, x = 0$).
- Apply boundary condition at $y = \pm c$ (v = 0). Prevent rigid body motion in x-direction. Not needed if symmetry will be applied.
- Compare stress results at specified locations of the FEA with theoretical results obtained by Eq. (6.17).

Chapter 7
Transfer of Thermal Analysis Results to the Structural Model

Abstract In general, the thermal model has a lower resolution than the structural FE model used for linear thermoelastic analysis. Besides a difference in mesh resolution, the methods used for the thermal and structural analyses differ as well. This complicates the transfer, also referred to as temperature mapping, of the calculated temperatures of the thermal model as temperature loading on the FE model for thermoelastic response calculation. This chapter discusses four different systematic methods for temperature mapping.

7.1 The Interface Problem

The relevant aspect of thermal and structural modelling for thermoelastic predictions has been discussed in Chaps. 5 and 6. The importance of consistency of the thermal and structural models in terms of level of detail and representation of important elements in both models is discussed in Chap. 4, "Modelling for thermoelastic".

The subject of this chapter is the transfer of temperatures resulting from the thermal analysis in as much as possible physically consistent and accurate way [43] to the structural analysis model for computing the thermoelastic responses. The challenge in this interface problem consists of two aspects. The first aspect is related to the difference in simulation methods between thermal and structural analysis. The second aspect is related to the logistics of handling the large amount of data involved in the process of thermoelastic analysis. The problem related to the second aspect can only be solved by adequate automation of the analysis chain. This aspect is already discussed in Sect. 4.7, "need for automation of the analysis chain". This chapter is focussing on the possible methods to handle the first aspect of the interface problem related to the difference in modelling between thermal and structural analysis.

The spacecraft thermal models are based on the thermal lumped parameter method, and the structural models are based on the finite element method. Due to the difference in methods, the corresponding models make use of different entity types as is explained in Sect. 7.2 in which the difference between a thermal lumped parameter node and a finite element node is explained.

S. Appel and J. Wijker, *Simulation of Thermoelastic Behaviour of Spacecraft Structures*, Springer Aerospace Technology, https://doi.org/10.1007/978-3-030-78999-2_7

Besides difference in method, there is in many cases a significant difference in mesh resolution between the two models. It may happen that the ratio of the number of thermal nodes in the TMM to the number nodes in the FE model is 1:1000, [50].

7.2 Thermal Lumped Parameter Node Versus Finite Element Node

It is good to realise that an important difference between the finite element method and the lumped parameter method lays in the definition of the "nodes" for which both methods try to solve the temperatures. In the lumped parameter method, a node represents the properties of a volume or an amount of material. In the finite element method, the node is nothing else than a discrete geometrical entity with no volume or mass. Thanks to the finite elements, a certain portion of the volume and material represented by connected elements can be associated with the finite element nodes [61].

In the lumped parameter method, pairs of nodes are connected with each other through conductors. The most frequently used types of conductors are linear and radiative, called GL and GR in ESATAN-TMS. In the finite element method, elements connect the nodes and define in the thermal domain conductive relations between two up to sometimes 20 finite element nodes.

In terms of results, the difference may be more clear. The thermal node temperature is the average temperature of the volume and material represented by the thermal node. On the other hand, the temperature at a finite element node is the temperature at a discrete location in space. Both entities, in their corresponding methods, represent the unknowns to be solved from the numerical formulations of the heat transfer equations.

Figure 7.1 illustrates schematically the difference in mesh and node definition for the thermal lumped parameter model and the finite element model.

For structural application of the finite element method, and specifically the thermoelastic application, the temperature load has to be assigned to the FE nodes[1] of the finite element model. Since both the type of nodes and the use by the two models are different, it shows the problem to transfer temperatures from the thermal nodes of a lumped parameter model to those of a finite element model.

7.3 Building Correspondence Between Models

Before any temperature mapping method can be applied, it is essential that the correspondence between the entities in the thermal model and the structural model is

[1] For some applications, it may be favourable to specify temperature loading by assigning temperatures to the elements.

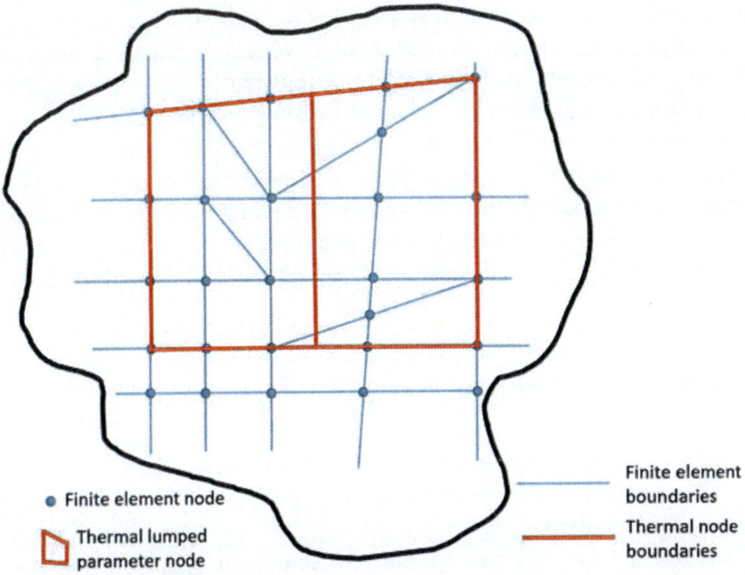

Fig. 7.1 Thermal lumped parameter nodes versus finite element nodes

available. The process in which this data is generated is referred to as overlap detection [9]. Despite the fact that this process can, to a large extent, be automated, this is the most time-consuming and labour-intensive part of the thermoelastic analysis process. This process benefits from good coordination between the engineer developing the thermal model and the person responsible for the development of the structural model. This increases the chances that meshes in both models are aligned that reduce the ambiguity in the correspondence relations.

The correspondence data or overlap data need to be generated only once and only need to be reviewed and updated when either the structural model or the thermal model is changed.

Correspondence between the thermal and structural model typically appears in two ways, depending on the temperature mapping method that is to be applied. The difference between these two ways of defining the correspondence data or overlap data lays in the way a thermal node is geometrically represented in the context of the temperature mapping process. The geometrical representations of a thermal node used by different temperature mapping methods are as follows:

- A discrete point in the geometric centre of the thermal node
- A surface or volume representing the thermal node.

There can be different definitions of the geometric centre, which are not discussed here.

In the first way of geometrical representation, a thermal node, it is implicitly assumed that the average temperature of the thermal node occurs at the geometric centre. In general, this is not correct but can be a reasonable approximation. The error related to this will reduce by increasing the mesh resolution.

In the next section, how each of the discussed temperature mapping methods is using the geometrical representation of the thermal node will be explained.

Temperature mapping methods that consider the thermal node as a volume or surface typically define the correspondence as the relation between the finite elements of the structural model and the thermal node of the thermal model. To indicate that an element corresponds to a thermal node is also expressed as "the element is overlapped by a thermal node".

7.4 Temperature Mapping Methods

The temperature mapping step is one of the steps in the thermoelastic analysis chain which is, like the thermal and structural analysis, a source of errors that need to be understood and minimised.

Four temperature mapping methods are discussed in this chapter as follows [58]:

- Geometric temperature interpolation method
- Centre-point prescribed temperature (CPPT) method
- Patch-wise temperature application method
- Prescribed average temperatures (PAT) method.

To evaluate these mapping methods, the following performance criteria are considered:

- Average mapped thermal node temperature. The weighted average temperature of the finite element (FE) nodes corresponding to the thermal node shall reproduce the thermal node temperature as good as possible.
- The displacement and stress fields, computed with the mapped temperature field, shall reproduce as good as possible reference results, determined with detailed thermal and structural models.

7.4.1 Geometric Temperature Interpolation Method

General-purpose preprocessors of finite element software (such as FEMAP, MSC Patran) offer the possibility to specify loads and boundary conditions as a field that is defined by values as a function of the position (spatial field). The coordinates of the geometric centres of the thermal nodes in combination with the thermal node temperatures are used to construct the temperature field in this geometric temperature interpolation method. The temperature loads on the finite elements nodes are then defined by linear or higher-order interpolation between the geometric centres of the thermal nodes that construct the temperature field.

Another geometrical interpolation method is the inverse distance weighting method (IDW) [28]. The FE model nodal temperatures T_i^s can be expressed in weighted sum of associated thermal node temperatures T_j^t

$$T_i^s = \sum_{j=1}^{N} W_{ij} T_j^t, \; i = 1, 2 \ldots \tag{7.1}$$

where N is the number of selected thermal node temperatures T_j^t and the weight factor (power distance weight) W_{ij} is given by

$$W_{ij} = \frac{\frac{1}{d_{ij}^k}}{\sum_{j=1}^{N} \frac{1}{d_{ij}^k}}, \; i = 1, 2 \ldots \tag{7.2}$$

in which d_{ij} is the Euclidean distance between the location of FE node i and the centre of thermal nodes j. The power k allows for tuning the weighting factors W_{ij}. The sum of the weighting factors W_{ij} shall be

$$\sum_{j=1}^{N} W_{ij} = 1, \quad i = 1, 2 \ldots \tag{7.3}$$

When $k = 0$, the weighting factor is $W_{ij} = 1/N$. In case the centre of thermal node j coincides with the location of FE node i, implying $d_{ij} \rightarrow 0$, the weighting factor will be $W_{ij} \rightarrow 1$ and (with Eq. 7.3) $W_{im} \rightarrow 0 (m \neq j)$.

The weighting factors W_{ij} can be considered to form the matrix $[W]$. With the vector of thermal node temperatures $\left(T^t\right)$ and the vector of FE node temperatures (T^s), Eq. (7.1) can be written in matrix form as

$$\left(T^s\right) = [W] \left(T^t\right) \tag{7.4}$$

The expression of Eq. (7.4) can be used to transform every temperature field in the form of thermal node temperatures into temperatures mapped at FE-nodes. This allows for efficient generation of multiple mapped temperature fields for instance originating from a thermal transient analysis.

These pure geometrical methods have the disadvantage that no physical and configuration information from the thermal and structural models is used. Discontinuities in the structure or variations in the conductivity are not taken into account during the generation of the geometrically interpolated temperature field. The user has to provide the information on the physical configuration of the structure to the interpolation process. This could be done by applying the method per region of the structure (e.g. separate panels). However, the transition of temperature fields over the interface of connected regions cannot be captured in this way.

In general, the geometrical interpolation methods do not have any constraints or rules ensuring that the average of the temperature, mapped to the entities in the finite element model corresponding to a thermal node, is equal to the thermal node temperature.

The main points in this temperature mapping method are as follows:

(+) Basic linear geometric interpolation is available in most finite element pre-processor tools.
(−) The average temperature of the thermal nodes is not necessarily ensured in the mapped temperature field.
(−) The interpolation method is based on geometry only ignoring any physical and configuration information such as interfaces and changes in heat flow path due to variation in conduction.
(+) A smooth temperature field results from this method.

7.4.2 Centre-Point Prescribed Temperature Method

Another method is the generation of the interpolated temperature field through a finite element (FE)-based conductive steady-state simulation. For this method, the structural mechanical FE model needs to be converted to a conductive FE model. As boundary conditions for the conductive steady-state problem are the thermal node temperatures assigned to the FE nodes near the geometric centre of the thermal nodes as prescribed temperatures. For this reason, the method is called centre-point prescribed temperatures (CPPT) method. The temperature field that is the solution of this simulation is the mapped temperature field.

The method has the advantage that it takes into account the conductive relations in the finite element model. This allows the method to include implicitly all discontinuities in the structure or variations in the conductivity. When interfaces between structural parts are conductively represented, this allows to capture the effect of these interfaces in the mapped temperature fields.

In MSC Nastran, the temperatures can be prescribed with SPCD entries, and SOL 153 can be used for thermal steady-state analysis. This method is potentially more correct than geometric interpolation methods. Since it uses the conduction matrix as physical information for interpolation, the generated temperature field takes into account the variation of and interruption in conduction due to geometry, material and interfaces.

This method can, through the conduction relations in the FE model, that often have a higher resolution and complement the temperature field provided by the thermal analysis. Parts of the FE model that are not included in the thermal model will obtain temperatures via the conduction relations. This may look as a convenient feature, but need to be considered with care, since the thermal model did not provide all the required information.

A known issue with application of boundary conditions, such as forces, displacements and temperatures at single FE nodes, is the mesh density dependency. This problem is therefore also applicable to the CPPT method. In other words, this method may produce a different temperature field on the FE model when the mesh resolution is changed.

Like the geometrical interpolation methods, the average temperature of finite element model items corresponding to a thermal node is not ensured to be equal to the thermal node temperature.

The main points of this temperature mapping method are as follows:

(−) Extra effort is required in preparing a conduction FE model (converted structural FE model).
(+) Conduction involved in the temperature interpolation providing a physical basis to the interpolation process and can respect discontinuities in the structure.
(−) The average temperature of the thermal nodes is not necessarily ensured in the structural FE model.
(+) The interpolation is based on the conductive relations (conduction matrix) of the complete FE model, therefore not the full FE models need to covered by thermal nodes.
(−) Important sensitivity to mesh resolution at the location of prescribed temperatures.
(+) A smooth temperature field results from this method.

7.4.3 Patch-Wise Temperature Application Method

One method that is also often applied in industry is, what is here referred to as the patch-wise temperature application method. In this method, a thermal node temperature is applied as an uniform temperature to a group of finite elements that correspond to the thermal node. The plot of this temperature field looks like a patch work. This is the reason for the name chosen for this method: "patch-wise temperature application method". In the following, the acronym PWT is used to refer to this method.

The thermal lumped parameter method assumes a constant temperature in the volume or surface area represented by the thermal node, thus lumping all properties including the temperature into a single node. The thermal lumped parameter method uses this single temperature that represents the average temperature of the thermal node volume. The PWT method applies the constant temperature assumption to the mapped temperature field. The result is a temperature field that has temperature steps at the edges of the thermal nodes.

This method is simple but it does not contain any attempt to approximate a physical, continuous and temperature field for the FE model.

The PWT method may be adequate in the case when the mesh resolution of the thermal and structural model is of the same order of magnitude, e.g. one or two finite elements corresponding to one thermal node. In the past, this was rarely the case. Nowadays, thermal meshes tend to get denser although typically they still are significantly coarser in comparison with FE meshes.

The key points of this mapping method are as follows:

(+) The average temperature of the thermal nodes is ensured in the mapped temperature field.
(−) No interpolation or extrapolation is performed. Therefore, a complete overlap needs to be implemented in order to assign a temperature to each element in the FE model.
(−) A characteristic discontinuous temperature distribution is the result of this method.

7.4.4 Prescribed Average Temperature Method

The prescribed average temperature (PAT) method [13] is based on the intrinsic assumption of the lumped parameter method, i.e. the temperature of the lumped parameter thermal node is the average temperature of the volume represented by the thermal node. In addition, the conduction relations between the finite element nodes in the form of the conduction matrix are used to produce a physically consistent temperature field.

The PAT method is the average temperature assumption translated into linear constraint relations between the thermal node and FE node temperatures. These constraint relations, together with the conductive properties (conduction matrix) of the FE model, form the linear system that has to be solved to generate an interpolated temperature field that satisfies both the average temperature of the thermal nodes and the physical conductive relations between the finite element nodes.

One could consider that the PAT method combines the strong points from the CPPT and the PWT method.

Like the CPPT method, this method can, through the conduction relations in the FE model, that often have a higher resolution and complement the temperature field provided by the thermal analysis. Parts of the FE model that are not included in the thermal model will obtain temperatures via the conduction relations. This may look as a convenient feature, but need to be considered with care, since the thermal model did not provide all the required information. The missing information of the temperature field is basically generated by the temperature mapping method under the assumption that no heat is injected in those areas of the FE model.

Due to the method of defining the linear constraint relations that ensures the average temperature relations, the method has a very low sensitivity for FE mesh resolution changes.

The PAT method will be discussed in detail in Chap. 8.

The main points of this mapping method are as follows:

(−) Extra effort is required in preparing a conduction FE model (converted structural FE model).
(+) Conduction involved in the temperature interpolation providing a physical basis to the interpolation process and can respect discontinuities in the structure.
(+) The average temperature of the thermal nodes is ensured in the temperature field mapped to the structural FE model.
(+) The interpolation is based on the conductive relations (conduction matrix) of the complete FE model, therefore not the full FE model need to covered by thermal nodes.
(+) Very low sensitivity to FE mesh resolution changes.
(+) A smooth temperature field results from this method.

7.5 Comparing Mapping Methods on a 1-D Problem

In this section, the four different temperature mapping methods are applied to a one-dimensional problem. Thanks to the simplicity of the problem, it is able to illustrate the behaviour of these methods.

7.5.1 *One-Dimensional Model Description*

An Al-alloy strip with thickness $t = 10$ mm, length $L = 1$ m and width $b = 0.25$ m is represented by a FE model consisting of 1-D rod conduction elements. The model contains 20 rod elements and 21 FE nodes. The strip, thermal node temperatures and FE model are shown in Fig. 7.2.

The conductivity is $k = 205$ W/m/°C, the CTE is $\alpha = 2.4 \times 10^{-5}$ m/m/°C, and Young's modulus is $E = 70. \times 10^9$ Pa.

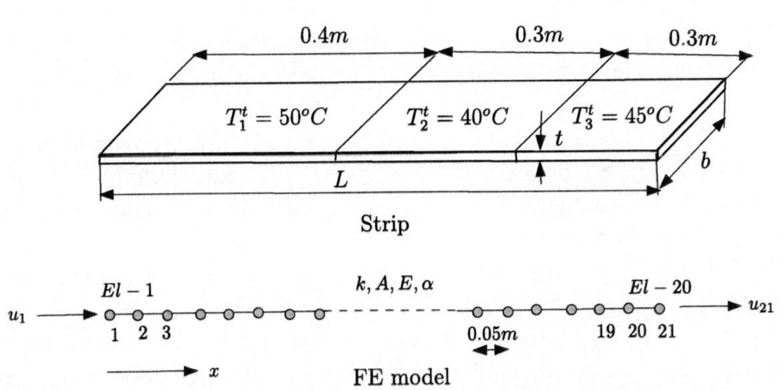

Fig. 7.2 Al-alloy strip, thermal nodes and corresponding FE model

Table 7.1 Thermal node, overlapped elements and FE nodes

Thermal node	Temperature (°C)	FE model rod elements	FE model nodes	Midpoint
1	50	1–8	1–9	5
2	40	9–14	9–15	12
3	45	15–20	15–21	18

The correspondence between thermal nodes and finite elements and nodes is presented in Table 7.1

7.5.2 *Temperature Mapping Results*

The temperature distribution in the FE model T_k^s, $k = 1, 2, \ldots, 21$ is calculated applying the four discussed temperature mapping methods using the overlap relations defined in Table 7.1.

After that the displacement field u_k, $k = 1, 2, \ldots, 21$ and element stress field σ_k, $k = 1, 2, \ldots, 20$ are calculated with $u_1 = u_{21} = 0$. The reference temperature is $T_{ref} = 0$ °C.

The resulting temperature fields in the rod from the four interpolation methods are shown in Fig. 7.3.

For the geometric temperature interpolation, a second-order inverse distance weighting (IDW) method is used ($k = 2$ in Eq. (7.2)). The temperature distribution is shown in graph (a) of Fig. 7.3. It can be noted that the FE node temperature at the centre of the thermal node is equal to the corresponding thermal node temperature. Also it can be observed that the temperatures at the edges are reduced.

The conduction relations cause the CPPT interpolated temperatures in the rod to be constructed from piece-wise linear temperature fields between the known temperatures at the centre of the thermal nodes that are coinciding with the FE model nodes 5, 14 and 18. From the centre of the thermal nodes to the ends of the rod, the produced mapped temperature field is constant. The results are shown in graph (b) of Fig. 7.3.

In case of the PWT interpolation method, the overlaid elements by a thermal node get assigned the same temperature as the corresponding thermal node. The temperature field consists of straight horizontal lines that indicate the constant temperature in the zones of the thermal nodes. This temperature distribution is shown in graph (c) of Fig. 7.3. This figure also shows the steps in temperature level at the edges of the thermal node.

In graph (d) of Fig. 7.3, the temperature distribution produced by the PAT methods is presented. It must be noted that the vertical axis of this graph is covering a larger range. Due to the constraint to ensure that the average temperature is reproduced in the

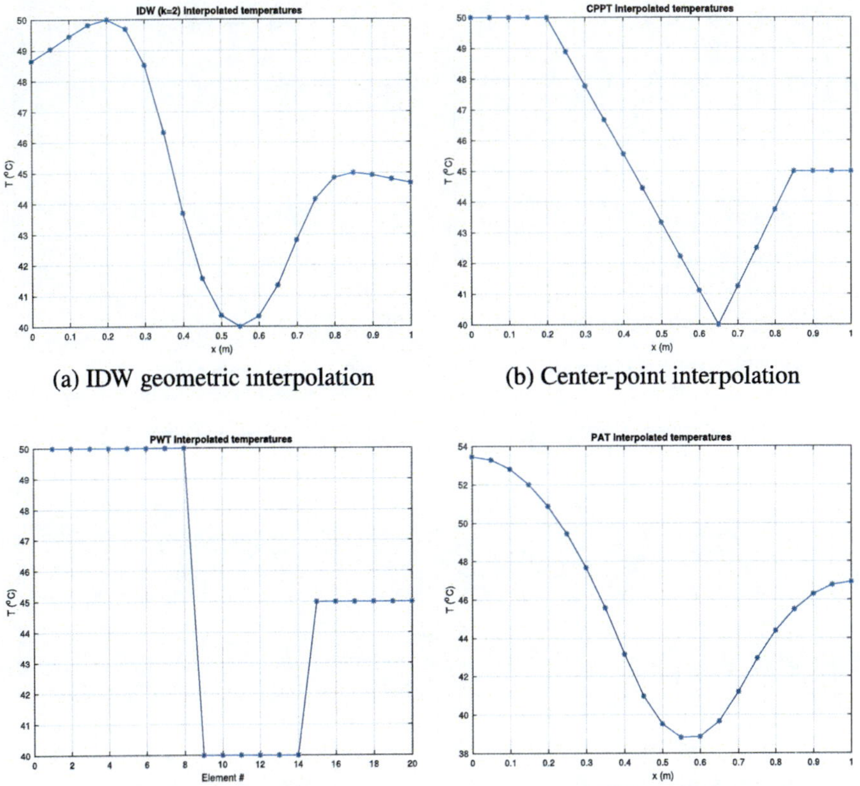

(a) IDW geometric interpolation
(b) Center-point interpolation
(c) Patch-wise interpolation
(d) PAT interpolation

Fig. 7.3 Interpolated temperature fields in rod

FE node temperatures, FE node temperatures lower and higher than the thermal node temperature are produced. Different from the other methods, the PAT temperature field increases towards the ends of the rod.

Like the result of the IDW method, the temperature field produced with the PAT method is smooth, but the IDW results are not respecting the average temperature of the thermal node.

7.5.3 Thermoelastic Responses

The calculated displacement fields in the rod constrained at the FE nodes 1 and 21 at both ends of the rod are shown for the four interpolation methods in Fig. 7.4. The displacements fields computed with the four temperature mapping methods show a number of similarities, but also some important differences.

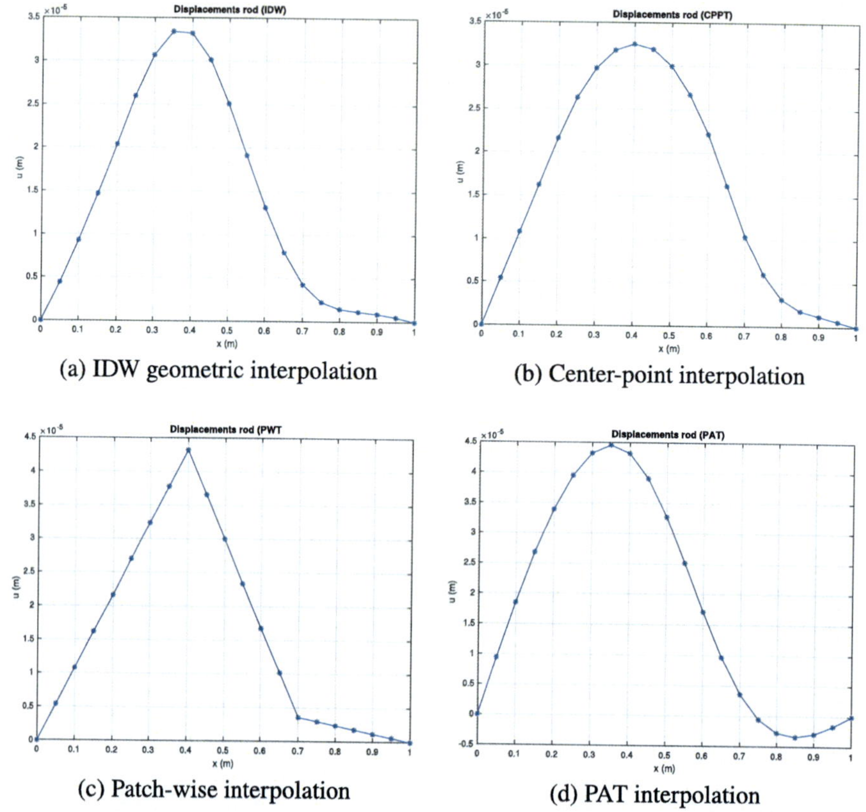

(a) IDW geometric interpolation

(b) Center-point interpolation

(c) Patch-wise interpolation

(d) PAT interpolation

Fig. 7.4 Displacement fields in rod, end nodes 1 and 21 are constrained

Except for the patch-wise temperature mapping method, all resulting displacement fields are rather smooth. The PWT temperature field results in a piece-wise linear displacement field with a higher peak in the displacements than produced with the IDW and CPPT temperature fields.

Only the displacement field, resulting from the temperature field produced with the PAT method, leads to negative displacement values, i.e. FE nodes are moving in the opposite direction. These negative displacements can be observed at the right end of graph (d) in Fig. 7.4.

Compared to the other interpolation methods, the PAT interpolated temperatures show higher maximum temperatures at the left and right ends of the rod and a lower minimum temperature at the mid of the rod. These over and undershoots compared to the other methods cause corresponding different thermal strain levels resulting in the negative displacement values. When the temperature of thermal node 2 is set to $T_2^t = 48$ °C, the over and undershoot of the temperature field are reduces displacement field becomes more flat, and negative displacement values disappear (results are not shown).

Table 7.2 Maximum displacement u_{max}

Method	Maximum displacement u_{max} (m)
Geometric interpolation	3.3419×10^{-5}
Centre-point prescribed temperature	3.2533×10^{-5}
Patch-wise temperatures	4.3200×10^{-5}
Prescribed average temperatures	4.4553×10^{-5}

Table 7.3 Mean temperatures, estimated and FE model stresses

Method	Mean temperature	Estimated Stress	FE model Stress
	T_{mean} (°C)	$\sigma_{xx} = -\alpha E T_{mean}$ (MPa)	σ_{xx} (MPa)
Geometric interpolation	44.731	−7.6002	−7.7638
Centre-point prescribed temperature	45.595	−7.6600	−7.5882
Patch-wise temperatures	45.500	−7.6440	−7.6440
Prescribed average temperatures	45.724	−7.6816	−7.6440

The maximum displacements, produced with the temperature fields with each temperature interpolation method, are presented in Table 7.2.

The stress in the rod for each temperature mapping method is presented in Table 7.3. Note that equilibrium requires a constant stress over the length of the rod, despite a varying displacement field. Table 7.3 also includes for reference the stress derived from the average temperature for each temperature mapping method.

It is remarkable that two temperature interpolation methods, PWT and PAT, obtain the same build-up stress. This is thanks to the fact these two methods both respect the average thermal node temperature and therefore inject the same amount of thermal energy in the model.

Suppose now that the displacement constraint at node 21 is removed. The rod can then expand stress free. The calculated displacement fields in the rod for the four interpolation methods are shown in Fig. 7.5.

The displacements for FE node 21 u_{21} are presented in Table 7.4. The geometric temperature interpolation shows a lowest displacement. The other three interpolation methods show equal displacements at node 21. Both the IDW and CPPT methods do not respect the average temperature of the thermal node. This implies that both methods have the potential to impose a different amount of thermal strain energy in the structure. Based on the common limitation of not respecting average thermal node temperature, it would have been expected that also the CPPT would have shown a difference in the produced displacement at FE node 21. The difference can be found in comparing the graphs (a) and (b) of Fig. 7.3 with temperature fields produced by

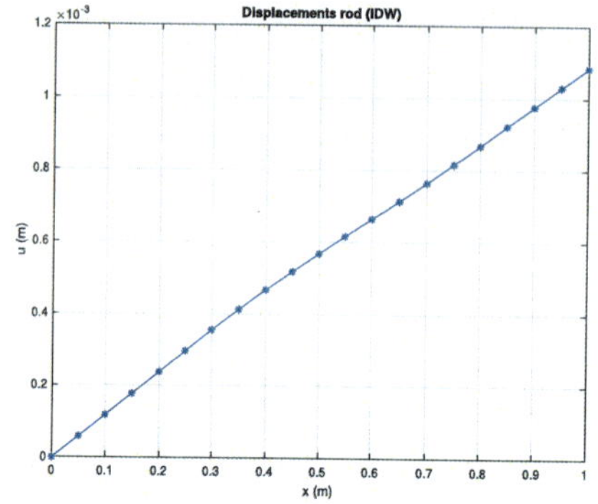

(a) IDW geometric interpolation

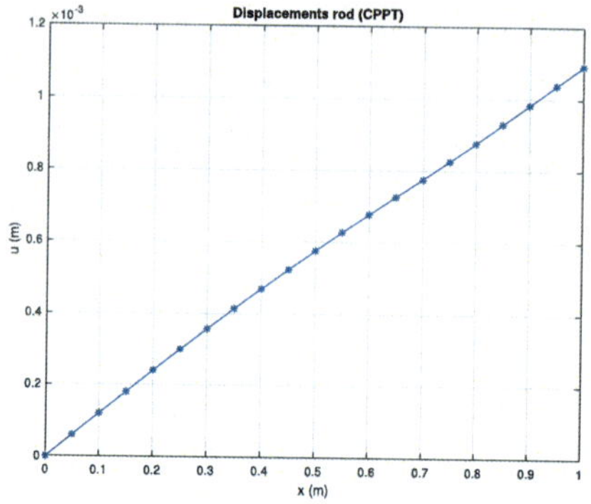

(b) Center-point interpolation

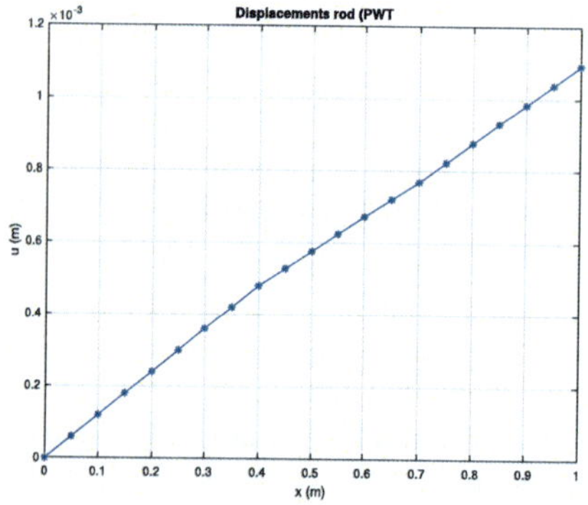

(c) Patch-wise interpolation

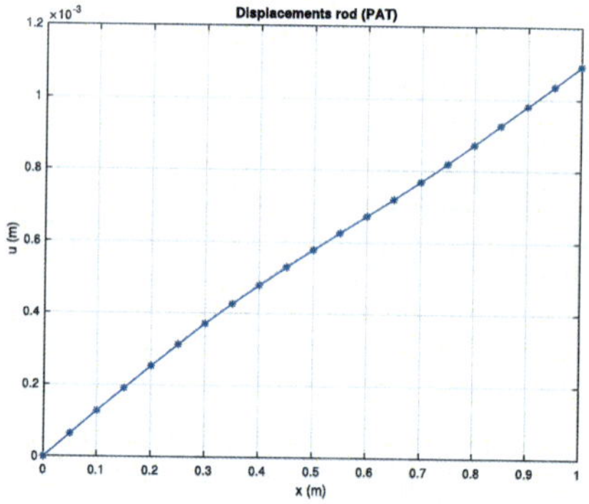

(d) PAT interpolation

Fig. 7.5 Displacement fields in rod, node 21 not constrained

Table 7.4 Maximum displacement u_{max}

Method	displacement u_{21} (m)
Geometric interpolation	1.0840×10^{-3}
Centre-point prescribed temperature	1.0920×10^{-3}
Patch-wise temperatures	1.0920×10^{-3}
Prescribed average temperatures	1.0920×10^{-3}

the two methods. The CPPT has clearly higher temperatures at both ends of the rod. This seems to be sufficient to produce the higher displacement compared to the IDW temperature field.

7.5.4 Conclusion of One-Dimensional Problem

Although this simple example has its limitations in terms of being representative for real-life structures, it shows nicely a number of strong and weak points of the different temperature mapping methods.

In this problem, there are no reference results provided to compare the four temperature methods against. This makes it not possible to declare, based on the results, one of the four temperature mapping methods the most adequate one.

Nevertheless, the PAT methods looks to be the method that does not have the limitations of the other three methods and is producing physically consistent results.

7.6 Benchmarking of Temperature Mapping Methods on a Two-Dimensional Problem

This section describes a comparison on a more complex structure compared to the model in the previous section. This model is two dimensional and includes interfaces between two parts of the structure. The description of this benchmark is an extended version of the paper [58], presented at the ECSSMET 2018.

7.6.1 Geometry, Mesh and Boundary Conditions

The benchmark model, for comparing results between the four different methods presented, consists of an assembly of two plates of each 1×0.5 m. The two plates are connected with each other at four locations over a length 0.05 m. A schematic representation is given in Fig. 7.6.

The plates have a thickness of $t = 2$ mm, and the material has a conductivity of $k = 167$ W/m/K. Young's modulus and Poisson's ratio are set to $E = 70$ GPa and $\nu = 0.3$, respectively. The CTE used is $\alpha = 23.2 \times 10^{-6}$ m/m/K.

The FE mesh consists of linear four node elements (MSC Nastran CQUAD4). The mesh has 40 elements along the short side of each of the two plates and 80 elements along the long side.

For the thermal model, each plate is meshed with five thermal nodes along the short side and ten thermal nodes along the long side. Note that therefore each of the red and the dark blue boundary condition patches in Fig. 7.6 represents a single thermal node. An overlay picture of the structural mesh and the thermal nodes is given in Fig. 7.7.

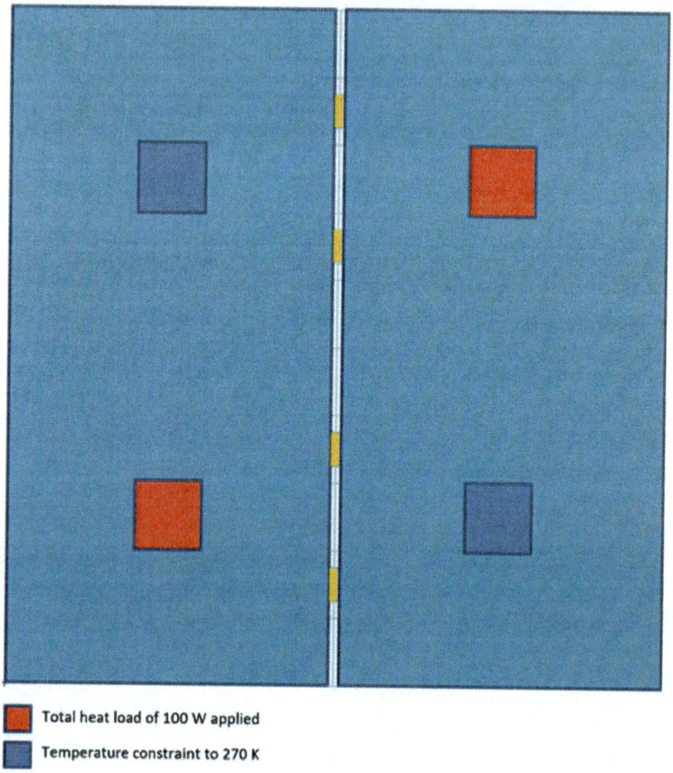

Fig. 7.6 Benchmark geometry and boundary conditions

7.6.2 Temperature Field to Be Mapped

The thermal model, as described in the previous section, has the thermal nodes connected with conductors generated with the PAT method [59] and described in Sect. 9.3.2. The resulting temperature field is shown in Fig. 7.8.

The temperature field from the thermal analysis with ESATAN-TMS is to be mapped to the FE model with the four described methods. The geometrical mapping method applied here is the field interpolation method implemented in MSC Patran.

7.6.3 Reference Temperature, Displacement and Stress

For the comparison of the four methods of mapping temperature fields on a FE model, a reference is needed. The FE method is known to have a good formulation of the conductive problem, especially when it comes down to representing 2-D and

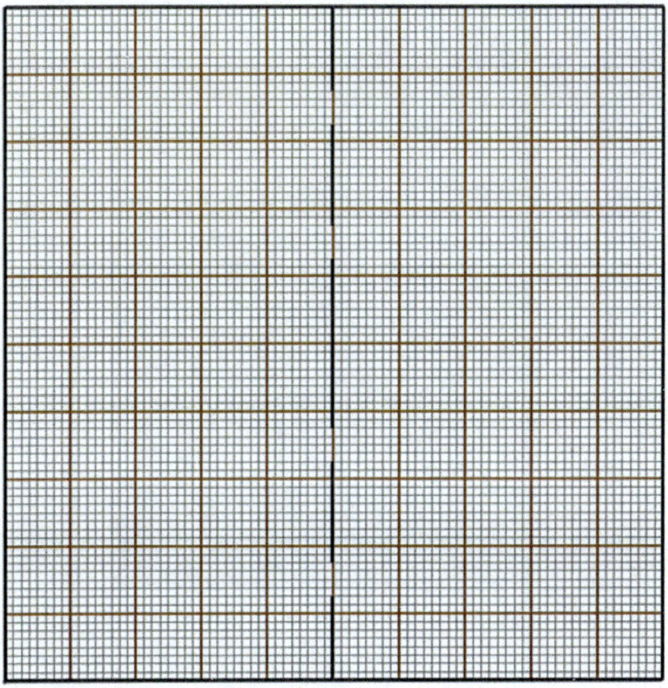

Fig. 7.7 Thermal and FE mesh shown in overlay. The FE mesh is shown in grey, thermal mesh in brown and the edges of the plates are shown in black

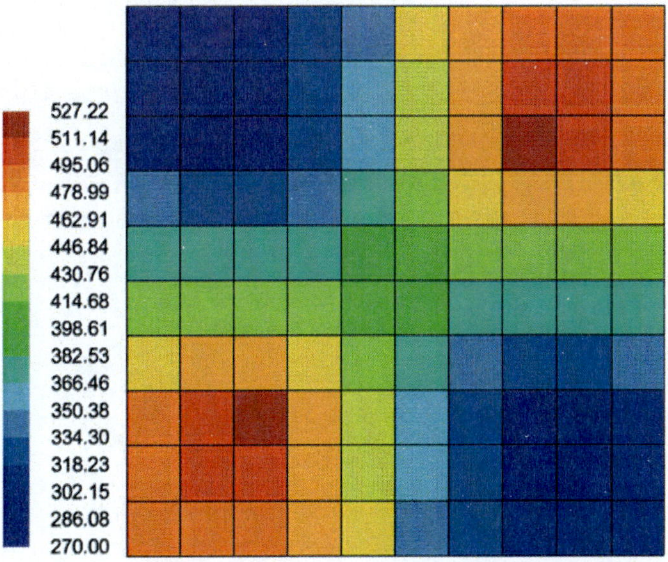

Fig. 7.8 Temperature field from the thermal model that is to be mapped to the FE mesh

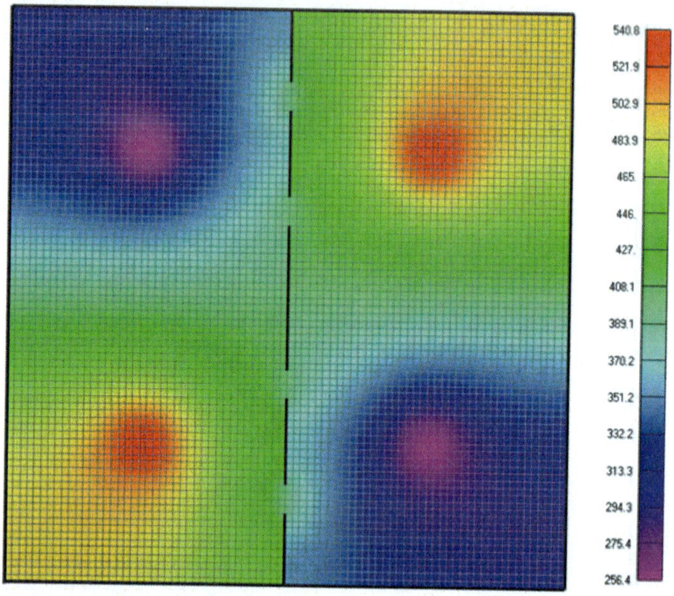

Fig. 7.9 Reference temperature field

3-D temperature fields. The FE mesh as is presented in Fig. 7.7 is used to produce a reference temperature field. To guarantee good comparison, it is important that the temperature boundary conditions applied to the FE model are consistent to those applied to the ESATAN-TMS model. For that purpose, a multi point constraint is implemented to enforce an average temperature of 270 K on the FE nodes corresponding to the thermal nodes that have these prescribed temperatures. The resulting reference temperature field is shown in Fig. 7.9.

For the thermoelastic response calculation, iso-static boundary conditions are applied. The node in the middle of the FE mesh on the right side of the slit is constrained in 1, 2 and 3 directions, the middle node on the outer right side of the FE mesh is constrained in 2 and 3 directions and the node in the middle of the bottom side to the right of slit is constrained in 3 direction only. This results in the displacement field as shown in Fig. 7.10. The corresponding von Mises stresses are shown in Fig. 7.11.

7.7 Comparing Performances of Mapping Methods

7.7.1 Performance Criteria for the Mapping Methods

In order to be able to quantify the performances of the four temperature mapping methods, the two performance criteria, defined in Sect. 7.1, are here recalled as follows:

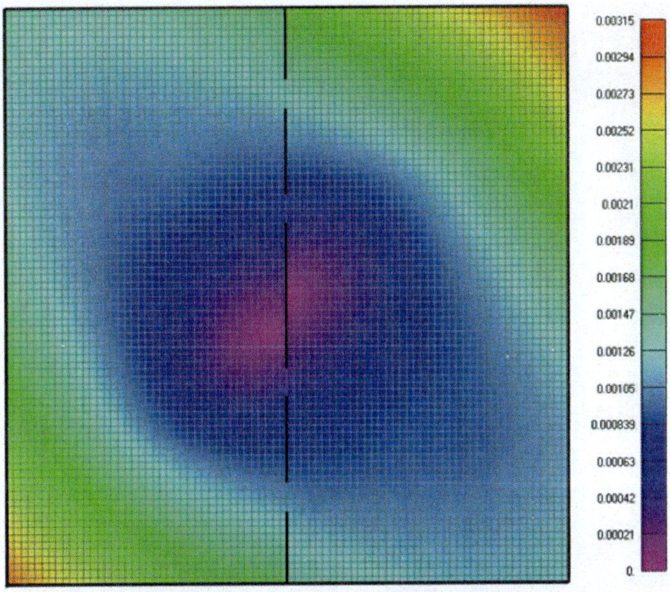

Fig. 7.10 Thermoelastic displacement field based on the reference temperature field

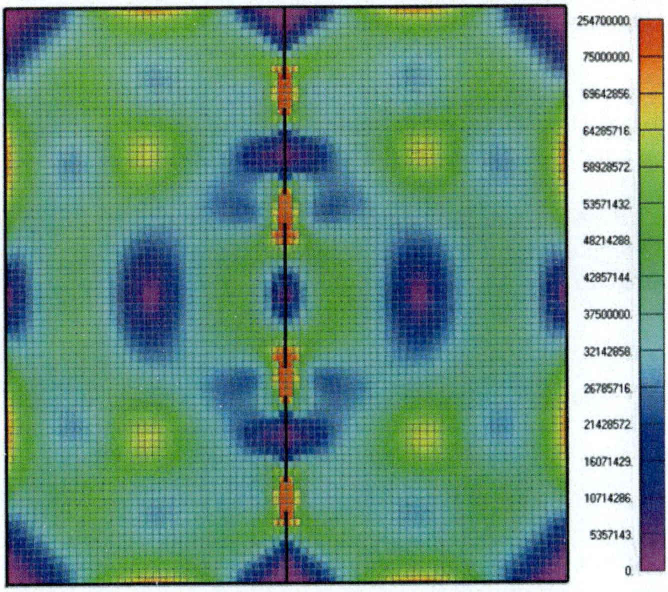

Fig. 7.11 Thermoelastic von Mises stress field based on the reference temperature field

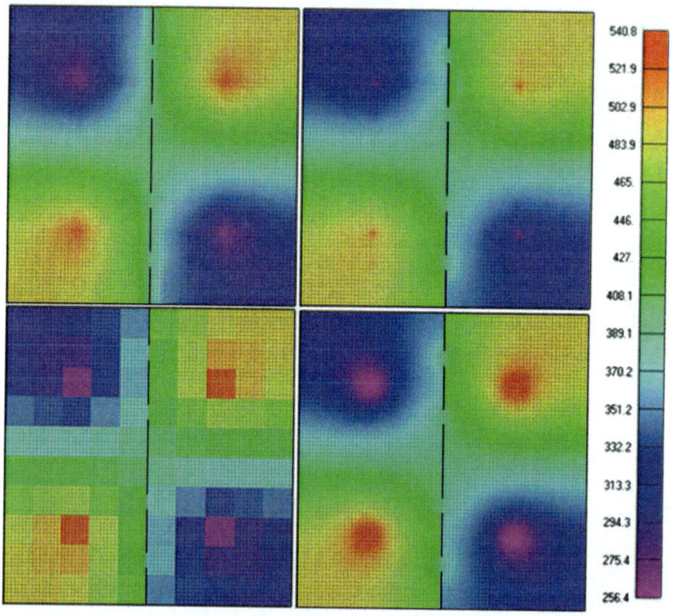

Fig. 7.12 Mapped temperature fields generated with different mapping methods. Top left: Geometrical interpolation. Top right: CPPT mapping. Bottom left: Patch-wise temperature application. Bottom right: PAT mapping

- Average mapped thermal node temperature: The weighted average temperature of the FE nodes corresponding to the thermal node shall reproduce the thermal node temperature as good as possible.
- The displacement field and the stress field computed from the mapped temperature field shall reproduce that of a detailed thermal and thermoelastic analysis done with the reference FE model, as good as possible.

In the following sections, these criteria are evaluated for the four mapping methods.

7.7.2 *Qualitative Comparison of the Mapped Temperature Fields*

For the purpose of direct comparison, Fig. 7.12 shows the temperature fields produced with all four mapping methods in one figure at reduced size.

Upon inspection of Fig. 7.12 and comparison with the reference temperature field in Fig. 7.9, the following is observed:

- The temperature field produced through the geometric interpolation method ignores the interruption of the structure, since it has no information to take this into account. It could be decided to do the temperature interpolation per panel, but in that case, the physical gradients in the temperature field near the connections between the panels would then not be taken into account.
- Both the CPPT and the PAT method, are able to simulate, near the interfaces, the interaction between the temperature fields of the two panels.
- It can be noted that the temperature field produced by the CPPT method shows a sharp spot at the centre of the thermal nodes that are dissipating the heat.
- The temperature field of the patch-wise temperature application method clearly shows the patches with constant temperature, identical to the thermal model results. This temperature field is clearly not a physical temperature field.
- The PAT method reproduces the reference temperature field very well.

7.7.3 Average Temperature Comparison

Both the patch-wise temperature application method and the PAT method inherently secure that the average temperature of the FE nodes corresponding to the thermal node is equal to the thermal node temperature. No further assessment for these two methods is therefore needed.

In Figs. 7.13 and 7.14, the differences between the thermal node temperatures and the average per thermal node area for the geometrical interpolation method and the CPPT method are presented. A positive difference means that the thermal node temperature is higher than the computed average of the interpolated FE node temperatures in each thermal node area. For both these two methods, the average temperature of the FE nodes, corresponding to a thermal node, deviates strongly from the associated thermal node temperature.

The strongest deviations from the thermal node temperatures are observed in the areas of the highest temperature peaks. This is inherent to the way of applying the temperature in those methods through applying the ESATAN calculated temperature to the middle of the thermal node region in the FE model. The peak temperature in those regions, after interpolation, will be equal to the ESATAN calculated thermal node temperature, which actually should represent the region's average temperature. This results in a corresponding deviation in those regions from the required average temperatures as calculated in the thermal analysis.

For the geometrical interpolation method, a mismatch due to not taking into account the boundary between the two plates is observed as well. This can be noted clearly at both the top and bottom centres of the Fig. 7.13.

Earlier a remark was made that the mapped temperature fields produced with the CPPT method have a sensitivity for the mesh density. Trials with coarser FE meshes (not shown in in this book) demonstrate that the mesh density dependence also shows up in the form of difference in the average temperature of the FE nodes corresponding to a thermal node for different mesh densities.

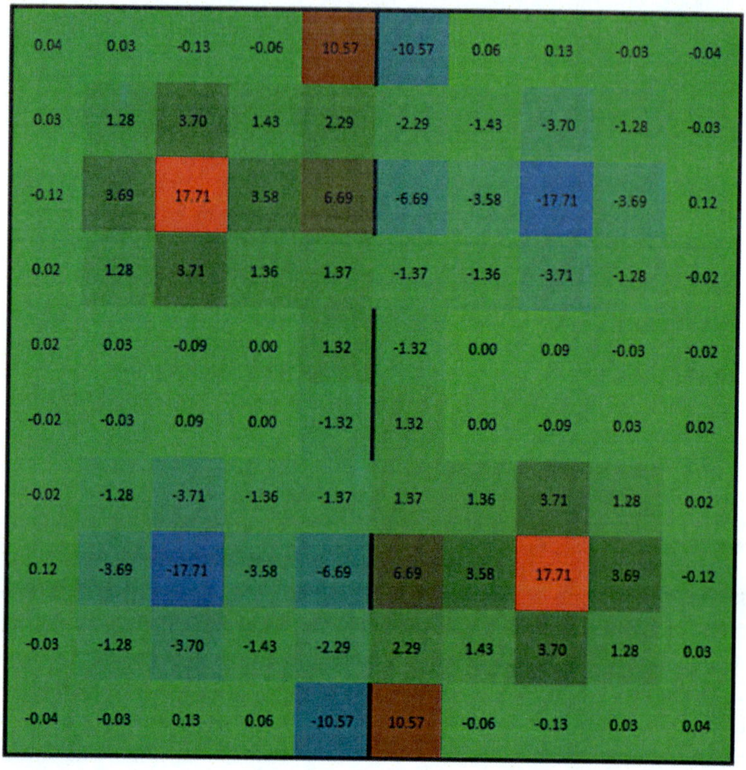

Fig. 7.13 Difference between thermal node temperature and average FE node temperature per thermal node area for the geometrical interpolation method

The geometrical interpolation method appears not to be dependent on FE mesh density.

7.7.4 Displacement Comparison

The four mapped temperature fields are used to compute the corresponding thermoelastic deformation with a static FE analysis. It turns out for this benchmark case that, except for the case with the PAT method-based temperature field, the displacement fields show only small differences relative to the reference displacement field. The differences in displacements relative to the reference displacement field are presented in Fig. 7.15 for the different mapping methods.

An error up to ca. 0.07 mm relative to a displacement of about 2.0 mm in the same region is observed with the CPPT mapping method. The geometrical interpolation method shows small peaks with errors up to ca. 0.09 mm in the corners of the slit

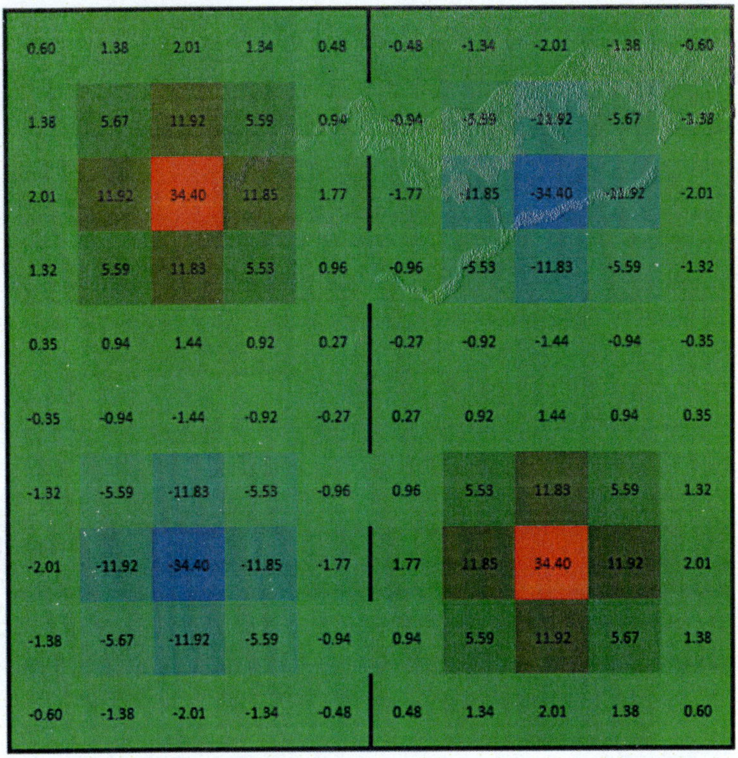

Fig. 7.14 Difference between thermal node temperature and average FE node temperature per thermal node area for the CPPT method

region. The patch-wise temperature application method shows a maximum error between 0.025–0.04 mm at the bottom side of the slit region and in the lower left quarter of the plate.

The displacement field based on the mapped temperatures produced by the PAT method nearly exactly coincides with the reference case. The differences are therefore shown in Fig. 7.16, with a smaller scale. Looking at the magnitude of the displacement error, the authors believe that the difference is a numerical truncation error in the temperature data originating from the use of a different thermal solver, ESATAN-TMS instead of MSC Nastran and having to perform an interpolation step from ESATAN-TMS to MSC Nastran.

7.7.5 *Stress Comparison*

Figure 7.17 shows the von Mises stress fields resulting from the temperature fields produced with the different temperature mapping methods. The comparison with the

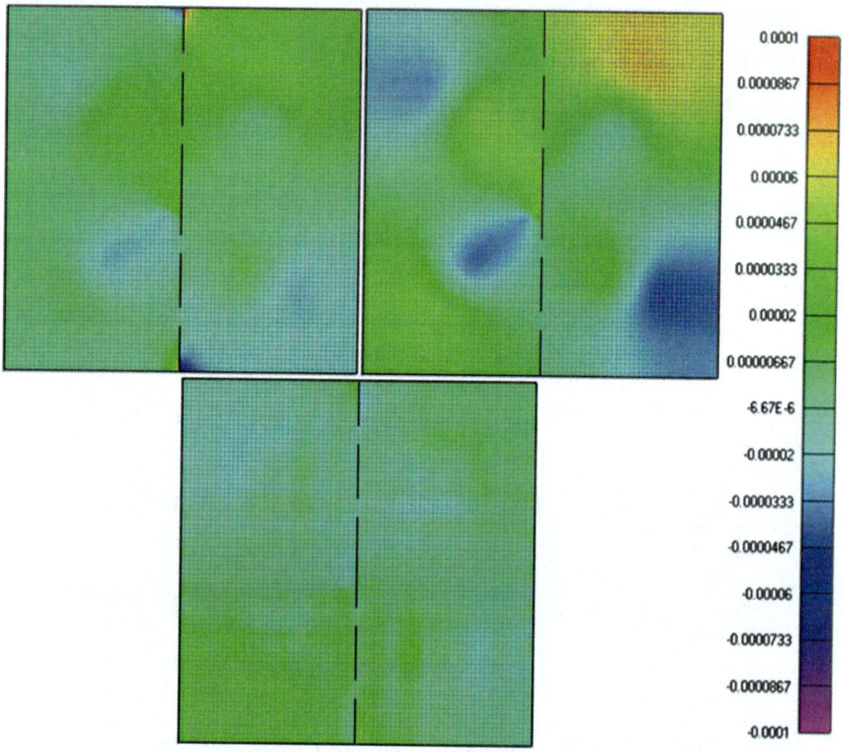

Fig. 7.15 Difference, relative to the reference case, for the total displacement field based on the temperature field generated with the different mapping methods. Top Left: Geometrical interpolation. Top right: CPPT mapping. Bottom: Patch-wise temperature application

reference stress field is presented in Fig. 7.18 as contour plots showing the distribution of the stress differences relative to the reference temperature field in Fig. 7.11 for each temperature mapping method. In Fig. 7.18, the stress difference plot for the PAT method is not provided, because the stress difference values for that method are much smaller and require a different scale. Therefore, these are shown in Fig. 7.19.

Like the displacements, the highest difference in the stress fields shows up in the regions with high gradients that occur near the boundary conditions and around the interfaces between the panels. One observation that can be made is that, even away from boundary conditions and interfaces, where the stress level is typically in the order of 50 MPa, the differences in stress levels range from 8 to 18 MPa. Closer to the boundary conditions, the differences get up to 28 MPa.

For the geometrical interpolation method (top left in Fig. 7.18), as expected, the highest differences show up near the interface and the gap between the two panels. These features were ignored by this interpolation method.

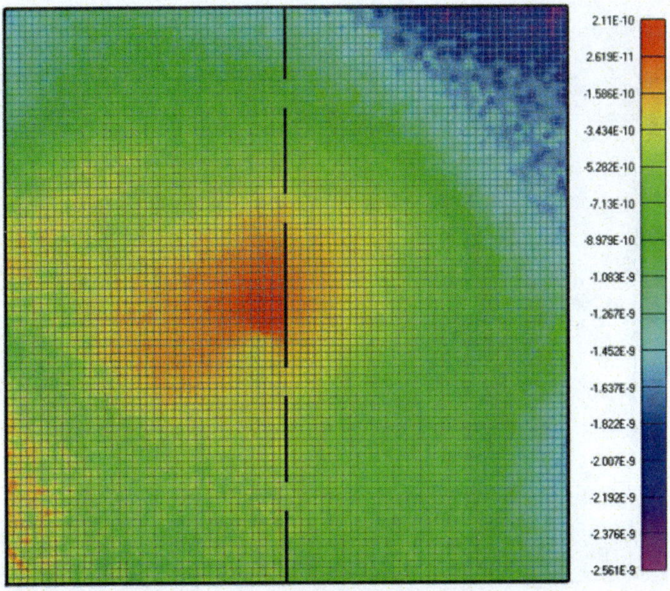

Fig. 7.16 Difference, relative to the reference case, for the total displacement field based on the temperature field generated with the PAT method plotted with a small scale to show the small numerical error

The temperature field from the CPPT method shows also in the stress difference the characteristic local peaks caused by the discrete temperature prescription (top right in Fig. 7.18).

Finally, in the stress field produced with the temperature field generated with the PWT method (bottom picture in Fig. 7.18), the thermal node edges can be noticed. At these thermal node edges, the temperature field has discrete steps with corresponding high-temperature gradients that are translated in locally higher stress levels.

Again the PAT methods nearly coincides with the reference stress field. Like for the displacements, the differences in stress level are four orders of magnitude smaller than the reference stress field: maximum difference around 300 Pa versus a typical stress level 50 MPa. This implies that with the temperature field based on PAT method, the reference stress field is very well reproduced, except for numerical noise.

7.7.6 Concluding the 2-D Benchmark Model

The results produced by all mapping methods, except the prescribed average temperature method, appear to deviate from the reference results.

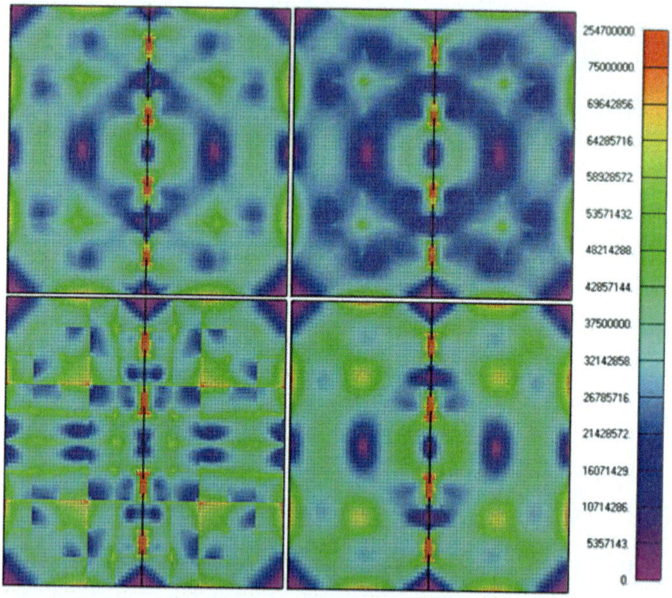

Fig. 7.17 von Mises stress based on the temperature field generated with the different mapping methods. Top left: Geometrical interpolation. Top right: CPPT mapping. Bottom left: Patch-wise temperature application. Bottom right: PAT mapping

A concern is that the displacements and the stresses produced by the other mapping methods contain significant errors and could also be non-conservative. The main cause of the deviation is that these mapping methods do not reproduce the temperature field as determined by the thermal analysis.

The results obtained with this benchmark model conclude with a preference for applying the PAT method, when accurate thermoelastic mapping results are required.

7.8 Summary Temperature Mapping/Interpolation Methods

In this chapter, four temperature mapping methods were described. A qualitative comparison of these methods is listed here, in which the positive or strong points of the method are indicated with a "(+)", and weaker points or drawbacks are indicated with "(−)" as follows:

Geometrical interpolation

(+) Basic linear geometric interpolation is available in most finite element preprocessor tools.

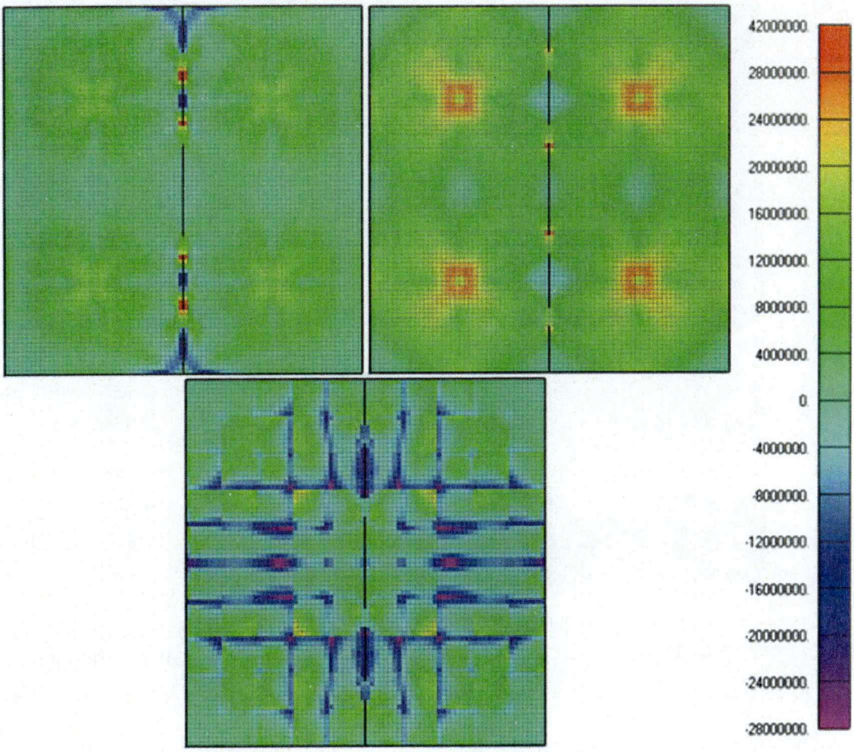

Fig. 7.18 Difference, relative to the reference case, for von Mises stress based on the temperature field generated with the different mapping methods. Top left: Geometrical interpolation. Top right: CPPT mapping. Bottom: Patch-wise temperature application

(−) The average temperature of the thermal nodes is not necessarily maintained in the mapped temperature field.
(−) The interpolation method is based on geometry only ignoring any physical and configuration information such as interfaces and changes in heat flow path due to variation in conduction.
(+) A smooth temperature field results from this method.

Centre-Point Prescribed Temperature Method (CPPT)

(−) Extra effort is required in preparing a conduction FE model (converted structural FE model).
(+) Conduction involved in the temperature interpolation providing a physical basis to the interpolation process and can respect discontinuities in the structure.
(−) The average temperature of the thermal nodes is not necessarily maintained in the structural FE model.

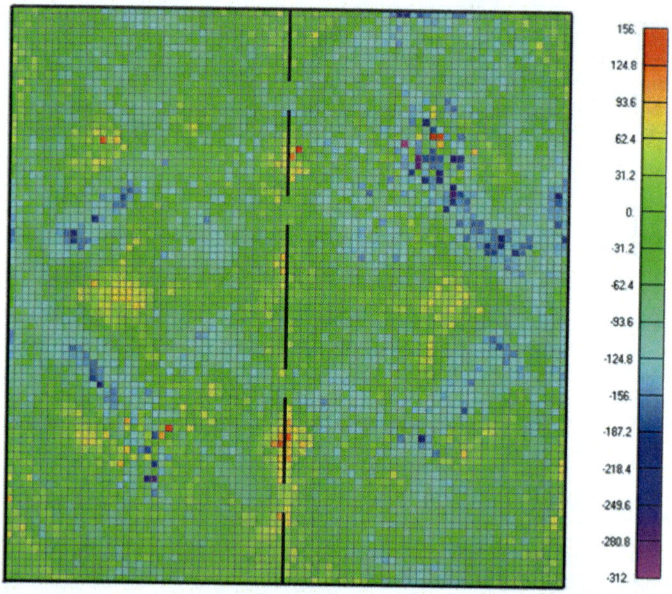

Fig. 7.19 Difference, relative to the reference case, for von Mises stress based on the temperature field generated with the PAT method plotted with a small scale to show the small numerical error

(+) The interpolation is based on the conductive relations (conduction matrix) of the complete FE model, therefore not the full FE model need to covered by thermal nodes.
(−) Important sensitivity to mesh resolution at the location of prescribed temperatures.
(+) A smooth temperature field results from this method.

Patch-Wise Temperature Application Method (PWT)

(+) The average temperature of the thermal nodes is assured in the mapped temperature field.
(−) No interpolation is performed. Therefore, a complete overlap needs to be implemented in order to assign a temperature to each element in the FE model.
(−) A characteristic discontinuous temperature distribution is the result of this method.

Prescribed Average Temperature method (PAT)

(−) Extra effort is required in preparing a conduction FE model (converted structural FE model).
(+) Conduction involved in the temperature interpolation providing a physical basis to the interpolation process and can respect discontinuities in the structure.
(+) The average temperature of the thermal nodes is maintained in the temperature field mapped to structural FE model.

(+) The interpolation is based on the conductive relations (conduction matrix) of the complete FE model, therefore not the full FE model need to covered by thermal nodes.
(+) Very low sensitivity to FE mesh resolution changes.
(+) A smooth temperature field results from this method.

Two example cases were presented. The PAT method appears to be the most adequate method for temperature mapping. For that reason, the method will be described in detail in the next chapter.

Problems

7.1 An Al-alloy strip with length $L = 1$ m. The number of nodes is 21. The strip, thermal nodes and associated thermal node temperatures and FE model are shown in Fig. 7.20. Calculate the temperatures in the FE nodes $T_j^s,\ j = 1, 2, \ldots, 21$ using the IDW interpolation method taking all thermal nodes into account using 1 and 5 for the power k of Eq. 7.2. The correspondence between FE nodes and thermal nodes must be obtained from Table 7.5. Compare the IDW interpolation results with earlier calculated interpolation results obtained with the PAT method.

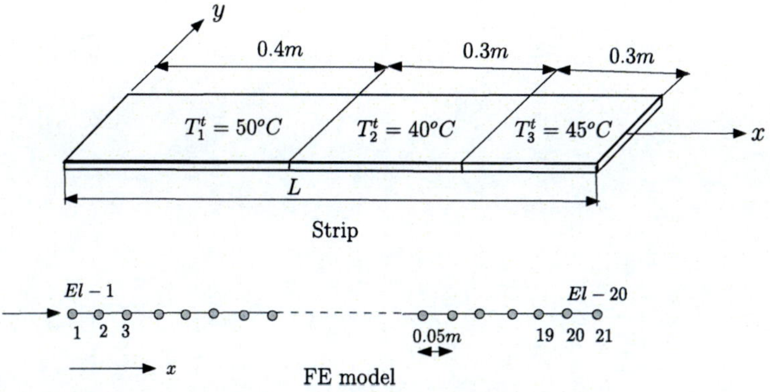

Fig. 7.20 Al-alloy strip, thermal nodes and corresponding FE model

Table 7.5 Isothermal node, overlaid elements, nodes

Thermal node	Temperature (°C)	FE model rod elements	FE model nodes	Midpoint
1	50	1–8	1–9	5
2	40	9–14	9–15	12
3	45	15–20	15–21	18

Chapter 8
Prescribed Average Temperature Method

Abstract The system equations of the prescribed average temperature (PAT) mapping method are explained. The method allows to transfer the temperature field as computed with the lumped parameter thermal analysis to the finite element model while respecting the assumptions of the lumped parameter method.

8.1 Introduction

In Chap. 7, four temperature mapping methods were described, and the performance of these methods were compared:

- Geometric temperature interpolation methods
- Centre-point prescribed temperature (CPPT) method
- Patch-wise temperature application method
- Prescribed average temperature (PAT) method.

Chapter 7 concluded with indicating advantages of the PAT method compared to the other methods. For that reason, this chapter is dedicated to the PAT method. The fundamentals of the method are described and illustrated with examples.

The PAT method has a useful spin-off, which is the generation of linear conductors for thermal lumped parameter models. The generation of these PAT-based conductors is explained in Sect. 9.3 which uses the principles described in the current chapter.

8.2 Relating Thermal Nodes and FEM Nodes

The objective of each of the mapping procedures discussed in this book is to relate the thermal node temperatures to the finite element node temperatures. This problem is schematically presented in Fig. 8.1.

The basis of the PAT method is a finite element conduction model of the structure. This conduction model is derived from the structural finite element model.

S. Appel and J. Wijker, *Simulation of Thermoelastic Behaviour of Spacecraft Structures*, Springer Aerospace Technology, https://doi.org/10.1007/978-3-030-78999-2_8

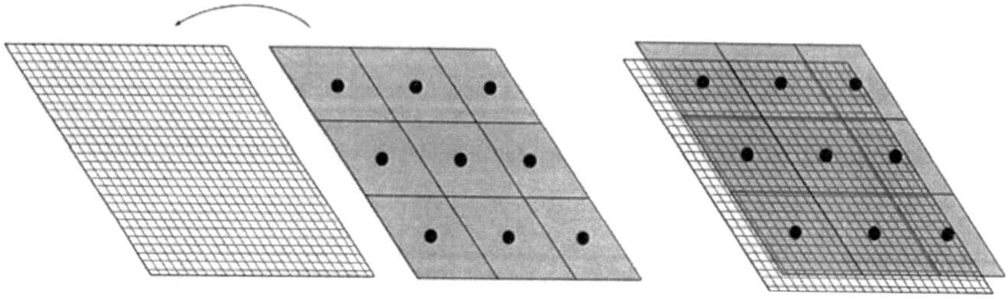

Fig. 8.1 Thermal mesh overlap onto FE model

The principle is to relate the temperature of a thermal node of the thermal model to the average temperature of the finite element nodes covered by the thermal node. This constraint relation is not enough to describe the temperature distribution over the finite element model in an unique way. For that reason, the constraints are complemented with a conduction balance to achieve an unique temperature distribution.

If the temperatures are known in all the FE nodes, then, with the use of the element shape functions, the temperature field over the complete body can be written as

$$T(x, y, z) = \sum_{i=1}^{n_s} \Psi_i(x, y, z) T_i^s, \tag{8.1}$$

with

T Temperature field over the finite element model as a function of position in the model (x, y, z)
T_i^s Temperature at finite element node i
Ψ_i Shape function value associated with finite element node i
n_s Number of finite element nodes in the model.

In this way, the temperature is defined over the complete body as function of the position in the body.

It must be noted that the shape function value Ψ_i is composed of the contributions of the shape functions of all finite elements connected to finite element node i. Assuming that there are n_e elements connected to FE node i, this could be written as

$$\Psi_i(x, y, z) = \sum_{j=1}^{n_e} \psi_{ij}^e(x, y, z), \tag{8.2}$$

with ψ_{ij}^e being the shape function of finite element j associated to FE node i.

The lumped parameter thermal analysis provides only one temperature per thermal node. As stated above, the principle of the PAT method is based on the assumption of the lumped parameter method that the temperature of the thermal node is the average temperatures of the volume represented by the thermal node. This can be written as

$$T_k^t = \frac{\int_{V_k} T(x, y, z) dV}{V_k}, \tag{8.3}$$

where

T_k^t Temperature of thermal node k
T Temperature field over the finite element model as a function of position (x, y, z) in the model
V_k Volume spanned by thermal node k.

Substitution of Eq. (8.1) in Eq. (8.3) gives

$$T_k^t = \frac{1}{V_k} \int\limits_{V_k} \sum_{i=1}^{n_s} \Psi_i(x, y, z) T_i^s dV = \sum_{i=1}^{n_s} \frac{1}{V_k} \int\limits_{V_k} \Psi_i(x, y, z) T_i^s dV = \sum_{i=1}^{n_s} a_{ki} T_i^s, \tag{8.4}$$

and

$$a_{ki} = \frac{1}{V_k} \int\limits_{V_k} \Psi_i(x, y, z) dV \tag{8.5}$$

The most essential relation that forms the basis of the PAT method is included in Eq. (8.4) and is worth to be expressed explicitly.

$$T_k^t = \sum_{i=1}^{n_s} a_{ki} T_i^s \tag{8.6}$$

In matrix–vector notation this can be written as

$$\left(T^t\right) = A \left(T^s\right) \tag{8.7}$$

where

$\left(T^t\right)$ Vector with thermal node temperatures
A The A-matrix, generally rectangular, containing the values a_{ki}
$\left(T^s\right)$ Vector with temperatures at finite element nodes.

With the characteristic of the element shape functions, it can be shown that the volume associated with FE node i is

$$v_{ki} = \int\limits_{V_k} \Psi_i(x, y, z) dV. \tag{8.8}$$

It must be noted that Ψ_i only has the contributions from the elements overlapped by thermal node k (see Eq. (8.2)). For that reason, the integral in Eq. (8.8) is limited to the volumetric bounds of the thermal node k.

The total volume of the thermal node is then

$$V_k = \sum_{i=1}^{n_s} \int_{V_k} \Psi_i(x, y, z) dV = \sum_{i=1}^{n_s} v_{ki}. \tag{8.9}$$

Combining Eq. (8.9) with Eq. (8.5) results in the characteristic of a row in the A-matrix as follows:

$$\sum_{i=1}^{n_s} \frac{v_{ki}}{V_k} = \sum_{i=1}^{n_s} a_{ki} = 1 \tag{8.10}$$

In other words, the coefficients in each row in the A-matrix represent the fraction of the total volume of the thermal node associated with each overlapped finite element node. The sum of these fractions should add up to 1 to obtain the complete volume.

The thermal node temperatures, (T^t), are obtained from a lumped parameter thermal analysis, and the coefficients of the matrix A are derived from the finite element model as described in the next section. Normally, (T^s) cannot be obtained from Eq. (8.7), because the matrix A is usually not square and thus cannot be inverted. Therefore, a conduction balance is added to the model to achieve a unique temperature distribution.

PAT relation for a one-dimensional problem

A one-dimensional rod is considered with a total length of $3L$ consisting of two parts: Rod 1 with a length of $2L$ and cross section of A and Rod 2 with a length of L and with cross section of $3A$ (see Fig. 8.2). The two rod elements correspond to a single thermal node from the thermal model. In this example, the weighting factors for the finite element node temperatures a_{ik} of Eq. (8.3) are going to be determined. The PAT relation between the thermal node temperature T_1^t and the FE node temperatures T_1^s, T_2^s and T_3^s is going to be derived.

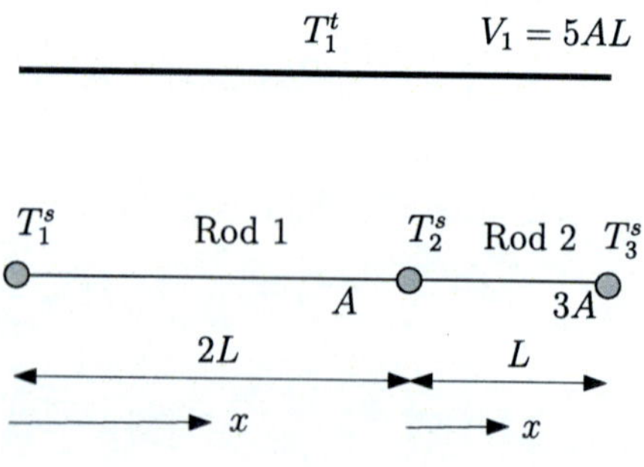

Fig. 8.2 FE model with two rod elements overlapped by one thermal node

For the elements, the following shape functions are chosen

$$(\psi^e)^T = \begin{pmatrix} \psi_1^e \\ \psi_2^e \end{pmatrix} = \begin{pmatrix} 1 - \frac{x}{a} \\ \frac{x}{a} \end{pmatrix} \tag{8.11}$$

Using Eq. (8.11), together with previous equations, the volume of the two elements associated with the thermal node 1 can be determined and assigned to the FE nodes

$$\begin{aligned} v_{11} &= \tfrac{1}{2}(2AL) = AL \\ v_{12} &= \tfrac{1}{2}(2AL) + \tfrac{1}{2}(3AL) = \tfrac{5}{2}AL \\ v_{13} &= \tfrac{1}{2}(3AL) = \tfrac{3}{2}AL \end{aligned} \tag{8.12}$$

with v_{1i} being the volume associated with FE node i that is in the domain of thermal node 1.

With Eq. (8.9), the total volume of the thermal node in 1 becomes

$$V_1 = v_{11} + v_{12} + v_{13} = \left(1 + \tfrac{5}{2} + \tfrac{3}{2}\right) AL = 5AL \tag{8.13}$$

The weighting factors a_{1j} then become

$$\begin{aligned} a_{11} &= \frac{v_{11}}{V_1} = \tfrac{1}{5} \\ a_{12} &= \frac{v_{12}}{V_1} = \tfrac{1}{2} \\ a_{13} &= \frac{v_{13}}{V_1} = \tfrac{3}{10}. \end{aligned} \tag{8.14}$$

The thermal node temperature T_1^t can now be expressed in FE model node temperatures

$$T_1^t = a_{11}T_1^s + a_{12}T_2^s + a_{13}T_3^s = \tfrac{2}{10}T_1^s + \tfrac{5}{10}T_2^s + \tfrac{3}{10}T_3^s \tag{8.15}$$

The sum of the elements in row $i = 1$ of the $[A]$-matrix $\sum_{j=1}^{3} a_{1j} = 1$.

8.3 Creation of Consistent Values of A-Matrix Coefficients with a Finite Element Code

The expressions for the weighting factors for a_{ki} in Eq. (8.5), collected in the A-matrix, as shown in Eq. (8.7), make use of the shape functions that are used by the finite element formulations. Unfortunately, the shape functions that are implemented in a FEM tool are in general not accessible for the users of the finite element software.

However, each finite element code is the calculation of the integral of Eq. (8.5) part of the standard procedure to convert element volumetric loads to nodal loads. The nodal loads are in most cases part of the right-hand side load vector of the linear equations to be solved (see vector (R_Q) of Eq. (B.33)).

It is assumed that it is known which elements of the finite element model correspond to which thermal node of the thermal model. The A-matrix coefficients for a thermal node can then be determined by applying an unit volumetric heat load (1 W/m^3) to all the elements that correspond to that thermal node. These artificial element heat loads are then converted to the FE nodal heat load vector (R_Q) with the use of Eq. (B.28). The heat flux q_{ki} for specifically FE node i overlapped by thermal node k can then be written as

$$q_{ki} = \int_{V_k} Q\Psi_i(x, y, z)dV, \tag{8.16}$$

with Q being the unit volumetric heat load.

Considering the fact that $Q = 1$ W/m^3, this equation can be reduced to

$$q_{ki} = \int_{V_k} \Psi_i(x, y, z)dV. \tag{8.17}$$

With Eq. (8.17), the expression Eq. (8.9) for the volume of the finite element nodes corresponding to the thermal node k can be written as

$$V_k = \sum_{i=1}^{n_s} q_{ki}. \tag{8.18}$$

Equation (8.18) shows that the sum of the nodal fluxes, derived from a unit volumetric heat flux, forms the volume of the thermal node. This implies that the values q_{ki} can be considered as the amount of volume associated with finite element node i that is overlapped by thermal node k.

$$v_{ki} = q_{ki} = \int_{V_k} \Psi_i(x, y, z)dV, \tag{8.19}$$

with v_{ki} being the amount of volume associated with finite element node i that is overlapped by thermal node k.

Equation (8.18) can then be rewritten to show that the volume of the thermal node is the sum of the volumes associated with the finite element nodes that are overlapped by a thermal node.

$$V_k = \sum_{i=1}^{n_s} v_{ki}. \tag{8.20}$$

The A-matrix coefficients a_{ki} can be obtained by normalising the nodal volumes with the total volume of the thermal node

$$a_{ki} = \frac{v_{ki}}{V_k} \tag{8.21}$$

With the above equations, the procedure for obtaining the A-matrix coefficients a_{ki} for a thermal node with the help of a finite element tool is as follows:

1. Apply a unit volumetric heat flux to the elements overlapped by the thermal node
2. Obtain the nodal heat loads (= volume associated with the FE node v_{ki})
3. Sum all the nodal heat loads (= total volume of the thermal node V_k)
4. Divide each volume associated with a FE node by the total volume of the thermal node ($= v_{ki}/V_k = a_{ki}$).

Generating the A-matrix for rectangular elements

In Fig. 8.3, a rectangular conduction finite element is shown. The constant thickness of that element is t, and the artificial internal heat source is $Q = 1$ W.

The linear shape functions for this kind of elements can be chosen to be

$$\left(\psi^e\right)^T = \begin{pmatrix} \psi_1^e(x, y) \\ \psi_2^e(x, y) \\ \psi_3^e(x, y) \\ \psi_4^e(x, y) \end{pmatrix} = \begin{pmatrix} 1 - \frac{x}{a} - \frac{y}{b} + \frac{xy}{ab} \\ \frac{x}{a} - \frac{xy}{ab} \\ \frac{xy}{ab} \\ \frac{y}{b} - \frac{xy}{ab} \end{pmatrix}. \tag{8.22}$$

Using Eq. (8.19), the volume associated with each node of the element is

$$v_{ki} = \int\limits_V \psi_i^e dV = t \int\limits_{A_e} \psi_i^e dA = \frac{abt}{4}, \tag{8.23}$$

with A_e being the surface area of the rectangular element.

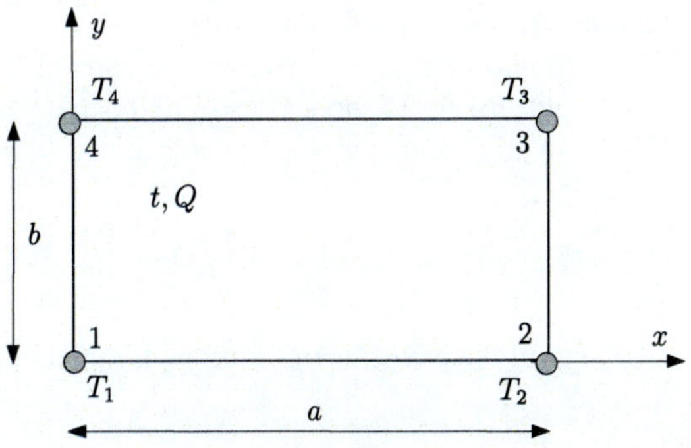

Fig. 8.3 Rectangular conduction element

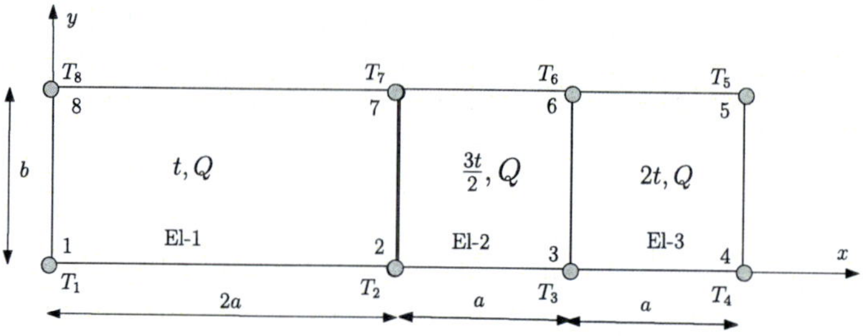

Fig. 8.4 Plate model with three rectangular conduction elements

Table 8.1 Volumes of the three finite elements

Element #	Volume
EL-1	$2abt$
EL-2	$\frac{3}{2}abt$
EL-3	$2abt$

For this simple rectangular-shaped element, the consistent approach with the use of shape functions results in a quarter of the volume of the element being associated with each of the FE nodes.

Considered is now the FE model shown in Fig. 8.4 with three conduction elements. The elements El-1 and El-2 are overlapped by thermal node 1 with temperature T_1^t, and El-3 is overlapped by thermal node 2 with temperature T_2^t (Table 8.1).

Since thermal node 1 is spanned by elements El-1 and El-2, the total volume associated with thermal node 1 is therefore the sum of both thermal elements

$$V_1 = 2abt + \tfrac{3}{2}abt = \tfrac{7}{2}abt \tag{8.24}$$

Elements El-1 and El2 contribute both to the volumes associated with the finite element nodes 1, 2, 3, 6, 7 and 8. Each of the two elements provides one quarter to each of its connected FE nodes. Table 8.2 shows the contributions of each element to the volume associated with each of the connected FE nodes and finally the resulting weighting factors a_{ki} for thermal node 1.

Thermal node 2 is only overlapping element EL-3. As a result for thermal node 2, the weights a_{2i} for FE nodes 3, 4, 5 and 6 are $\frac{1}{4}$.

The A-matrix for this problem then becomes

$$[A] = \begin{bmatrix} \frac{1}{7} & \frac{1}{4} & \frac{3}{28} & 0 & 0 & \frac{3}{28} & \frac{1}{4} & \frac{1}{7} \\ 0 & \frac{1}{4} & \frac{1}{4} & 0 & 0 & \frac{1}{4} & \frac{1}{4} & 0 \end{bmatrix} \tag{8.25}$$

The sum of the elements of each row is equal to one.

Table 8.2 Determining a_{ki} for thermal node i

	Volume	Volume	Total volume	
FE node i	Contribution from EL-1	Contribution from EL-2	v_{ki}	a_{ki}
1	$\frac{1}{2}abt$	0	$\frac{1}{2}abt$	$\frac{1}{7}$
2	$\frac{1}{2}abt$	$\frac{3}{8}abt$	$\frac{7}{8}abt$	$\frac{1}{4}$
3	0	$\frac{3}{8}abt$	$\frac{3}{8}abt$	$\frac{3}{28}$
6	0	$\frac{3}{8}abt$	$\frac{3}{8}abt$	$\frac{3}{28}$
7	$\frac{1}{2}abt$	$\frac{3}{8}abt$	$\frac{7}{8}abt$	$\frac{1}{4}$
8	$\frac{1}{2}abt$	0	$\frac{1}{2}abt$	$\frac{1}{7}$

8.4 Coupling TMM to the FE Model

The coupling of the two mathematical models, TMM with FE model, is illustrated in Fig. 8.5. The coupling is achieved by the constraint equation Eq. (8.7).

With the TMM, the vector with thermal node temperatures $\left(T^t\right)$ is determined. Potentially, also external heat fluxes $\left(Q^t\right)$ may be applied to the thermal nodes.

The temperatures (T^s) at the FE model nodes are coupled to the temperatures of the TMM $\left(T^t\right)$ by means of the $[A]$-matrix. No applied heat flows (Q^s) are assumed in the FE model. The conduction matrix $[K_c]$ of the FE model is utilised to lead the internal heat flow due to the prescribed temperatures $\left(T^t\right)$ and is functioning as a temperature interpolation matrix.

The derivation of the system equations of the PAT method is started from the stationary heat diffusion equation in three dimensions that can be written as (see also Sect. B.2)

$$\nabla.k\nabla T(\mathbf{x}) + Q(\mathbf{x}) = 0, \tag{8.26}$$

where k is the thermal conductivity (W/m K), T (K) the temperature distribution, the Q (W/m^3) the heat source and ∇ the nabla operator.

Fig. 8.5 Coupling TMM with FE model

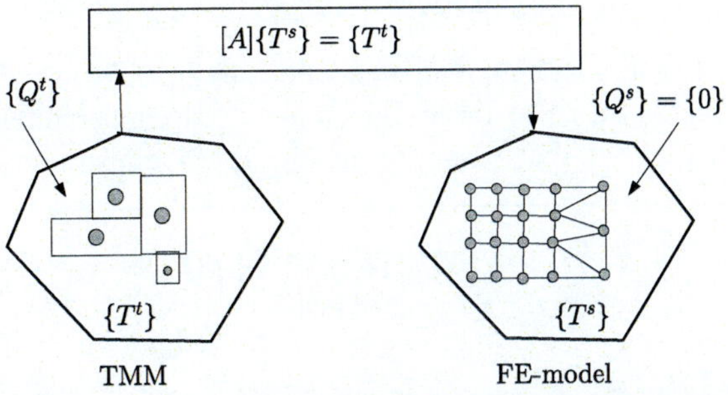

In accordance with the calculus of variation, the approximate solution of Eq. (8.26) can be obtained with an assumed function minimising the thermal functional $J(T)$ [36, 51]

$$J(T) = \frac{1}{2}\int_V \left[k\left(\frac{\partial T}{\partial x}\right)^2 + k\left(\frac{\partial T}{\partial y}\right)^2 + k\left(\frac{\partial T}{\partial x}\right)^2 - QT \right] dV. \tag{8.27}$$

In accordance with Eq. (8.27), the thermal functional $J(T^s, \Lambda)$ can be written as follows:

$$J(T^s, \Lambda) = \tfrac{1}{2}\left(T^s\right)^T [K_c]\left(T^s\right) - \left(T^s\right)^T\left(Q^s\right) + \left(\left(T^s\right)^T [A]^T - \left(T^t\right)^T\right)(\Lambda), \tag{8.28}$$

where

- $\frac{1}{2}\left(T^s\right)^T [K_c]\left(T^s\right)$ represents the internal thermal energy in the FE model,
- $-\left(T^s\right)^T\left(Q^s\right)$ represents the thermal work done by the external heat flow on the thermal model, and finally
- $\left(\left(T^s\right)^T [A]^T - \left(T^t\right)^T\right)(\Lambda)$ the constraint equations between the temperatures of the TMM and the FE model are introduced in the thermal functional. (Λ) are the Lagrange multipliers.

When $\left(T^t\right)$ is known and the temperatures in the FE model (T^s) are approximated by minimising the thermal functional $J\left(\left(T^s\right)\right)$ such that

$$\delta J\left(\left(T^s\right), (\Lambda)\right)) = \frac{\partial J}{\partial\left(T^s\right)}\delta\left(T^s\right) + \frac{\partial J}{\partial(\Lambda)}\delta(\Lambda) = 0. \tag{8.29}$$

The consequences of Eq. (8.29) are

$$\begin{aligned} \frac{\partial J}{\partial\left(T^s\right)} &= \left[[K_c]\,[A]^T\right]\begin{pmatrix}\left(T^s\right)\\(\Lambda)\end{pmatrix} = \left(Q^s\right) = (0), \\ \frac{\partial J}{\partial(\Lambda)} &= [A]\left(T^s\right) - \left(T^t\right) = (0) \end{aligned} \tag{8.30}$$

Combining both matrix equations in Eq. (8.30) will give the interpolation matrix equation to solve the weighted averaged temperatures (T^s) when the conduction matrix $[K_c]$, the $[A]$-matrix and the TMM temperatures $\left(T^t\right)$ are known.

$$\begin{bmatrix}[K_c] & [A^T]\\ [A] & [0]\end{bmatrix}\begin{pmatrix}\left(T^s\right)\\(\Lambda)\end{pmatrix} = \begin{pmatrix}(0)\\ \left(T^t\right)\end{pmatrix}. \tag{8.31}$$

In the following, this equation will be referred to as "PAT equation".

The solution of Eq. (8.31) provides the temperatures (T^s) of the FE model nodes. Each element of the vector of Lagrange multipliers (Λ) represents the total heat flow to the corresponding thermal node coming from the other thermal nodes. This heat flow originates from the boundary conditions that PAT method imposes to the finite element model by prescribing the average thermal node temperature. The sum of the Lagrange multipliers is zero, indicating that the mapped temperature field is in balance with the thermal node temperatures.

One-dimensional example

The TMM consists of two thermal nodes with temperatures $T_1^t = 300,\ T_2^t = 320$ K. The FE model consists of three elements and four nodes. Thermal node 1 overlaps finite element El-1 with the FE nodes 1 and 2, and thermal node 2 overlaps the finite elements El-2 and El-3 with the FE nodes 2, 3 and 4. This is illustrated in Fig. 8.6. The following properties are used for the finite elements: $A = 0.005$ m^2, $L = 0.5$ m and the thermal conductivity $k = 237$ W/m/K.

The conduction matrix $[K_c]_{el}$ for a structural line element with two nodes is given by (see also Sect. B.3)

$$[K_c]_{el} = \frac{A_{el}k_{el}}{L_{el}} \begin{bmatrix} 1 & -1 \\ -1 & 1 \end{bmatrix} \tag{8.32}$$

After assembly of all element conduction matrices, the overall conduction matrix $[K_c]$ is, corresponding to the four finite element nodes,

$$[K_c] = \begin{bmatrix} 0.5925 & -0.5925 & 0 & 0 \\ -0.5925 & 4.1475 & -3.5550 & 0 \\ 0 & -3.5550 & 13.0350 & -9.4800 \\ 0 & 0 & -9.4800 & 9.4800 \end{bmatrix}. \tag{8.33}$$

Based on the overlap relations, the $[A]$-matrix is given by

$$[A] = \begin{bmatrix} 0.5 & 0.5 & 0.0 & 0.0 \\ 0.0 & 0.3 & 0.5 & 0.2 \end{bmatrix} \tag{8.34}$$

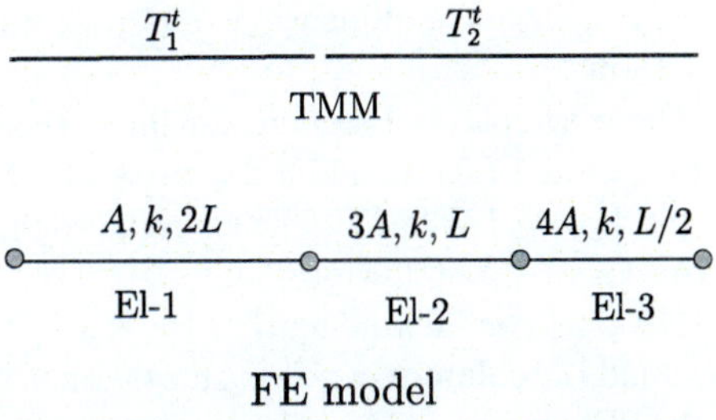

Fig. 8.6 TMM/FE model interpolation configuration

Using Eq. (8.31), the system to be solved becomes

$$\begin{bmatrix} 0.5925 & -0.5925 & 0 & 0 & 0.5 & 0 \\ -0.5925 & 4.1475 & -3.5550 & 0 & 0.5 & 0.3 \\ 0 & -3.5550 & 13.0350 & -9.4800 & 0 & 0.5 \\ 0 & 0 & -9.4800 & 9.4800 & 0 & 0.2 \\ 0.5 & 0.5 & 0.0 & 0.0 & 0 & 0 \\ 0.0 & 0.3 & 0.5 & 0.2 & 0 & 0 \end{bmatrix} \begin{pmatrix} (T^s) \\ (\Lambda) \end{pmatrix} = \begin{pmatrix} 0 \\ 0 \\ 0 \\ 0 \\ 300 \\ 320 \end{pmatrix}. \tag{8.35}$$

The solution of this equation is the nodal temperatures (T^s) (in K) in the FE model

$$(T^s) = \begin{pmatrix} 285.037 \\ 314.963 \\ 321.945 \\ 322.693 \end{pmatrix} \tag{8.36}$$

and the Lagrange multipliers Λ (in W)

$$(\Lambda) = \begin{pmatrix} 35.461 \\ -35.461 \end{pmatrix} \tag{8.37}$$

Example: Effect of non-uniform conductivity

This example is inspired by the presentation by ESA/ESTEC during a workshop on 7 February 2012. The presentation included a simple PAT temperature interpolation problem, which will be numerically repeated here in this example. No dimensions and values of the conduction properties were provided, and therefore, these are assumed in order to perform the calculations. The PAT analysis is done with the aid of the software package MATLAB®. The thermal mathematical model (TMM) consists of three thermal nodes 1, 2 and 3. The temperatures are $T_1^t,\ T_2^t,\ T_3^t = 50.0, 100.0, 0.00\,°\mathrm{C}$. The TMM is shown in Fig. 8.7.

This TMM completely covers the conduction FE model, which is shown in Fig. 8.8. The thermal nodes of the TMM are shown with red-dashed rectangles. The conduction FE model consists of 165 iso-parametric four-noded elements and 192 FE nodes. The dimensions are given in metres. The thickness of the plate is $t = 5$ mm.

The grey-coloured rectangular finite elements represent a low conduction strip with a width of 2 mm and a thickness of 0.5 mm. The isotropic conduction in this strip is $k_{gap} = 1$ W/m/°C. The isotropic conduction for all other elements is $k = 237$ W/m/°C.

The total conduction matrix $[K_c]$ is a 192×192 matrix, and the A-matrix has 3 rows and 192 columns. Knowing the thermal node temperatures (T^t), the temperature in the FEM nodes (T^s) can be evaluated using Eq. (8.31). The resulting interpolated

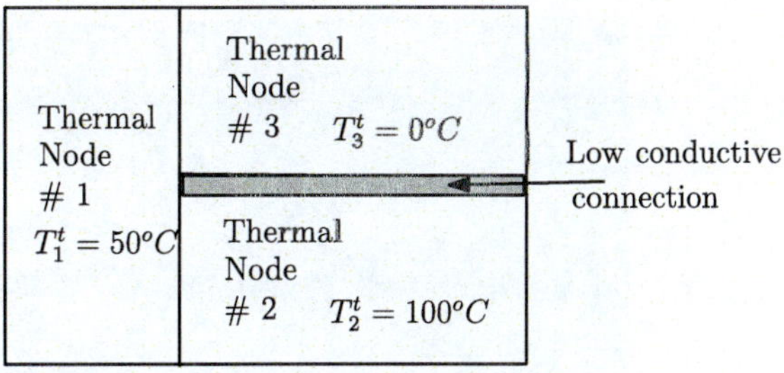

Fig. 8.7 Thermal mathematical model (TMM)

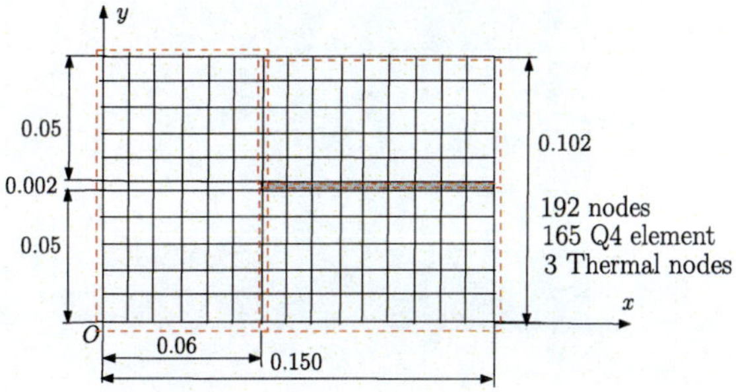

Fig. 8.8 Finite element conduction model

FE node temperatures are presented in a contour plot as shown in Fig. 8.9. The maximum temperature is $T_{max} = 115.15\,°\text{C}$, and the minimum temperature is $T_{min} = -15.452\,°\text{C}$. The mean temperature is $T_{mean} = 50.000\,°\text{C}$. The contour plot was generated with a temperature increment of $\Delta T = 5\,°\text{C}$.

The iso-temperature lines in Fig. 8.9 show that the heat is flowing around the low conductive strip. Thanks to the use of the conduction matrix as interpolation tool, the PAT method is able to take into account change or interruption of the conductive coupling between parts of the structure. The direction of the temperature gradient is normal to the iso-temperature lines and is all nicely oriented along the path around the low conductive strip. Temperature mapping methods that are not aware of the presence of this change of conductance, such as geometrical interpolation methods, will produce temperature fields with gradients normal to the strip, since the highest temperature difference between thermal nodes is found between the thermal nodes at the two sides of the strip.

Plate with hole

This example shows the application of the PAT interpolation method followed with the calculation of the displacements and von Mises stress σ_{VM},[1] [76].

[1] $\sigma_{VM} = \sqrt{\sigma_{xx}^2 + \sigma_{yy}^2 - \sigma_{xx}\sigma_{yy} + 3\tau_{xy}^2}$.

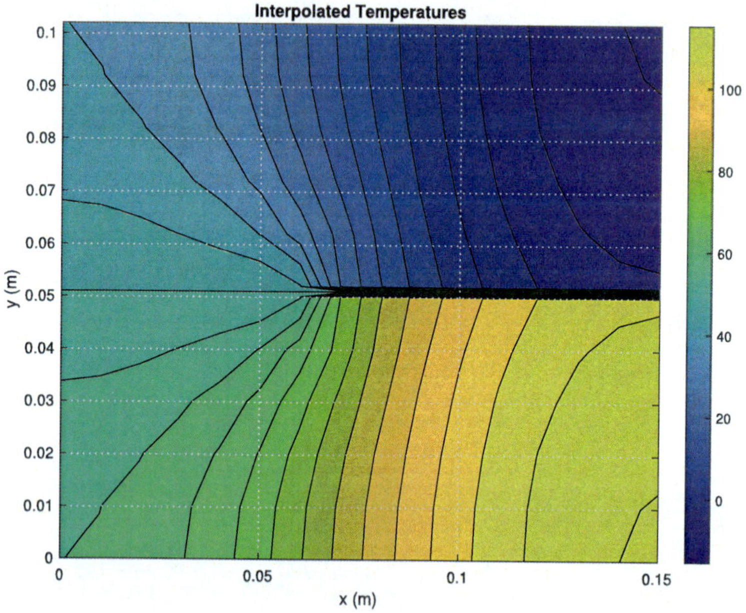

Fig. 8.9 PAT interpolated temperatures

At the left side in Fig. 8.10, a simply supported panel containing a hole and dimensions are shown, while at the right side the associated thermal mathematical model (TMM) and thermal nodes 1–6 are shown. The outer edges of the panel are simply supported.

The height of the panel is $2H = 500$ mm, the width is $W = 400$ mm, and the thickness is $t = 5$ mm. The diameter of the hole is $d = 110$ mm. The material properties of the panel are as follows: the CTE $\alpha = 24 \times 10^{-6}$ m/m/°C, the conductivity $k = 167$ W/m/°C, Young's modulus $E = 70$ GPa and Poisson's ratio $\nu = 0.33$. The reference temperature is $T_{ref} = 20\,°\text{C}$.

The temperatures of the thermal nodes in the TMM are T_i^t, $i = 1, 2 \ldots, 6$, 80, 60, 50, 40, 30 and 55 °C, respectively.

The following steps were performed:

- A conductive FE model has been derived from the structural model with the same mesh using quadrilateral and triangular plate (membrane) finite elements. The TMM and FE model are shown under (a) in Fig. 8.11. Both models are presented on top of each other with the finite elements shown in different colours indicating the different thermal nodes that are overlapping the elements. The structural FE model mesh is much more detailed than the TMM detailed.
- Based on the correspondence between finite elements and thermal nodes, the A-matrix has been assembled.
- The PAT equation is solved to obtain the FE model nodal temperatures. The interpolated nodal temperatures are shown under (b) in Fig. 8.11. The resulting

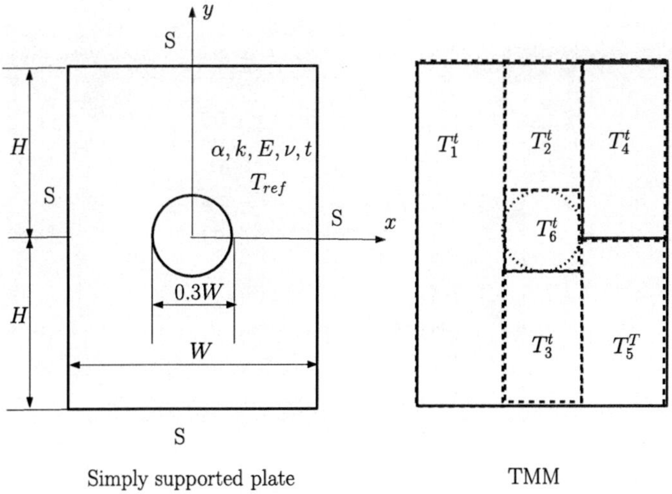

Fig. 8.10 Simply supported panel with hole

temperature field is nicely respecting the constraint of the hole in the plate. The interpolation is complementing the low resolution of the TMM.

- Through the application of mapped temperature field as thermoelastic loading on the structural FE model, displacements and von Mises stresses are calculated and shown in Fig. 8.12.

This example shows that the PAT method can construct a physically consistent temperature distribution on the finite element model. It must be noted that despite the fact that results look physically realistic, the method is producing complementary information that is not provided by the TMM with the rather low resolution. For reliable thermoelastic analysis, it is therefore recommended to increase the mesh resolution of the thermal model in order to constrain better the PAT method.

8.5 Evaluating PAT Method Results

The temperature fields of the thermal model that is mapped to the finite element model may show temperatures lower and higher than the temperatures of the thermal nodes. The level of under or overshoot may be a relevant aspect to be evaluated. In this section, it is briefly explained what is causing the under and overshoots and how this could be reduced if there is a need to.

The PAT method is based on constraints that assure that the temperatures of the thermal nodes are equal to the average temperature of the finite element nodes

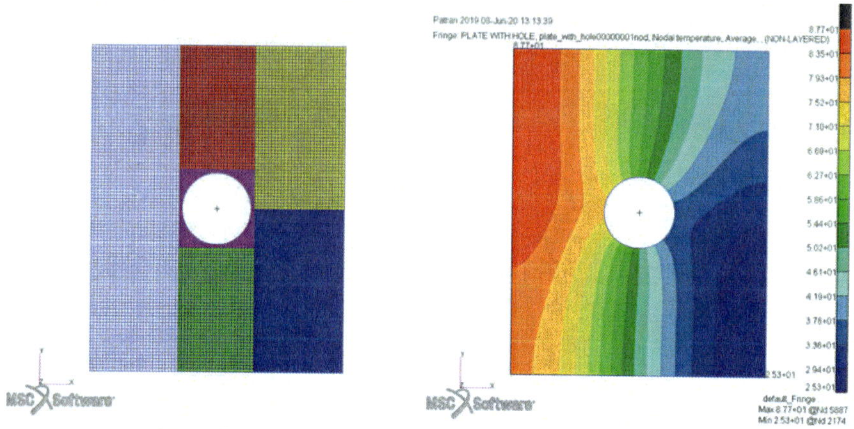

(a) TMM, FE model (b) Temperature field

Fig. 8.11 TMM, FE model and temperature field

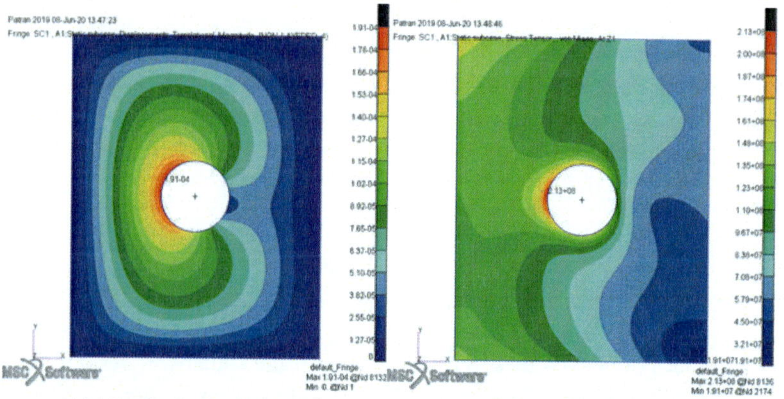

(a) Magnitude displacements (b) von Mises stress

Fig. 8.12 Displacement field and von Misses stress distribution

corresponding to each thermal node. In general, the structure does not have a uniform temperature, which implies that the temperature of a thermal node is different from the temperatures of the surrounding thermal nodes. With the help of the conduction matrix, a smooth continuous temperature field is constructed over the FE nodes, causing the temperature of the FE nodes in each thermal node being lower and higher than the thermal node temperature. Overall the PAT constraint makes that the average of the FE node temperature is equal to the corresponding thermal node temperature.

The fact that FE node temperatures are higher or lower than the thermal node temperature cannot be avoided. However, when the amount of under or overshoot is becoming large, the temperature at the FE mesh is not sufficiently constrained due to too low resolution of the mesh of the thermal model. Increasing the thermal mesh resolution will reduce the under and overshoot and will at the same time capture better the temperature gradients in the temperature field. Higher mesh resolution in general will improve the quality of the thermal analysis results with corresponding increase of quality in the subsequent steps of the thermoelastic analysis chain.

8.6 Mathematical Models Checks for PAT Method

8.6.1 Introduction

The conduction matrix $[K_c]$ and the A-matrix form important inputs to the interpolation matrix Eq. (8.31). The quality of both matrices has to be checked by well-defined health checks. The checks on both matrices can be divided into the following:

- Checks on the conduction matrix
  - Transition of the structural FE model into a conduction FE model
  - Disconnected parts in the conduction FE model
  - Zero heat flux for uniform temperatures
  - Zero conductive functional
- Checks on A-matrix
  - Disconnected parts
  - Normalised A-matrix in accordance with Eq. (8.10).

8.6.2 Conduction FE Model Health Check

8.6.2.1 Conduction FE Model Obtained from the Structural FE Model

In many cases, a FE model developed for structural application is used for thermal purpose. The conduction matrix from the conduction FE model is used for a conduction-based temperature mapping process, like the CPPT and the PAT methods. The conduction FE model is also used to compute linear conductors with one of the methods described in Chap. 9.

When a structural FE model has been the basis for the thermal conductive finite element model, the model might have been developed for the calculation of modal characteristics and responses in time and frequency domain [93, 94]. For the transition from the structural analysis domain of the model to the thermal domain, some points have to be kept an eye on.

The nodes in a structural FE model represent in general six degrees of freedom: three translations and three rotations. In the thermal domain, the nodes in the FE model only have one degree of freedom, being the temperature. It might be required that simplified representations, of for instance connections, need to be implemented as well in the thermal conductive model. These connections may be modelled in the structural model with multi point constraints, like rigid body elements, or with spring elements that make reference to degrees of freedom that do not exist any more in the thermal domain. It is also possible that some element types are not supported in the thermal domain or have a different function.

In many FE codes, the elastic structural elements have a thermal counterpart. In some codes, like MSC Nastran, the element definition and geometric properties remain unchanged. Nevertheless, always the structural materials have to be replaced with thermal material properties.

Also when a thermal model is built from scratch, there may be chances on mistakes that could jeopardise the quality of the model. It is therefore as good practice to perform a few basic checks on the model. This can be done as follows.

8.6.2.2 Check on Disconnected Parts of the Model

Considered is the bare thermal finite element model. This is the model without prescribed boundary temperatures and applied heat loads. When the temperature at only one of the FE nodes is prescribed to a unit temperature, then the solution of the system shall produce the same unit temperature at all the FE nodes, when all nodes are properly connected.

In case there are conductively isolated parts in the model, then it depends on the solver whether or not this approach will produce a temperature field. In the case of a nonlinear steady-state solver, the parts that are not connected to the node with the prescribed temperature will remain at the initial temperature.

When a linear solver is used, it will most likely identify that the system is singular. This information is then also an indication that there are disconnected parts in the model.

In case it is intended that parts are conductively disconnected from each other, then the temperature shall be prescribed at each of the independent parts. This shall then result into uniform temperatures in each of the parts at the level of the prescribed temperature.

8.6.2.3 Check on Zero Heat Flux for a Uniform Temperature

The check in the previous section should produce a uniform temperature. This would mean that there are no thermal gradients. Basically, all terms in Eq. (B.1) in Appendix B must be equal to zero. The element fluxes produced with the finite element analysis should show zero fluxes in all three directions for each element.

If all temperatures in the model obtain the set temperature T_s, the global set temperature vector becomes

$$(T_s) = \begin{pmatrix} T_s \\ T_s \\ \dots \\ T_s \end{pmatrix} = T_s \begin{pmatrix} 1 \\ 1 \\ \dots \\ 1 \end{pmatrix} = T_s\,(U)\,. \tag{8.38}$$

In such a case, the linear conductive equilibrium equation from Eq. (B.56) has to produce a zero right-hand side flux vector, because all the terms on a row of the conduction matrix must add up to zero, like is shown in Eq. (8.39)

$$[K_c]\,(T_s) = T_s[K_c]\,(U) = \left(R_Q\right) = (0)\,, \tag{8.39}$$

8.6.2.4 Check on Zero Conductive Functional

This check can only be performed if there is access to the conduction matrix and matrix operations can be performed. In MSC Nastran, this is possible through direct matrix abstraction programming (DMAP).

The conductive functional can be defined as

$$J = \frac{1}{2}\,(T)^T\,[K_c]\,(T)\,. \tag{8.40}$$

Substitution of the uniform temperature from Eq. (8.38) into Eq. (8.40) with the support from Eqs. (8.39), (8.40) shall then produce a zero value for the functional if the terms of each row in the conduction matrix add up nicely to zero.

$$J = \frac{1}{2}\,(T_s)^T\,[K_c]\,(T_s) = \frac{1}{2}T_s^2\,(U)^T\,[K_c]\,(U) = 0 \tag{8.41}$$

8.6.3 Checking A-Matrix Input to the PAT Method

For the functioning of the PAT method, the quality of two matrices is of importance: the conduction matrix and the A-matrix. In this section, it is explained how the quality of the two matrices can be verified by making use of the solution of the PAT equation Eq. (8.31).

In previous sections, already four methods to check the quality of the conduction FE model are discussed, with one of them being the check on disconnected parts of the model. It is also possible by solving Eq. (8.31) to identify whether disconnected parts are present in the conduction finite element model. For this purpose, the following steps have to be followed:

- Define in the overlap or correspondence relations for a single thermal node that is corresponding to one finite element of the conduction FE model and determine the corresponding A-matrix coefficients. It will be a single row A-matrix, since there is only one artificial thermal node involved.

- Assign a temperature to the single thermal node (e.g. a temperature of 1.0 °C).
- Solve Eq. (8.31).
- Check that all finite element nodes in the model have a temperature of 1.0 °C.

If the temperature throughout the whole model is 1.0 °C, it means that, via the conduction relations in the FE model, the prescribed temperature at one element of the FE model could be extended to the complete model. If the results show groups of finite element nodes with a temperature different from 1.0 °C, then the mentioned extension of the temperature field could not be achieved to the complete model and indicates a conductively disconnected part in the FE model. Depending on the solver, used to solve PAT equation Eq. (8.31), the produced temperature of disconnected FE nodes can be 0.0 or the solver is not able to produce a solution, because of singularity. If the latter occurs, then this is also a sign that there are disconnected parts in the model.

It may be the case that two or more parts are intentionally not connected via conduction. In that case, above test could be repeated for each disconnected part. An additional test could be to introduce for each disconnected part an individual thermal node and assign different temperatures to each of them. The solution of the PAT equation Eq. (8.31) should then lead to different uniform temperatures for each of the disconnected parts. The temperatures of each of the disconnected parts must then be the same as the temperature of the associated thermal node.

Check on disconnected parts

Two conduction FE models are verified (see Fig. 8.13)

- FE model 1: Complete FE model with seven FE nodes that are all connected to each other through six conduction rod elements.
- FE model 2: The same as FE model 1, but with the rod element between FE nodes 4 and 5 removed. FE model 2 consists therefore of two disconnected parts.

The thermal node TN_1 is associated with both models as indicated in Fig. 8.13. The temperature of thermal node TN_1 is $T_1^t = 1$. The thermal node TN_1 is associated with the first rod element connecting FE nodes 1 and 2. The material properties of the rod elements are such that $Ak/L = 1$, A is area of the cross section, k is conductivity, and L is the length of rod element. The A-matrix in this problem consists of a single row

$$[A] = \begin{bmatrix} 0.5 \; 0.5 \; 0. \; 0. \; 0. \; 0. \; 0. \end{bmatrix} \tag{8.42}$$

With the use of the PAT equation Eq. (8.31), the temperatures at the FE nodes (T^s) for FE model 1 then becomes

$$\left(T^s\right)^T = \left(1. \; 1. \; 1. \; 1. \; 1. \; 1. \; 1.\right), \tag{8.43}$$

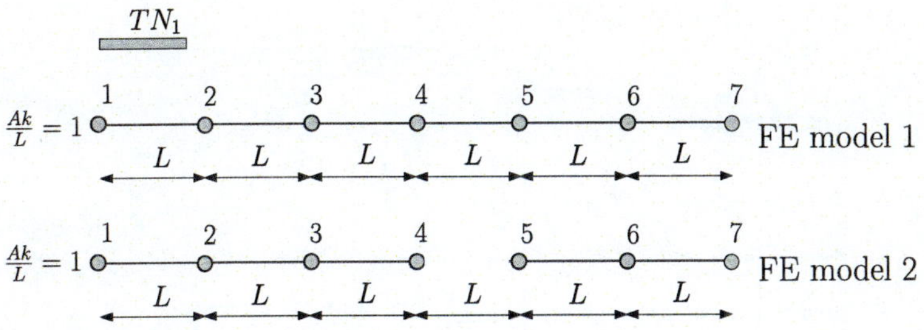

Fig. 8.13 Thermal node TN_1 and FE models 1 and 2

and for FE model 2

$$\left(T^s\right)^T = \left(1.\ 1.\ 1.\ 1.\ 0.\ 0.\ 0.\ \right). \tag{8.44}$$

The constant uniform temperature found with FE model 1 tells that no unconnected parts are in that FE model. The zero temperatures produced with FE model 2 tell that an unconnected part is present in that model.

Above check on the PAT method focussed on parts in the model that were unintentionally conductively disconnected. In case there is a problem in the weighting factors in the A-matrix, then this may also lead to problems in the solution of the PAT equation. Provided that the sum of the terms of each row of the A-matrix add up to 1.0, the solution should be physically consistent.

This consistency may be checked by assigning the same temperature to each thermal node overlapping the FE model. Typically, one could use 1.0 °C as a temperature for each thermal node. If the A-matrix has no problems, the solution of the PAT equation Eq. (8.31) should give a perfectly uniform temperature for all FE nodes over the complete FE model.

A-matrix check example

The thermal models consists of three thermal nodes TN_1, TN_2, TN_3, and the conduction FE model consists of seven nodes and six rod elements. Both models are illustrated in Fig. 8.14.

The material properties of the rod elements are such that $Ak/L = 1$, A is area of the cross section, k is conductivity, and L is the length of each rod element.

All thermal nodes are assigned the same temperature of 1.0

$$\left(T^t\right) = \begin{pmatrix} T_1^t \\ T_2^t \\ T_3^t \end{pmatrix} = \begin{pmatrix} 1.0 \\ 1.0 \\ 1.0 \end{pmatrix}. \tag{8.45}$$

In Table 8.3, the A-matrix is shown for this problem in which the sum of A-matrix elements on each row is equal to 1.0. Each row in that matrix corresponds to a thermal

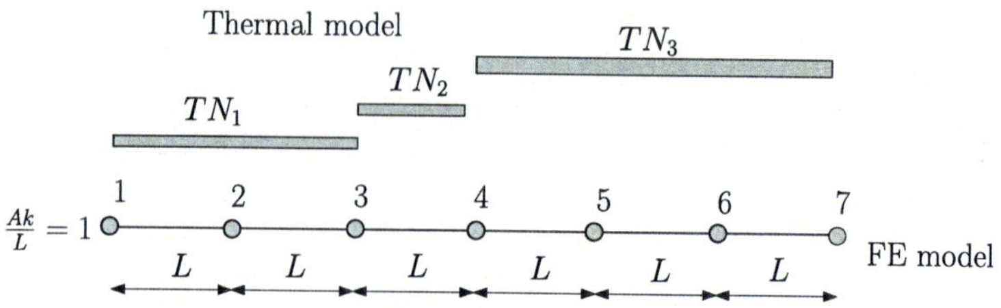

Fig. 8.14 Thermal model and associated FE model

Table 8.3 Correct A-matrix check

A-matrix	(T^s)
$\begin{bmatrix} 0.25 & 0.50 & 0.25 & 0. & 0. & 0. & 0. \\ 0. & 0. & 0.5 & 0.5 & 0. & 0. & 0. \\ 0. & 0. & 0. & 0.1667 & 0.3333 & 0.3333 & 0.1667 \end{bmatrix}$	$\begin{pmatrix} 1.0 \\ 1.0 \\ 1.0 \\ 1.0 \\ 1.0 \\ 1.0 \\ 1.0 \end{pmatrix}$

Table 8.4 Incorrect A-matrix

A-matrix	(T^s)
$\begin{bmatrix} 0.25 & 0.50 & 0.25 & 0. & 0. & 0. & 0. \\ 0. & 0. & 1.0 & 1.0 & 0. & 0. & 0. \\ 0. & 0. & 0. & 0.1667 & 0.3333 & 0.3333 & 0.1667 \end{bmatrix}$	$\begin{pmatrix} 1.0494 \\ 0.8784 \\ 0.5751 \\ 0.5470 \\ 0.8267 \\ 1.0480 \\ 1.1661 \end{pmatrix}$

node. Since the example model has three thermal nodes, there are three rows in the A-matrix. The resulting vector with temperatures at the FE nodes is shown in the same table and shows that the temperatures at each FE node are equal to 1.0. This A-matrix clearly has no problems.

In Table 8.4 however, not all the terms of the A-matrix are adding up to 1.0. In the second row of the A-matrix, the coefficients add up to 2.0. The consequence is that this leads to a non-uniform temperature field at the FE nodes, despite the uniform temperatures at the thermal nodes. Clearly, this A-matrix has problems and is not passing the test.

There may be cases that the PAT equation Eq. (8.31) cannot be solved. A possible cause is that the PAT relations overconstrain the temperature field that is to be produced for the FE model. This means that through the PAT relations of Eq. (8.6), two or more thermal nodes are prescribing, potentially different, average temperatures for the same FE nodes. This may occur when a mistake is made in the definition of the correspondence between thermal nodes and finite elements in which a number of finite elements are overlapped by more than one thermal node.

This problem may also occur when the number of finite elements corresponding to a thermal node is small. The consequence may be that too many FE nodes, connected to the overlapped elements, are shared by more than one thermal node. In other words, there are no or just a small number of FE nodes that are corresponding with only one thermal node. In general, this problem can be prevented by assuring that the mesh resolution of the finite element model is at least a factor of two higher than the resolution of the thermal model.

8.7 Effect of Incomplete Correspondence

The PAT method is quite forgiving when it comes to sensitivity to incomplete overlap of correspondence relations. This section, with the help of a simple example, investigated what the effect could be of incomplete correspondence.

In Fig. 8.15, the FE model of a Al-alloy rectangular plate with dimensions 1×0.5 m^2 is shown. The thickness of the plate is $t = 5$ mm, and the conductivity is $k = 237$ W/m°C. The number of iso-parametric Q4 conduction elements is 50, and the number of nodes is 60.

The first 25 Q4 elements are overlaid by thermal node 1 with a temperature of $T_1^t = 60\,°$C, and the Q4 elements 26–50 are covered by thermal node 2 with a temperature of $T_2^t = 40\,°$C. So in this initial case, all 50 elements are covered by thermal nodes. The thermal nodes are indicated by red-dashed/dotted squares. The Q4 elements 1, 25, 26 and 50 are shown in Fig. 8.15.

Fig. 8.15 FE model plate

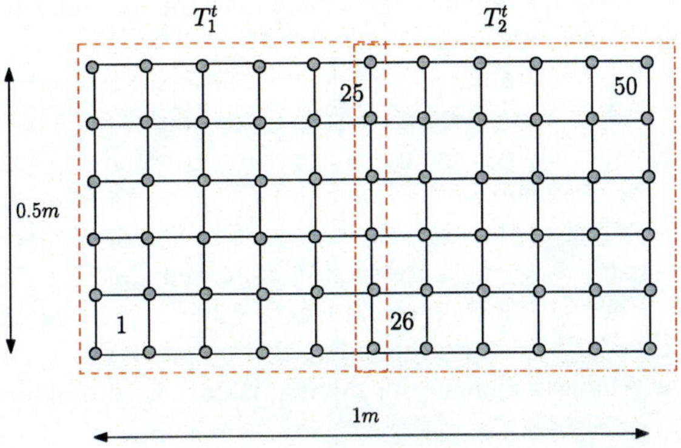

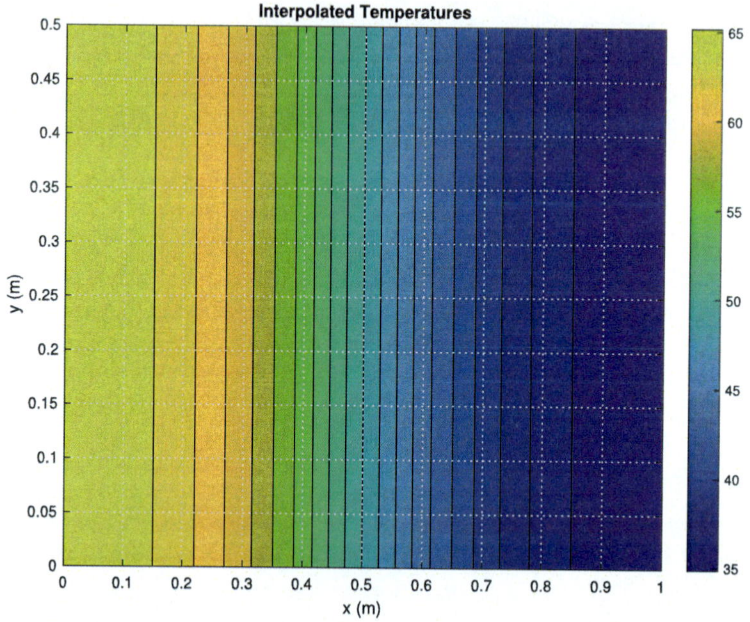

Fig. 8.16 PAT interpolated temperatures, all elements overlaid

The contour plot of the interpolated temperatures is presented in Fig. 8.16. The calculated average, maximum and minimum temperatures are, respectively, 50 °C, 65.1515 °C and 34.8485 °C. All iso-temperature lines are vertical, indicating a one-dimensional variation of temperature.

Suppose the Q4 conduction elements 1 and 25 are not overlaid by thermal node T_1^t and elements 26 and 50 are not covered by thermal node T_2^t. The results of the interpolation analysis are shown in Fig. 8.17. In this figure, a change can be observed that iso-temperature lines that were nice vertical in Fig. 8.16 now bend. Although the effect at temperature level does not seem to be dramatic, the thermomechanical problem is now changed from a 1-D to 2-D problem.

The calculated averaged, maximum and minimum temperatures are, respectively, 50 °C, 65.1496 °C and 34.8504 °C.

The rows of the A-matrices for these two problems are graphically visualised in Figs. 8.18 and 8.19. Figure 8.18 corresponds to the configuration in which all finite elements are overlapped by the two thermal nodes. Each of thermal nodes has half of the elements covered. Each of the curves represents a row in the A-matrix with the A-matrix element value versus the FE node number. The blue dots represent the A-matrix elements corresponding thermal node 1 and the red ones to thermal node 2. Where the dots represent a zero value, it indicates that there is no overlap of the FE nodes by the corresponding thermal node. For instance for FE node number 40, the A-matrix element for thermal node 1 is zero, indicating that FE node 40 is not

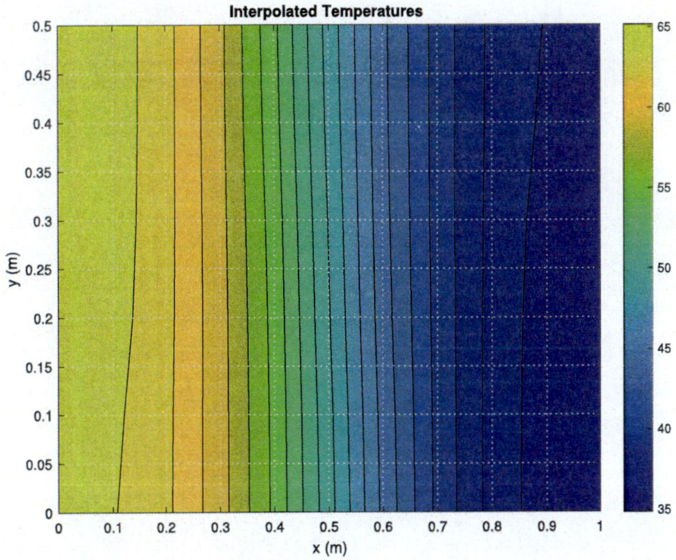

Fig. 8.17 PAT interpolated temperatures, elements 1, 25, 26 and 50 are not overlaid

overlapped by thermal node 1. This FE node appears to be overlapped by thermal node 2, since the A-matrix element value for this thermal node is not equal to zero.

Three different nonzero A-matrix values can be identified: the highest values of 0.04 correspond to FE nodes that are located away from the edges of the thermal nodes. The FE nodes have contributions to their weight from four elements. The lowest nonzero values of 0.01 correspond to the FE nodes at the corners of the thermal nodes with contributions of only one element. The in-between values of 0.02 correspond to the edges away of the corners which is the result of the contribution of two elements.

In Fig. 8.19, the profile of the A-matrix is visualised for the case in which a few of the finite elements are left out of the overlap relations. Comparing this figure with Fig. 8.18, the effect on the coverage of the PAT relations can be observed. Due to the fact that the fraction of the total weight now has to be distributed over less FE nodes, the A-matrix elements are now a bit higher. Also can be observed that there are more FE nodes with a weight of zero.

Missing four elements in the overlay of the two thermal nodes have a small influence on the average, maximum and minimum temperatures. However, the temperature distribution changes which may have an effect on the thermoelastic responses. It is therefore relevant to take well care of the construction of the overlap relations. In this problem, all finite elements were of the same size. One should realise that when thermal nodes correspond to elements of different size, especially the largest element should not be left out. The effect of omission of the larger elements will be more significant.

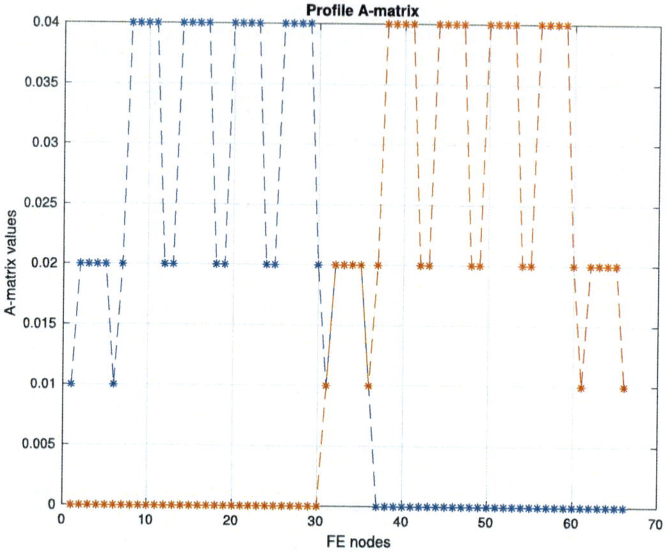

Fig. 8.18 Profile A-matrix, all elements overlaid

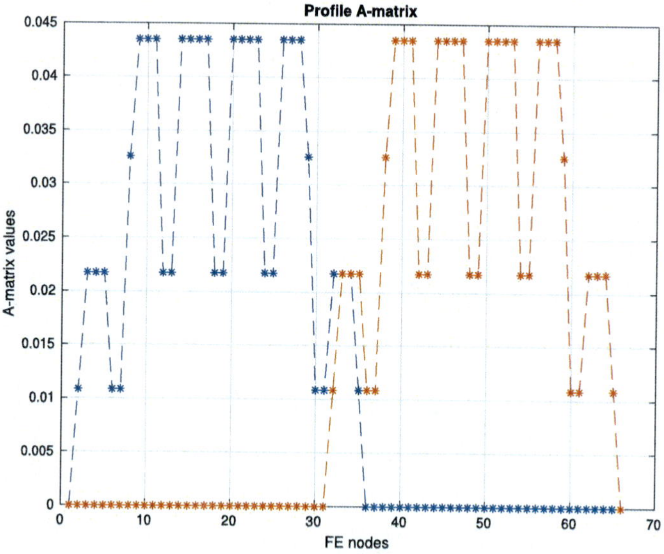

Fig. 8.19 Profile A-matrix, conduction elements 1, 25, 26 and 50 are not overlaid

Problems

8.1 A solar array consists of a solar panel and a yoke to keep the panel out of the shadow of the spacecraft. The configuration and dimensions of the solar array wing are shown in Fig. 8.20. The yoke shall be very firmly coupled to the panel using either MPC equations (see Appendix D) or with scalar elements with a large conduction. The pairs of nodes 12 and 14 and 10 and 13 shall be coupled (see dashed ellipses).

- The model of the solar array wing is a conduction FE model (one DOF per node).

The solar panel is made of a sandwich construction with Al-alloy face sheets and a thickness $t = 0.5$ mm. The core height is $h = 20$ mm. The conductivity of the core is $k = 10$ W/m/°C. The yoke is made of an Al-alloy tube structure with a thin-walled square cross section with thickness $t = 1$ mm and height and width $h = b = 50$ mm. The conductivity of the Al-alloy is $k = 200$ W/m/°C.

- Perform a PAT interpolation calculation, $TN_1 = 50, TN_2 = 40$ and $TN_3 = 30$ °C.
- Discuss the results of the interpolated temperatures of nodes 2, 5, 8 and 11.

8.2 This problem is based on an article of the Korean Aerospace Research Institute [64] and personal communications with Prof. Dr. Jae Hyuk Lim of the Chonbuk National University. Perform a PAT analysis on the phase-change memory (PCRAM) consisting of five materials. The configuration of the PCRAM is shown in Fig. 8.21. Assume an uniform thickness of the PCRAM $t = 50$ nm ($1\text{ nm} = 10^{-9}$ m). The conductivity of the five materials, the TMM estimated temperatures and the (prescribed) number of iso-Q4 conduction elements are provided in Table 8.5. Perform the following assignments:

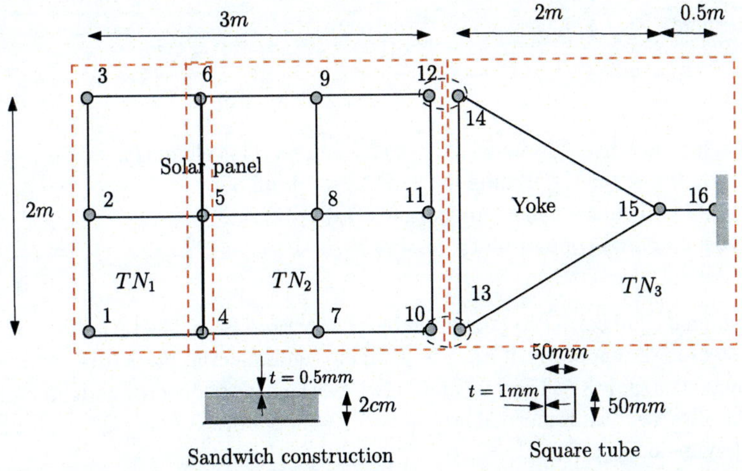

Fig. 8.20 Solar array wing

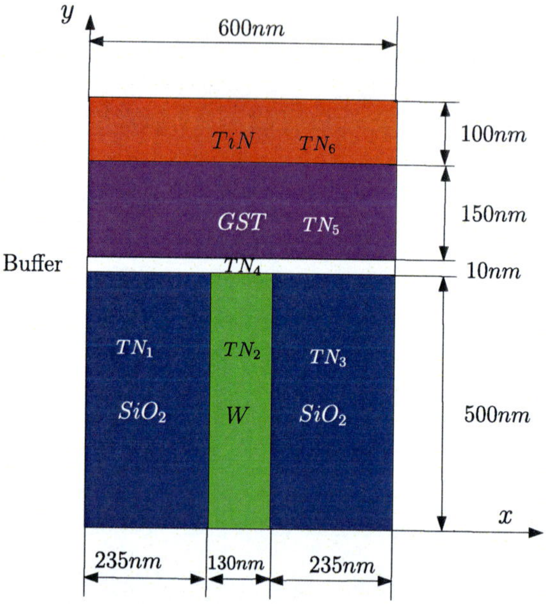

Fig. 8.21 PCRAM configuration [64]

Table 8.5 Material properties [64], TMM estimated temperatures and (minimum prescribed) number of quadrilateral (Q4) elements

Material	Conductivity (W/m°C)	Temp. T (°C)	Q4 elements
W (green)	178	1000	5×20
Tin (red)	13	200	23×4
GST (purple)	0.5	500	23×6
SiO_2 (blue)	1.4	750	$2 \times 9 \times 20$
Buffer (white)	17.4	600	23×1

- Create a conduction FE model using quadrilateral (Q4) elements.
- Generate the A-matrix relating the TMM and FE model.
- Calculate PAT interpolated temperatures (Eq. 8.31).
- Create a coloured contour plot.

8.3 A hinged rigid bar AB is supported by a copper rod BD and an Al-alloy rod CE. The system is shown in Fig. 8.22. The geometry and material properties of the rods are provided in Table 8.6. Thermal node TN 1 overlays the copper rod and thermal TN 2 the Al-alloy rod. Both thermal nodes have a temperature $\Delta T = -25\,°C$. Perform the following assignments:

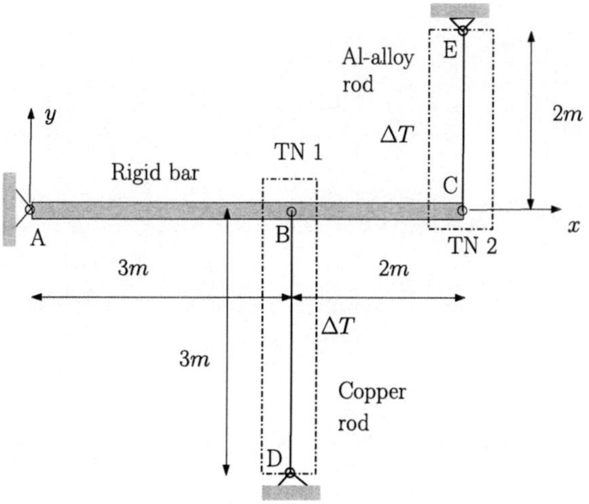

Fig. 8.22 Rigid bar supported by rods

Table 8.6 Geometry and material properties of rods

Parameters	Dimensions	Copper	Al-allow
L	m	3	2
A	mm^2	1500	1200
E	GPa	120	70
α	m/m/°C	16.8×10^{-6}	24.0×10^{-6}
k	W/m/°C	385	237

- Create a conduction FE model of the given system, 11 nodes and 10 rod conductive finite elements per supporting rod (total 20 elements and 22 nodes). Couple point B and point C with a scalar conductive element (k = 10) to create a conductive coupling between the two supporting rods.
- Carry out the PAT interpolation. Temperature field will be used for the subsequent thermoelastic analysis. Thermal node temperatures are equal to $\Delta T = -25\,°\text{C}$.
- Create a structural FE model of the given system, 11 nodes and 10 rod structural finite elements per supporting rod (total 20 elements and 22 nodes). The rigid bar will be incorporated using a MPC equation $[U_m]\,(u) = (0)$ (see Appendix D).
- Carry out thermoelastic FEA and obtain stresses in both copper rod and Al-alloy rod.

Chapter 9
Generation of Linear Conductors for Lumped Parameter Thermal Models

Abstract The generation of linear conductors for a thermal lumped parameter model is in many cases a time-consuming task. The finite element method can help to automate this process and in addition increase the quality of these conductors especially for complex-shaped parts. Three methods for the generation of linear conductors between pairs of the thermal nodes are discussed first. One of these three is the "Far Field" method implemented in ESATAN-TMS. The other two methods are based on static reduction of the finite element conduction matrix and a steady-state solution technique, respectively. The PAT method provides relations between thermal nodes and finite element nodes consistent with the lumped parameter assumption. The system equation used for the temperature mapping with the PAT method is exploited to determine a complete conductive network between in principle all conductively connected nodes in the thermal model.

9.1 Need for Automated Conductor Generation

The traditional implementation of the lumped parameter method requires the user to provide the linear conductors and radiative conductors (GL and GR in ESATAN). Radiative solvers exist for many years to support the user with the generation of radiative conductors. In Europe ESARAD, today embedded in ESATAN-TMS, and SYSTEMA-THERMICA [5] are the most frequently used tools and are already around for many years. However, for a long time, there was no functionality implemented in the lumped parameter thermal tools to support the users with the generation of linear conductors. The thermal engineers had to develop their own methods to generate these conductors.

For most of these methods, two thermal nodes are considered in isolation from the other thermal nodes. Considered are then two thermal nodes that are for instance part of a thermal mesh on a surface. These nodes are presented in Fig. 9.1 in which the geometric centres are marked with solid dots. For the determination of the conductor value between these two thermal nodes, the length L of the heat flow path between the two thermal nodes is taken as the distance between the geometric centres of these thermal nodes.

S. Appel and J. Wijker, *Simulation of Thermoelastic Behaviour of Spacecraft Structures*, Springer Aerospace Technology, https://doi.org/10.1007/978-3-030-78999-2_9

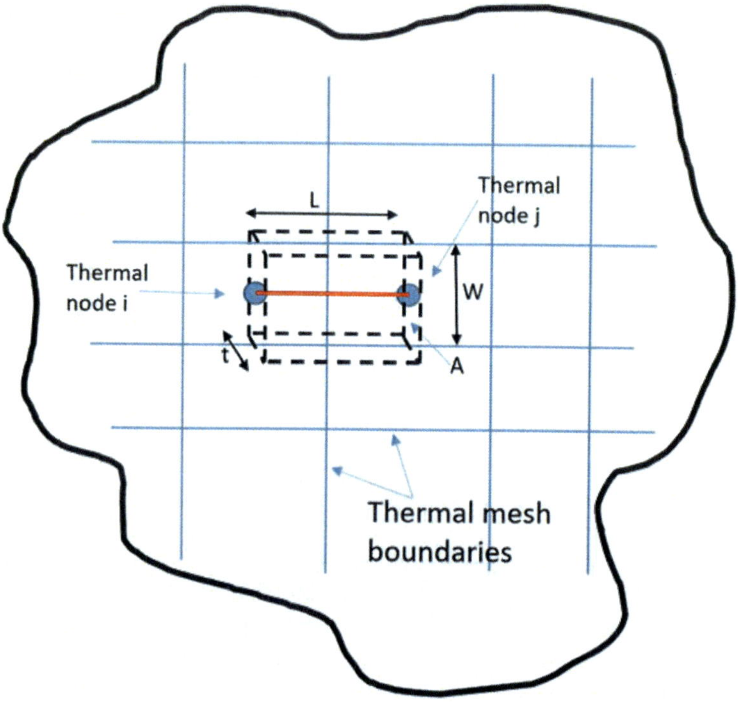

Fig. 9.1 Basis for calculation of linear conductor with considering two thermal nodes in isolation

Since the thermal node temperature in the lumped parameter method represents the average temperature of the material represented by the thermal node (see also Sects. 5.4.2 and 7.2), it is assumed that the these geometric centres coincide with the location where the temperature inside the thermal node equals to the average temperature. This is an approximation that may not always be correct.

The heat flows from thermal node i to thermal node j through an effective cross section with dimension of W, being the width of the interface between the two thermal nodes, and the thickness t of the plate surface (in case of 2-D modelling). With k being the material conductivity, this results then in the following expression for the linear conductor GL between thermal node i and thermal node j:

$$GL(i, j) = \frac{kA}{L}. \quad (9.1)$$

Since some time, the geometric modellers associated with radiative analysis tools have linear conductor generators included. The method, that is implemented in ESATAN-TMS, is based on the "Far Field" method [11, 55, 88]. This method uses "under the hood" a finite element model generated on the geometry of the thermal nodes to generate conductors between thermal node pairs of the type presented in

Fig. 9.1. Thanks to the use of finite elements, in principle thermal nodes with any shape can be handled.

Despite the enhanced support for automatic conductor generation, many thermal engineers still prefer to generate manually, or with the aid of (often ad-hoc) custom made tools, the linear conductors.

As Fig. 9.1 clearly shows, the conductor is basically a scalar link. It has no directional information included and is also independent of the surrounding thermal mesh. These scalar conductors have proven to be quite adequate when there are discrete connections between two thermal nodes, and the effect of surrounding thermal nodes is not present or is very limited. For continuous structural systems, such as panels with 2-D temperature fields or even complex solid pieces with 3-D temperature fields, these conductors appear to underestimate the conductivity of the structural components [11, 59]. A rather high mesh resolution is needed to overcome this problem.

An other problem with the traditional methods to create linear conductors is that these have difficulties with structural components with complex geometries. Sometimes circular holes have to be approximated with a rectangular or square hole in order to allow for approximation of the conductor values. This limitation could in principle be removed with methods like the "Far Field" method.

Since most of the applied methods require quite some manual involvement of the thermal engineer, it is not easy to run a mesh convergence exercise as explained in Sect. 4.5. In case a series of spreadsheets are used for the generation of the linear conductors, it is often not trivial to implement these steps in an automated analysis chain. Also, running several jobs in an automated way to assess the uncertainties (see Chap. 10) becomes quite complicated and cumbersome.

This chapter aims to contribute to removing the above-explained limitations. The conductive modelling with the FE method is the basis for the presented approach to compute linear conductors in an automated way. As background information, a summary of the fundamentals of the finite element method for conductive modelling is provided in Appendix B.

Before moving to the methods for generation of linear conductors, it is important to first recall the fundamental difference between a lumped parameter thermal node and the finite element node as is explained in Sect. 7.2. With the understanding of this important difference, three methods are discussed for generating single linear conductors between two thermal nodes. One of these three is the "Far Field" method.

The explanation of the methods for generation of single conductors between thermal node pairs can be considered as an introduction to the conductor generation methods described in Sect. 9.3 based on the PAT method detailed in Chap. 8. These latter methods provide both a basis for automation and gives a more accurate conductive network for the thermal lumped parameter model.

The explained methods are illustrated by examples.

9.2 Calculation of a Single Linear Conductor with a Conduction FE Model

All mathematics that was explained in Appendix B is provided as background information. It is all applied "under the hood" of a finite element code. Users of finite element codes are normally not directly confronted with these equations, but these provide a good introduction to the content of this section.

Although the explanation in the previous section did only discuss a simple one-dimensional element, most finite element codes have two-dimensional and three-dimensional elements implemented. This allows to model complex geometries of for instance shape optimised machined parts. For these complex parts, it is not trivial to use the commonly applied method that is described in Sect. 9.1 to generate the linear conductors for the lumped parameter model.

This section tries to explain three methods that are based on the finite element method for the calculation of the GL (W/°C) between two lumped parameter nodes that are connected through a structure with a complex geometry. These methods are discussed in the following subsections and cover the calculation of the lumped parameter conductor through

- Reduction of the conduction matrix to the two lumped parameter nodes
- Solving a steady-state thermal analysis with constraints applied to two lumped parameter nodes
- Using the "Far Field" method.

A more generalised discussion of the first two methods is presented, not only for the computation of a single conductor between two thermal nodes, but for a complete conductive network of GLs for a full structure or larger structural components.

The "Far Field" method is also a finite element-based method and deserves a position in this list. The method is also implemented in ESATAN-TMS.

9.2.1 Calculation of a Conductor Through Reduction of the Conduction Matrix

Imagine a complex-shaped component, for instance, the fictitious machined bracket of which the design is shown in Fig. 9.2. The design is called fictitious, because it has not been used for any real structure and has been made up specifically for this explanation. Suppose that for the building of the thermal model, the coupling between two surfaces need to be known:

1. The inner cylinder wall of the lug eye
2. The contact surface with a panel on which the bracket is supposed to be mounted.

These are the interface surfaces with other structural components between which typically the conductor value is needed for the thermal model that is used for thermal

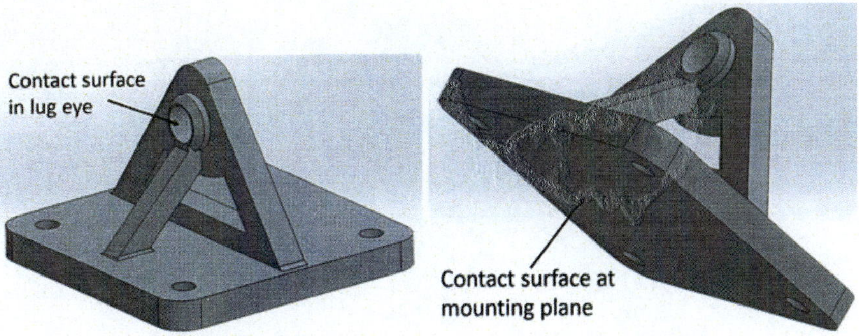

Fig. 9.2 Geometry of a fictitious machined bracket with contact surfaces for conductor calculation

control verification. In such a model, two nodes at each of the contact surfaces would suffice. In principle, from thermoelastic perspective, a temperature field in this bracket may cause this bracket to deform, and equipment attached to it may be tilted due to this deformation. When this tilting has the potential to cause that the alignment requirements for this equipment to be jeopardised, then it is important to capture the internal temperature field properly and be able to predict reliably the deformation of the bracket. Corresponding thermal discretisation has to be applied as well. In Sect. 4.6.3, recommendation is given on how to derive the most adequate level of detail in the modelling of features that also include brackets like this one.

Assuming that it is justified that for the application of this bracket in a thermal model, it is sufficient to represent the bracket with only two thermal nodes: one at each of the two contact surfaces indicated in Fig. 9.2. Note that this implies that here each of the thermal nodes represent a surface rather than a volume. Since no volume also means no mass and thus no thermal capacity, it would mean that no heat capacity is associated to these thermal nodes. A distribution of the heat capacity over the two thermal nodes can be generated, but this is not further discussed here.

With the help of the finite element method in combination with matrix manipulation methods that are available in several finite element codes, like MSC Nastran, this can be achieved.

First of all the geometry of the bracket in Fig. 9.2 has to be meshed with finite elements. The geometry is most suitable to be meshed with solid elements. To follow the geometry rather accurately, it is needed to use many elements. Since the model of this bracket is not integrated in full detail in a system-level model, the higher mesh resolution is not an issue. The material of the bracket is assumed to be aluminium.

In Fig. 9.3, the finite element model of the bracket is presented with two sets of so-called multipoint constraints. In each of these sets, all the FE nodes at the contact surface are connected to a dedicated FE node that represents the thermal node temperature of the thermal node representing this contact surface. With this approach, all the FE nodes at the complete contact surface will have the same temperature as the FE node representing the thermal node temperature. This is a simplified assumption, because in reality there will be no uniform temperature field at the bottom surface of

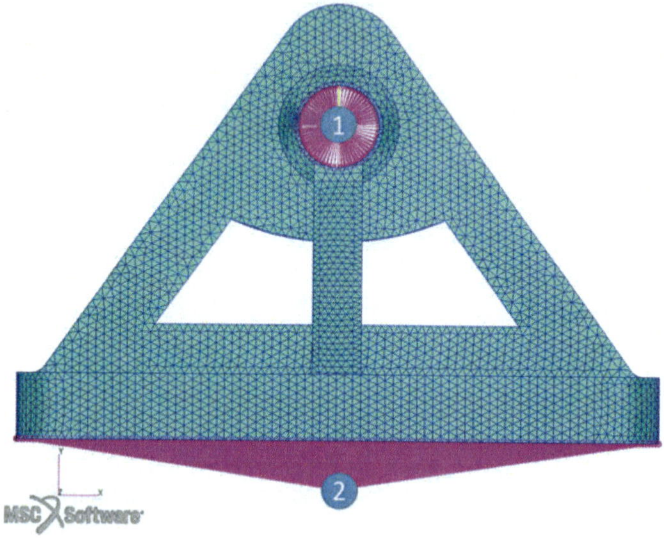

Fig. 9.3 Bracket meshed with constraints implemented for thermal node temperature representation

the bracket nor in the lug eye. In Sect. 9.3, a more sophisticated method is explained that allows to take into account an a-priori unknown temperature field inside a thermal node.

In Fig. 9.3, the two FE nodes, representing the two thermal nodes between which the conductor is going to be calculated, are marked with "1" for the lug eye interface thermal node and "2" for the bottom plane interface thermal node.

In the following derivation, two sets of FE nodes are identified:

- Boundary FE nodes: In this case, the FE nodes marked with "1" and "2" in Fig. 9.3. These represent the thermal nodes between which the conductors are to be calculated. Quantities related to these nodes are marked with subscript "b" (boundary or b-set).
- The interior FE nodes: These are all the remaining FE nodes. Quantities related to these nodes are marked with subscript "i" (interior or i-set).

The objective of this section is to provide an expression that eliminates all the interior nodal temperatures and reduces the conductive relation between node 1 and 2 to the form of

$$GL_{12}\,(T_1 - T_2) = \frac{1}{2}\,(Q_1 - Q_2)\,. \tag{9.2}$$

with Q_1 and Q_2 the heat flow rates (W) at the two thermal nodes, T_1 and T_2 the two thermal node temperatures and the GL_{12} the linear conductor (W/K) between the two thermal nodes.

This definition of FE node sets makes it possible to introduce the vector with all FE node temperatures partitioned in two sets or sub-vectors through

$$(T_g) = \begin{pmatrix} (T_b) \\ (T_i) \end{pmatrix}, \tag{9.3}$$

with

$$(T_b) = \begin{pmatrix} T_1 \\ T_2 \end{pmatrix} \tag{9.4}$$

being the vector with the temperatures of the boundary nodes.

In this section, instead of referring to the global conduction matrix with $[K_c]$, the same matrix is indicated with $[C]$. With the partitioning of the global nodal temperature vector, the steady-state conductive equilibrium equation can be written as

$$[C]\,(T) = (Q)\,, \begin{bmatrix} [C_{bb}] & [C_{bi}] \\ [C_{ib}] & [C_{ii}] \end{bmatrix} \begin{pmatrix} (T_b) \\ (T_i) \end{pmatrix} = \begin{pmatrix} (Q_b) \\ (0) \end{pmatrix} \tag{9.5}$$

Note that it is assumed that no heat load is applied to the interior of the model and therefore

$$(Q_i) = (0)\,. \tag{9.6}$$

Equation (9.5) can be split into two separate equations

$$\begin{aligned} [C_{bb}]\,(T_b) + [C_{bi}]\,(T_i) &= (Q_b)\,, \\ [C_{ib}]\,(T_b) + [C_{ii}]\,(T_i) &= (0) \end{aligned} \tag{9.7}$$

The second equation of Eq. (9.7) can be rewritten to

$$(T_i) = -[C_{ii}]^{-1}[C_{ib}]\,(T_b) \tag{9.8}$$

When this expression for (T_i) is substituted in the first equation of Eq. (9.7), then the following conductive relation for the boundary FE nodes is obtained

$$\begin{aligned} &\left([C_{bb}] - [C_{bi}][C_{ii}]^{-1}[C_{bi}]\right)(T_b) = (Q_b) \\ &\text{or} \\ &[\bar{C}_{bb}]\,(T_b) = (Q_b)\,. \end{aligned} \tag{9.9}$$

Using the vector with temperatures of the boundary nodes from Eqs. (9.4) and (9.9) can also be written as

$$\begin{bmatrix} \bar{C}_{bb,11} & \bar{C}_{bb,12} \\ \bar{C}_{bb,21} & \bar{C}_{bb,22} \end{bmatrix} \begin{pmatrix} T_1 \\ T_2 \end{pmatrix} = \begin{pmatrix} Q_1 \\ Q_2 \end{pmatrix}. \tag{9.10}$$

When this matrix expression is expanded, the following two equation is obtained

$$\begin{aligned} \bar{C}_{bb,11}T_1 + \bar{C}_{bb,12}T_2 &= Q_1 \\ \bar{C}_{bb,21}T_1 + \bar{C}_{bb,22}T_2 &= Q_2 \end{aligned}. \tag{9.11}$$

Because the system under consideration is still an unconstrained system (no temperatures are prescribed), the sum of the elements of the rows and columns of the reduced conduction matrix $[\bar{C}_{bb}]$ is zero that means

$$\bar{C}_{bb,11} = -\bar{C}_{bb,12} = -\bar{C}_{bb,21} = \bar{C}_{bb,22} = \hat{C}. \tag{9.12}$$

Substitution of Eq. (9.12) in Eq. (9.11) leads then to

$$\begin{aligned} \hat{C}T_1 - \hat{C}T_2 &= Q_1 \\ -\hat{C}T_1 + \hat{C}T_2 &= Q_2 \end{aligned} \tag{9.13}$$

When from the above expressions, the second equation is subtracted from the first, the following is obtained:

$$\begin{aligned} &2\hat{C}\,(T_1 - T_2) = Q_1 - Q_2 \\ &\text{or} \\ &\hat{C}\,(T_1 - T_2) = \frac{1}{2}\,(Q_1 - Q_2) \end{aligned} \tag{9.14}$$

Calling $\hat{C} = GL_{12}$ shows that the objective to reduce the conductive relation between node 1 and 2 to the form of Eq. (9.2) has been achieved. In general, it can be stated that the conductor between node 1 and 2 is

$$GL_{12} = -\bar{C}_{bb,12}. \tag{9.15}$$

The method applied in this section to reduce the size of the conduction matrix is also referred to as Guyan's or static reduction technique in mechanical vibrations [94].

Basic example of conductor calculation through static reduction

Considered is a finite element model consisting of two rod elements and three nodes 1, 2 and 3. Node 2 is the internal node. Nodes 1 and 3 are the two boundary nodes, which represent more or less the geometric centres of the thermal nodes. The conductor GL_{AB}, which is the conductive coupling between node 1 and 3, has to be calculated. The FE model and the conductor are shown in Fig. 9.4. The global FE model conduction matrix is

$$[C] = \frac{kA}{L}\begin{bmatrix} 1 & -1 & 0 \\ -1 & 3 & -2 \\ 0 & -2 & 2 \end{bmatrix} \tag{9.16}$$

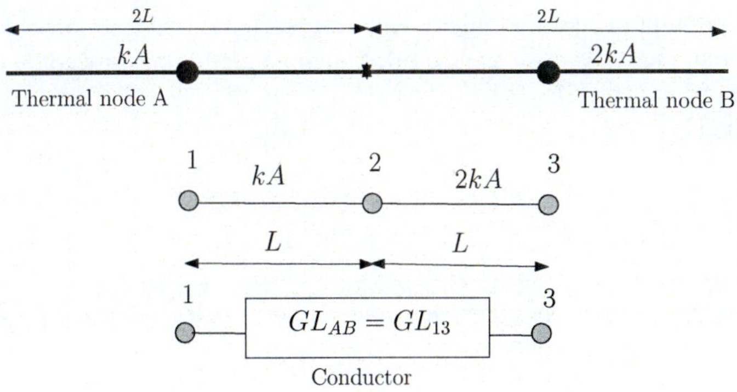

Fig. 9.4 Example FE model and conductor to be derived

This matrix is partitioned in line with Eq. (9.5) based on a b-set consisting of nodes 1 and 3 and an i-set with node 2:

$$[C_{bb}] = \frac{kA}{L}\begin{bmatrix}1 & 0\\ 0 & 2\end{bmatrix},\ [C_{bi}] = \frac{kA}{L}\begin{pmatrix}-1\\ -2\end{pmatrix},\ [C_{ib}] = [C_{bi}]^T,\ [C_{ii}] = \frac{3kA}{L} \quad (9.17)$$

The reduced conduction matrix $[\bar{C}_{bb}]$ can now be obtained with the expression of Eq. (9.9)

$$[\bar{C}_{bb}] = \frac{kA}{3L}\begin{bmatrix}2 & -2\\ -2 & 2\end{bmatrix}. \quad (9.18)$$

Finally, the conductor GL_{AB} can be calculated

$$GL_{AB} = GL_{13} = -\bar{C}_{bb,12} = \frac{2kA}{3L} \quad (9.19)$$

When the FE model in Fig. 9.4 is augmented with one rod element (ka, L) left of node 1 and one rod element $(2ka, L)$ right of node 3, the two thermal nodes are now completely represented by the FE model. The FE nodes 1 and 3 remain in (T_b). The calculated value of the GL_{AB} has not changed and is similar with the value found in Eq. (9.19).

Conductor calculation of bracket through static reduction

In this example, the approach with the use of MSC Nastran is explained to compute the conductor for the bracket that was presented in Fig. 9.2 and of which the model with the constraints to introduce the two thermal nodes was shown in Fig. 9.3.

To instruct MSC Nastran to follow the above static reduction procedure, the two nodes that are to be retained have to be assigned to the so-called A-set (analysis set).

Since matrix output is not a standard output, it has to be arranged through a small intervention in the solver with an ALTER. The input file for this procedure for MSC Nastran can be as shown in

```
NASTRAN HEAT=1
APP HEAT
TIME 1000
DIAG 8,13
SOL 101
COMPILE DMAP=PHASE1DR, SOUIN=MSCSOU, NOREF, NODECK, NOLIST
MALTER '$MALTER:AFTER SUPERELEMENT MATRIX AND LOAD REDUCTION TO A-SET,'
MATGPR GPLS,USET,SILS,KAA//'A' $
EXIT $
ENDALTER $
CEND
MPC=27376
SEALL = ALL
ECHO = none
BEGIN BULK
INCLUDE 'bracket.bdf'
ASET1,1,1000000,2000000
ENDDATA
```

Note that at the before last line, for pragmatic reasons, GRID IDs 1000000 and 2000000 are used for node 1 and 2 indicated in Fig. 9.3.

Below is a snippet from the output file from MSC Nastran showing the matrix KAA, the equivalent of the matrix $[\bar{C}]$ from Eq. (9.9).

```
0                    KAA
 POINT               VALUE        POINT              VALUE


 COLUMN                  1 (      1000000- S).
 1000000   S  2.96347E+00    2000000   S -2.96347E+00

 COLUMN                  2 (      2000000- S).
 1000000   S -2.96347E+00    2000000   S  2.96347E+00
```

In equation form, the MSC Nastran result can be written as

$$[\bar{C}_{bb}] = \begin{bmatrix} 2.96347 & -2.96347 \\ -2.96347 & 2.96347 \end{bmatrix}. \tag{9.20}$$

The conductor between node 1 and 2 then becomes

$$GL_{12} = -\bar{C}_{bb,12} = 2.96347 \tag{9.21}$$

9.2.2 *Conductor Calculation Through Steady-State Thermal Analysis*

In this section, the objective is the same as in the previous section: To obtain the representative conductor GL value between two locations at a component with a complex geometry. In the previous section, mathematics has been applied to produce an expression for the GL value. In this section, a more physical approach is followed through a steady-state thermal analysis.

Considered is again the finite element model of Fig. 9.3. The following thermal loading conditions is applied:

- A unit heat flow $Q_1 = 1$ W is applied to node 1.
- The temperature at node 2 is constrained to $T_2 = 0\,°\mathrm{C}$.

This case is analysed with a steady-state thermal analysis approach. The temperature T_1 is calculated with the conductive FE model.

With the computed value of T_1, the heat flow through the conductor GL_{12} W/°C is then

$$GL_{12}T_1 = Q_1 = 1 \tag{9.22}$$

The representative conductor GL_{12} can now be calculated

$$GL_{12} = \left.\frac{1}{T_1}\right|_{Q_1=1} \tag{9.23}$$

The conductor GL_{12} can also be obtained by constraining $T_1 = 0$ K and $Q_2 = 1$ W, and subsequently, T_2 is calculated.

Basic example of conductor calculation by solving the steady-state problem

Equation (9.16) shows the conduction matrix corresponding to the model from Fig. 9.4. The unconstrained heat equation of this model is

$$\frac{kA}{L}\begin{bmatrix} 1 & -1 & 0 \\ -1 & 3 & -2 \\ 0 & -2 & 2 \end{bmatrix}\begin{pmatrix} T_1 \\ T_2 \\ T_3 \end{pmatrix} = \begin{pmatrix} Q_1 \\ 0 \\ Q_3 \end{pmatrix} \tag{9.24}$$

When $T_3 = 0$ and $Q_1 = 1$, Eq. (9.24) is reduced to

$$\frac{kA}{L}\begin{bmatrix} 1 & -1 \\ -1 & 3 \end{bmatrix}\begin{pmatrix} T_1 \\ T_2 \end{pmatrix} = \begin{pmatrix} 1 \\ 0 \end{pmatrix} \tag{9.25}$$

The solution of this system is $T_1 = \frac{3}{2}\frac{L}{kA}$ (°C), $T_2 = \frac{1}{2}\frac{L}{kA}$ (°C). Node 3 is the heat sink, and the heat that is applied to node 1 is absorbed at node 3. This implies $Q_3 = -1$ (W).

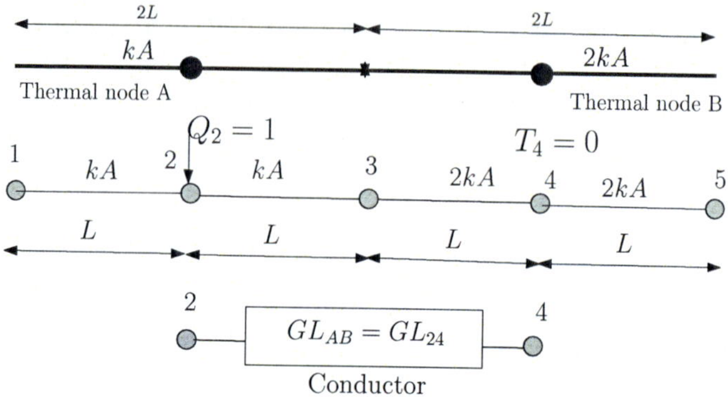

Fig. 9.5 Example FE model and conductor to be derived

With Eq. (9.23), GL_{13} can be obtained

$$GL_{13} = \frac{1}{T_1}\bigg|_{Q_1=1} = \frac{2}{3}\frac{kA}{L} \tag{9.26}$$

which is the same result, as expected, as was obtained in the example "Basic example of conductor calculation through static reduction" in the previous section.

The FE model in Fig. 9.4 will be expanded as illustrated in Fig. 9.5. The conduction FE models are now completely associated with the thermal model. The FE nodes 2 and 4 are common to the geometric centre of the thermal nodes.

The unconstrained heat equation of this model is

$$\frac{kA}{L}\begin{bmatrix} 1 & -1 & 0 & 0 & 0 \\ -1 & 2 & -1 & 0 & 0 \\ 0 & -1 & 3 & -2 & 0 \\ 0 & 0 & -2 & 4 & -2 \\ 0 & 0 & 0 & -2 & 2 \end{bmatrix}\begin{pmatrix} T_1 \\ T_2 \\ T_3 \\ T_4 \\ T_5 \end{pmatrix} = \begin{pmatrix} 0 \\ Q_1 \\ 0 \\ Q_4 \\ 0 \end{pmatrix}. \tag{9.27}$$

When $T_4 = 0$ and $Q_2 = 1$, Eq. (9.27) is reduced to

$$\frac{kA}{L}\begin{bmatrix} 1 & -1 & 0 & 0 \\ -1 & 2 & -1 & 0 \\ 0 & -1 & 3 & 0 \\ 0 & 0 & 0 & 2 \end{bmatrix}\begin{pmatrix} T_1 \\ T_2 \\ T_3 \\ T_5 \end{pmatrix} = \begin{pmatrix} 0 \\ 1 \\ 0 \\ 0 \end{pmatrix}. \tag{9.28}$$

The solution of this system Eq. (9.28) is

$$T = \begin{pmatrix} T_1 \\ T_2 \\ T_3 \\ T_5 \end{pmatrix} = \frac{L}{kA} \begin{pmatrix} \frac{3}{2} \\ \frac{3}{2} \\ \frac{1}{2} \\ 0 \\ 0 \end{pmatrix} \tag{9.29}$$

Node 4 is the heat sink, and the heat that is applied to node 1 is absorbed at node 4. This implies $Q_4 = -1$. With Eq. (9.23), GL_{24} can be obtained

$$GL_{24} = \frac{1}{T_2}\bigg|_{Q_2=1} = \frac{2}{3}\frac{kA}{L}, \tag{9.30}$$

The temperature gradient in the most left and right rod elements $dT/dx = 0$, which is equivalent to zero heat flux in those elements. These elements are in fact obsolete when the FE nodes 2 and 4 are common to the geometric centre of the thermal nodes (solid black dots in top part of Fig. 9.5).

Conductor calculation of bracket by solving the steady-state problem

The example from the previous section "Conductor calculation of bracket through static reduction" is now repeated with the method of the current section. The temperature of node 1 is set to 0 °C, and a unit heat flux is applied to node 2. The temperature that is obtained for node 2 by running the model with MSC Nastran is 3.374427×10^{-1} °C. This makes that the conductor becomes

$$GL_{12} = \frac{1}{3.374427 \times 10^{-1}} = 2.96346, \tag{9.31}$$

which is the same value as was obtained through the static reduction method in the previous section.

9.2.3 *Far Field Method for Generation of 1-D Linear Conductors*

In ESATAN-TMS, a method called "Far Field" method [11, 55, 88] is implemented to automatically compute the conductors between pairs of thermal nodes. The method uses the finite element method to simulate the temperature fields within the thermal nodes. Also, this method is considering two thermal nodes in isolation, but has an interesting physical basis in which heat is flowing from the "thermally most remote edge" of one thermal node via the interface with an other thermal node to the "thermally most remote edge" at this other thermal node. The "thermally most remote edges" form the basis to construct far field temperature fields in which heat is flowing through the two thermal nodes from one remote heat source to an other remote heat sink. The method is a good and physically well-justified alternative for the manual process as described in Sect. 9.1.

The two coupled thermal nodes and corresponding conductor GL_{AB} are presented in Fig. 9.6.

The principle of the "Far Field" method is schematically illustrated in Fig. 9.7. It shows how heat flows from the remote heat source at one edge of thermal node A, through thermal node A to its interface with thermal node B and then via thermal node B to the remote heat sink at the edge of thermal node B: from one far field to another.

The conductor between the two thermal nodes is assembled from the two conductors between each thermal node and their shared interface.

Each thermal node is overlaid with a conductive 3-D FE model that is used to simulate far field temperature fields, from which far field edges and average temperatures are obtained. This is illustrated in Fig. 9.8.

The calculation of each of the sub-conductors between a thermal node and the shared interface is done in two steps:

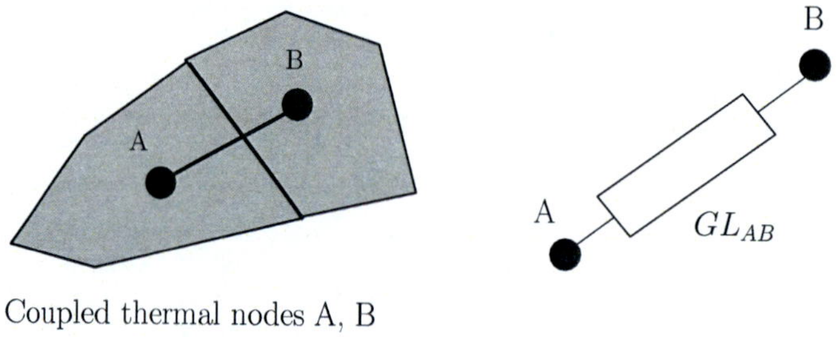

Fig. 9.6 Conductance between lumped parameter thermal nodes A and B

Fig. 9.7 1-D heat flow from one far field to the other far field

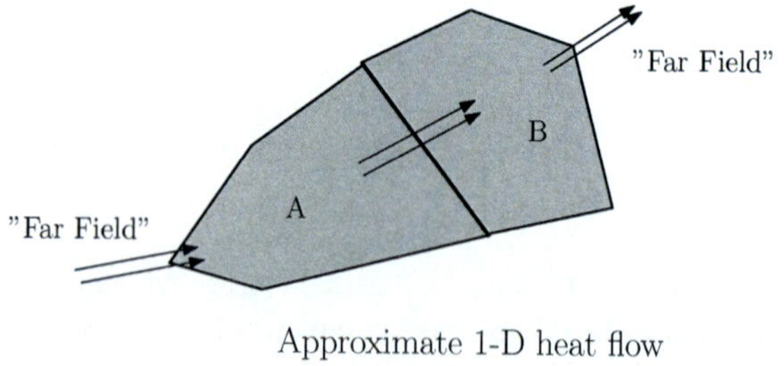

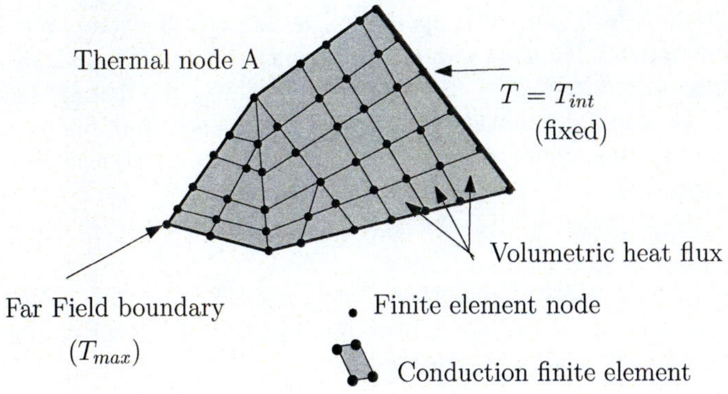

Fig. 9.8 Conductive FE model identifying far field boundary

Step 1: Identification the far field boundary nodes: This is achieved by imposing an uniform volumetric heat flux over the thermal node and calculating the resulting steady-state temperature field with the temperature fixed at the interface at an arbitrary level T_{int}. The rest of the boundary is assumed to be adiabatic.
The hottest points on the boundary form the far field boundary. A small temperature band below the observed maximum temperature T_{max} is applied, $T_{max}(1-\varepsilon) \leq T \leq T_{max}$, to prevent that only a single FE node is selected as the far field boundary. At least two connected nodes are needed to define the far field as being an edge of an element. Disjoint regions are allowed.
Step 2: Sub-conductor calculation: In this step, an arbitrary uniform area heat flux is applied (with total heat flow Q) to the far field boundary. The interface with the other thermal node is held at a fixed temperature T_{int}, and the remainder of the boundary of the thermal node is kept adiabatic. The temperature field is determined through a steady-state thermal analysis. This is the 1-D approximation of the far field heat flow. From this temperature distribution, the average temperature T_{avg} over the domain of the thermal node is calculated (see Fig. 9.9).

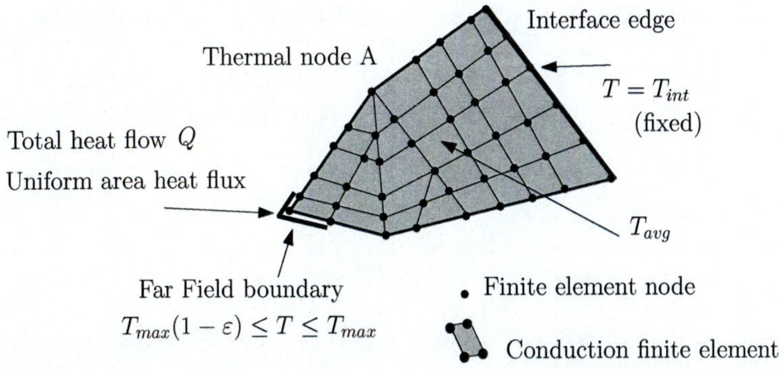

Fig. 9.9 Conductive FE model and average temperature

Since the total heat flow Q is applied to the far field boundary, this heat flow must flow through the thermal node to the interface with temperature T_{int}. The average temperature T_{avg} of the temperature field in the thermal node that is imposed through these boundary conditions is obtained from the finite element model through integration over the elements. With this 1-D, heat flow equation can be written as

$$Q = GL_A(T_{avg} - T_{int}), \tag{9.32}$$

where GL_A is the effective conductance between the thermal node and the interface with the neighbour thermal node. Inverting Eq. (9.32), the conductance from the thermal node to the interface is defined by

$$GL_A = \frac{Q}{(T_{avg} - T_{int})}. \tag{9.33}$$

The procedure is repeated for the second thermal node B, and two conductance values obtained, GL_A and GL_B, are combined using the reciprocal rule for conductors in series. The total conductance GL_{AB} between the two thermal nodes A and B is then given by

$$GL_{AB} = \frac{1}{\frac{1}{GL_A} + \frac{1}{GL_B}}. \tag{9.34}$$

Example "Far Field" method

Two thermal nodes A and B are coupled. The corresponding conductive FE models are illustrated in Fig. 9.10. The FE sub-model, consisting of the two rod elements connecting the FE nodes 1, 2 and 3, corresponds to thermal node A. The FE sub-model, consisting of two rod elements connecting the FE nodes 3, 4 and 5, corresponds to thermal node B. The cross-sectional area of thermal node B is twice that of thermal node A. The conductivity and length are equal for both thermal nodes.

The far field boundary points are in this example rather trivial, namely the FE nodes 1 and 5.

The heat equation of the conductive FE model corresponding with thermal node A is given by

$$\frac{kA}{L}\begin{bmatrix} 1 & -1 & 0 \\ -1 & 2 & -1 \\ 0 & -1 & 1 \end{bmatrix} \cdot \begin{pmatrix} T_1 \\ T_2 \\ T_3 = T_{int} \end{pmatrix} = \begin{pmatrix} Q \\ 0 \\ Q_3 \end{pmatrix}. \tag{9.35}$$

The solution of Eq. (9.35) is

$$T_1 = \frac{2QL}{kA} + T_{int},\ T_2 = \frac{QL}{kA} + T_{int},\ T_3 = T_{int}. \tag{9.36}$$

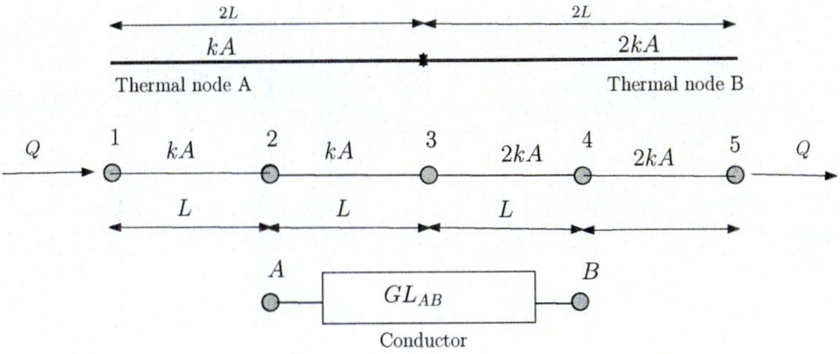

Fig. 9.10 FE model and thermal network

The average temperature of thermal node A is

$$T_{avg,A} = \frac{QL}{kA} + T_{int}, \tag{9.37}$$

and the conductance GL_A is then

$$GL_A = \frac{Q}{(T_{avg,A} - T_{int})} = \frac{kA}{L}. \tag{9.38}$$

The heat equation of the conductive FE model corresponding with thermal node B is given by

$$\frac{kA}{L}\begin{bmatrix} 2 & -2 & 0 \\ -2 & 4 & -2 \\ 0 & -2 & 2 \end{bmatrix} = \begin{pmatrix} T_3 = T_{int} \\ T_4 \\ T_6 \end{pmatrix}\begin{pmatrix} Q_3 \\ 0 \\ -Q \end{pmatrix}. \tag{9.39}$$

The solution of Eq. (9.39) is

$$T_3 = T_{int}.\ T_4 = \frac{-QL}{kA} + T_{int},\ T_5 = \frac{-QL}{2kA} + T_{int}. \tag{9.40}$$

The average temperature of thermal node B is

$$T_{avg,B} = \frac{-QL}{2kA} + T_{int}, \tag{9.41}$$

and the conductance GL_B is then

$$GL_B = \frac{Q}{(T_{int} - T_{avg,B})} = \frac{2kA}{L}. \tag{9.42}$$

The conductor GL_{AB} between the thermal nodes A and B can be calculated with

$$GL_{AB} = \frac{1}{\frac{1}{GL_A} + \frac{1}{GL_A}} = \frac{2kA}{3L}. \tag{9.43}$$

An alternative way to calculate the conductor GL_{AB} is done as follows

$$GL_{AB} = \frac{Q}{(T_{avg,A} - T_{avg,B})} = \frac{2kA}{3L}. \tag{9.44}$$

The value of the conductor GL_{AB} calculated with the "Far Field" method is equal to the value of the conductor GL_{13} as given in Eqs. (9.19), (9.26) and (9.30).

9.3 PAT-Based Methods for Generating TMM Conductors

In Sect. 9.2, different methods to compute the linear conductors (GLs) between two thermal nodes of the TMM are explained. In this section, two systematic methods to compute the conductors between all thermal nodes corresponding to a FE model are discussed. The two methods are both based on the PAT method.

The first method makes directly use of the properties of the solution of the PAT equation Eq. (8.31) and specifically the Lagrange multipliers Λ.

The second method requires extra operations on the sub-matrices of the PAT equation which include static reduction.

Both methods produce the same values for the GLs and are both based on the consistent PAT relation between the thermal nodes and the finite element model.

9.3.1 Extracting Conductors from Lagrange Multipliers Λ

In this method, the property of the Lagrange multipliers is exploited. The Lagrange multipliers (Λ) are part of the solution of the PAT equation Eq. (8.31) that is recalled here.

$$\begin{bmatrix} [K_c] & [A]^T \\ [A] & [0] \end{bmatrix} \begin{pmatrix} (T^s) \\ (\Lambda) \end{pmatrix} = \begin{pmatrix} 0 \\ (T^t) \end{pmatrix}$$

Each element of the vector of Lagrange multipliers represents the total heat flow to the corresponding thermal node coming from the other thermal nodes. This heat flow originates from the boundary conditions that the PAT method imposes to the finite element model by prescribing the average thermal node temperature. The sum of the Lagrange multipliers is zero, indicating that the mapped temperature field is in balance with the thermal node temperatures.

The definition of the linear conductor value states that it is equal to the heat flow between two thermal nodes when there is a unit temperature difference between these two thermal nodes. With this characteristic, all the conductors between thermal node k and the other thermal nodes can be computed by introducing a special vector of thermal node temperatures in which $T_k^t = 1.0$ and the other thermal node temperatures are set to $T_j^t = 0.0$ with $j \neq k$. This introduces a unit temperature difference between thermal node k and each of the other thermal nodes. The corresponding solution of the Lagrange multiplier vector (Λ) contains the GL values between thermal node j and thermal node k for $k \neq j$.

When the vector with thermal node temperatures with only one value equal to 1.0 and the others set to 0.0 is repeated for each thermal node, the unity matrix $[\tilde{T}^t]$ is obtained with thermal node temperatures.

$$[\tilde{T}^t] = \begin{bmatrix} 1\,0\,0 \cdots 0 \\ 0\,1\,0 \cdots 0 \\ 0\,0\,1 \cdots 0 \\ \vdots\,\vdots\,\vdots\,\ddots\,\vdots \\ 0\,0\,0 \cdots 1 \end{bmatrix} \tag{9.45}$$

When in the PAT equation Eq. (8.31) the vector $\left(T^t\right)$ is substituted with matrix $[\tilde{T}^t]$, the solution then become the matrix with FE node temperatures $[\tilde{T}^s]$ and the matrix with Lagrange multipliers $[\tilde{\Lambda}]$. The PAT equation is then rewritten to

$$\begin{bmatrix} [K_c] & [A]^T \\ [A] & [0] \end{bmatrix} \begin{bmatrix} [\tilde{T}^s] \\ [\tilde{\Lambda}] \end{bmatrix} = \begin{bmatrix} [0] \\ [\tilde{T}^t] \end{bmatrix} \tag{9.46}$$

The matrix with Lagrange multipliers $[\tilde{\Lambda}]$ is part of the solution of Eq. (9.46). The GL values can then directly be obtained from this matrix with

$$GL_{ij} = \tilde{\Lambda}_{ij} \quad i, j = 1, 2, \ldots, N, i \neq j, j > i. \tag{9.47}$$

with N being the number of thermal nodes.

This method can completely reuse the numerical infrastructure for solving the PAT equation and could therefore be considered as a sort of free side product of the PAT solution.

9.3.2 *Reduction of FE Model Conduction Matrix*

The method for generation of linear conductors based on the method summarised in [59] is described in detail in this section.

The starting point for the method is the constraint relation between the thermal nodes and the nodes of the finite element model from Eq. (8.7) and recalled here in a slightly different form

$$[A]\left(T^s\right) - \left(T^t\right) = (0) \tag{9.48}$$

Equation (9.48) can be considered to be a linear constraint relation between two sets of unknown temperatures coming from two different models. In many FE codes, this kind of linear constraint relations exists. In MSC Nastran, these relations are called multipoint constraints (MPC).

In MPC relations, there are both dependent and independent degrees of freedom, in which the dependent degrees of freedom can be expressed in terms of the independent degrees of freedom. In the context of this type of constraints, the independent unknowns are also referred to as master degrees of freedom or master nodes. The dependent unknowns are then referred to a slave degrees of freedom or slave nodes. The dependent unknown can be removed from the system through a mathematical elimination process that is applied and explained in this section.

Since the objective of the method is to express the conduction relation only in terms of the thermal node temperatures, the thermal node temperatures shall not be eliminated in the process and shall therefore be considered as part of the independent degrees of freedom. This implies that a number of FE node temperatures have to be selected as dependent degrees of freedom. For each constraint relations defined by Eq. (9.48), thus for each thermal node, one dependent FE node temperature has to be identified. See also Fig. 9.11 showing that for each thermal node a dependent (slave) node has to be selected.

As a consequence, the vector with FE node temperatures has to be split into two sets of unknowns.

$$\left(T^s\right) = \begin{pmatrix} \left(T^d\right) \\ \left(T^i\right) \end{pmatrix} \tag{9.49}$$

with $\left(T^d\right)$ being the vector with dependent FE node temperatures and $\left(T^i\right)$ the vector with independent FE node temperatures.

The complete vector of temperatures, including both FE node and thermal node temperatures, can now be written as

$$(T) = \begin{pmatrix} \left(T^d\right) \\ \left(T^i\right) \\ \left(T^t\right) \end{pmatrix} \tag{9.50}$$

In the first part of this section, the expression is derived that defines the relation between the dependent FE node temperatures and the independent FE node and thermal node temperatures. The independent temperatures $\left(\hat{T}\right)$ form the following subset of (T)

$$\left(\hat{T}\right) = \begin{pmatrix} \left(T^i\right) \\ \left(T^t\right) \end{pmatrix} \tag{9.51}$$

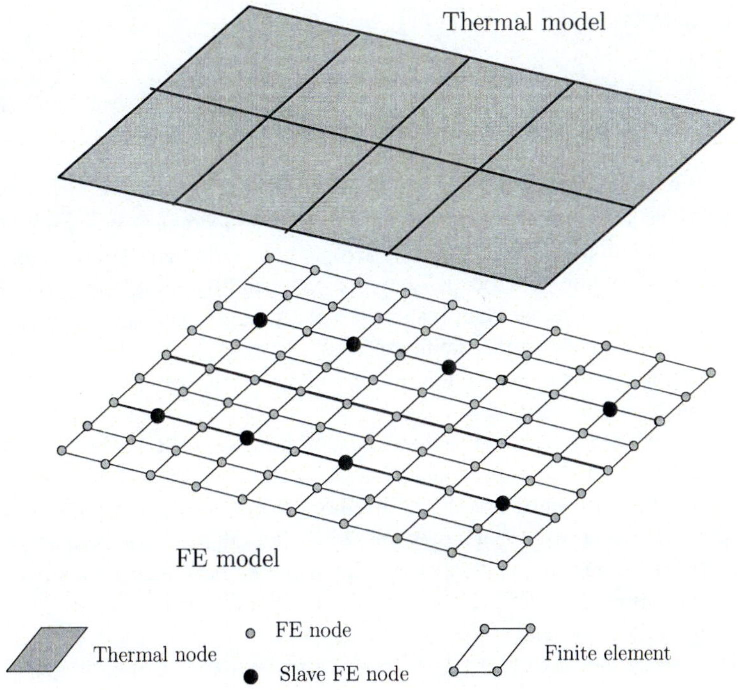

Fig. 9.11 Dependent (slave) FE nodes to be selected for each PAT relation

With Eq. (9.49), the FE node temperatures are partitioned into two sets of unknowns. In accordance with this, the A-matrix can be partitioned as well

$$[A] = \left[[A^{d,d}]\;[A^{d,i}]\right] \tag{9.52}$$

with $[A^{d,d}]$ being a square matrix corresponding to the dependent FE node temperatures. The number of rows and columns in this matrix is equal to the number of thermal nodes, which is equal to the number of a dependent FE nodes. The matrix $[A^{d,i}]$ is in general a rectangular matrix with the number of rows equal to the number of dependent FE node temperatures and the number of columns equal to the number of independent FE node temperatures.

Combining Eqs. (9.49) and (9.52), Eq. (9.48) can be written as

$$\left[[A^{d,d}]\;[A^{d,i}]\;-[I^{t,t}]\right]\begin{pmatrix}(T^d)\\(T^i)\\(T^t)\end{pmatrix} = (0) \tag{9.53}$$

Equation (9.53) can be rewritten into

$$[A^{d,d}]\left(T^d\right) = \left[-[A^{d,i}]\;[I^{t,t}]\right]\begin{pmatrix}(T^i)\\(T^t)\end{pmatrix} \tag{9.54}$$

With Eq. (9.54), the solution of $\left(T^d\right)$ can be formulated as

$$\left(T^d\right) = [A^{d,d}]^{-1}\left[-[A^{d,i}]\;[I^{t,t}]\right]\begin{pmatrix}(T^i)\\(T^t)\end{pmatrix} \tag{9.55}$$

This solution only exists when $[A^{d,d}]$ is not singular. Singularity of $[A^{d,d}]$ can occur when the selection of the dependent FE node temperatures is unfortunate or that the PAT relations overconstrain the temperature field. The latter means that through the PAT relations of Eq. (8.7) two or more thermal nodes are prescribing, potentially different, average temperatures for the same FE nodes. This may occur when a mistake is made in the definition of the correspondence between thermal nodes and finite elements in which a number of finite elements are overlapped by more than one thermal node.

Singularity may also occur when the number of finite elements corresponding to a thermal node is small. The consequence may be that too many FE nodes, connected to the overlapped elements, are shared by more than one thermal node. In other words, there are no or just a small number of FE nodes that are corresponding with only one thermal node. In general, this problem can be prevented by assuring that the mesh resolution of the finite element model is at least a factor of two higher than the resolution of the thermal model.

Assuming that the inverse of $[A^{d,d}]$ exists, then Eq. (9.55) can be used to write all FE node and thermal node temperature in terms of the independent FE node and thermal node temperatures.

$$\begin{pmatrix}(T^d)\\(T^i)\\(T^t)\end{pmatrix} = \begin{bmatrix}-[A^{d,d}]^{-1}[A^{d,i}] & [A^{d,d}]^{-1}\\ [I^{i,i}] & [0]\\ [0] & [I^{t,t}]\end{bmatrix}\begin{pmatrix}(T^i)\\(T^t)\end{pmatrix} \tag{9.56}$$

In Eq. (9.56), the matrices $[I^{i,i}]$ and $[I^{t,t}]$ are identity matrices of the size of the number of independent FE node temperature and the number of thermal node temperatures, respectively.

The matrix with sub-matrices in Eq. (9.56) can be referred to as the matrix $[M]$ as

$$[M] = \begin{bmatrix}-[A^{d,d}]^{-1}[A^{d,i}] & [A^{d,d}]^{-1}\\ [I^{i,i}] & [0]\\ [0] & [I^{t,t}]\end{bmatrix} = \begin{bmatrix}[M^{d,i}] & [M^{d,d}]\\ [I^{i,i}] & [0]\\ [0] & [I^{t,t}]\end{bmatrix} \tag{9.57}$$

With this, Eq. (9.56) can be rewritten to

$$\begin{pmatrix}(T^d)\\(T^i)\\(T^t)\end{pmatrix} = [M]\begin{pmatrix}(T^i)\\(T^t)\end{pmatrix} \tag{9.58}$$

Equation (9.58) is basically a reformulation of Eq. (9.48) implementing the partitioning of the FE node and thermal node temperatures as in Eq. (9.50).

With the expression of Eq. (9.58), it is possible to express the dependent temperatures as function of the independent temperatures and with that the first part of this section is completed. The next part of this section focusses on the use of this expression to eliminate the dependent temperatures from the system equations and reduce the conduction relations to a set corresponding to the thermal node temperatures.

Consider now the conduction matrix $[C]$ for the unknowns in the vector presented in Eq. (9.50), being the complete set of FE node temperatures and the thermal node temperatures

$$[C] = \begin{bmatrix} [C^{d,d}] & [C^{i,d}]^T & [0] \\ [C^{i,d}] & [C^{i,i}] & [0] \\ [0] & [0] & [0] \end{bmatrix} \tag{9.59}$$

In this matrix, the third row and column contain null-matrices. This means that the FE conduction matrix does not link the thermal node temperatures to the FE node temperatures. This linking is obtained with the help of the relation in Eq. (9.58). This would transform the conduction matrix of Eq. (9.59) to the set of FE node and thermal node temperatures $\left(\hat{T}\right)$ of Eq. (9.54) from which the dependent FE node temperature is eliminated

$$[\hat{C}] = [M]^T[C][M] = \begin{bmatrix} [\hat{C}^{i,i}] & [\hat{C}^{i,t}]^T \\ [\hat{C}^{i,t}] & [\hat{C}^{t,t}] \end{bmatrix} \tag{9.60}$$

Now, the dependent FE node temperatures are eliminated, the conductive link is established between the FE model and the thermal model with conduction matrix $[\hat{C}]$, and it is possible to apply static reduction [94] to extract the conduction matrix $[\tilde{C}]$ related to the thermal node temperatures, by eliminating the independent FE node temperatures

$$[\tilde{C}] = -[\hat{C}^{i,t}]^T[\hat{C}^{i,i}][\hat{C}^{i,t}] + [\hat{C}^{t,t}] \tag{9.61}$$

In a way similar to Eq. (9.47), the GL values between thermal node i and thermal node j can be extracted from the matrix $[\tilde{C}]$

$$GL_{ij} = -[\tilde{C}]_{ij} \quad i, j = 1, 2, \ldots, N, i \neq j, j > i. \tag{9.62}$$

with N being the number of thermal nodes.

This method is clearly more complicated than the method described in Sect. 9.3.1. However, the method in this section can also be applied with the MSC Nastran and does not require a dedicated PAT solver.

GL Generation for a one-dimensional model

In this example, the two methods for generating linear conductors are demonstrated with the help of a one-dimensional model.

An Al-alloy strip with thickness $t = 10$ mm, length $L = 1$ m and width $b = 0.25$ m is modelled as an 1-D rod in a conduction FE model. The conductivity is $k = 205$ W/m/°C. The number of rod elements is 20 and 21 nodes. The strip, thermal nodes and FE model are shown in Fig. 9.12.

The correspondence between the thermal nodes and the finite elements is presented in Table 9.1. Also, the FE node in the middle of each thermal node is indicated. This FE node is used as dependent FE node for the conductor generation method based on static reduction.

The conductor generation method, based on extraction of the Lagrange multiplier values, uses three temperature fields that form together the identity matrix

$$[\tilde{T}^t] = \begin{bmatrix} 1 & 0 & 0 \\ 0 & 1 & 0 \\ 0 & 0 & 1 \end{bmatrix} \tag{9.63}$$

In addition, a temperature field with all thermal node temperatures equal to 1 °C is applied to verify the correct definition of the A-matrix.

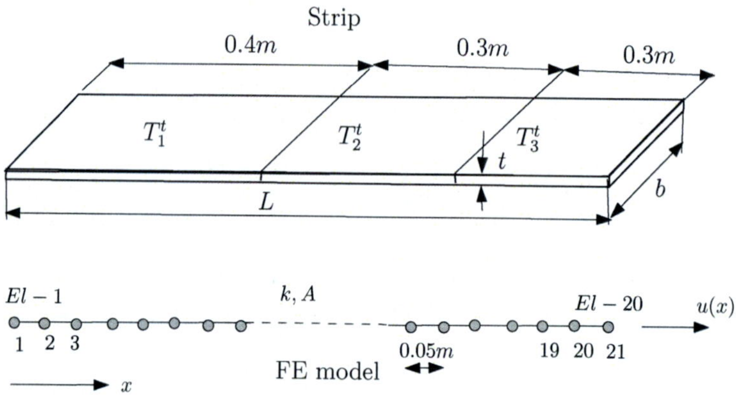

Fig. 9.12 Al-alloy strip, thermal nodes and corresponding FE model

Table 9.1 Thermal node, overlaid elements, nodes

Thermal node	FE model rod elements	FE model nodes	Midpoint
1	1–8	1–9	5
2	9–14	9–15	12
3	15–20	15–21	18

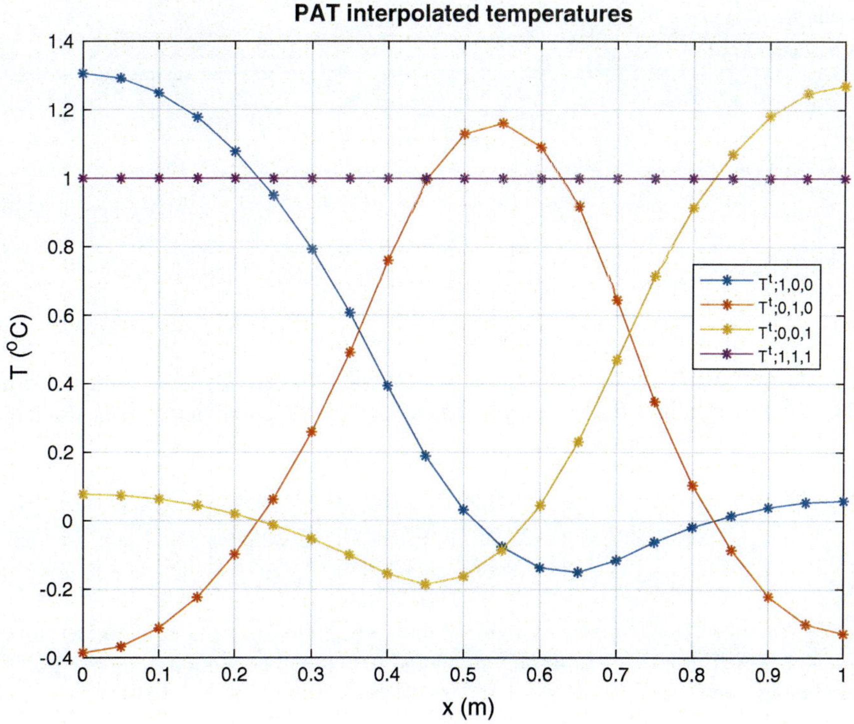

Fig. 9.13 PAT interpolated temperature fields in strip (rod)

The temperature distribution in the FE model is calculated with the prescribed average temperature method for the four sets of thermal node temperatures, and the results are presented in Fig. 9.13.

The matrix $[\tilde{\Lambda}]$ with Lagrange multipliers corresponding to the thermal node temperatures $[\tilde{T}^t]$ for this example becomes

$$[\tilde{\Lambda}] = \begin{bmatrix} -2.3383 & 2.9351 & -0.5968 \\ 2.9351 & -6.2647 & 3.3296 \\ -0.5968 & 2.9351 & -2.7328 \end{bmatrix} \tag{9.64}$$

It is good to highlight that for the case in which all thermal node temperature have the same unit temperature, all three Lagrange multipliers are equal to 0.0. This indicates that there is no heat flow between the thermal nodes, which should be the case for a uniform temperature field. It also is an indication that the A-matrix is not showing problems.

In the second part of this example, the conductors GL are calculated introducing multipoint constraints in combination with condensation of the conduction matrix. The conductors GL can be extracted from the final reduced conduction matrix $[\tilde{C}^{t,t}]$.

Table 9.2 Conductors between thermal nodes 1, 2, 3

Conductors W/°C		
	GL_{12}	GL_{13}
	2.9351	−0.59683
GL_{21}		GL_{23}
2.9351		3.3296
GL_{31}	GL_{32}	
−0.59683	2.9351	

The slave temperatures are the temperatures of nodes 5, 12 and 18 that coincide with the centres of the three thermal nodes. The reduced conduction matrix of the thermal model is given hereafter.

$$[\tilde{C}^{t,t}] = \begin{bmatrix} 2.33828 & -2.93511 & 0.59683 \\ -2.93511 & 6.26470 & -3.32959 \\ 0.59683 & -3.32959 & 2.73276 \end{bmatrix} \quad (9.65)$$

As could be expected from Eqs. (9.47) and (9.62), the following relation between the matrix with Lagrange multipliers $[\tilde{\Lambda}]$ and the reduced conduction matrix $[\tilde{C}^{t,t}]$ can be observed

$$[\tilde{C}^{t,t}] = -[\tilde{\Lambda}] \quad (9.66)$$

which illustrates that both methods lead to the same results.

The resulting conductors are presented in Table 9.2.

Comparing PAT-based conductors with classical hand-calculated conductors

Traditionally, the linear conductors for the TMM are calculated with method that is explained in Sect. 9.1 using the expression kA/L. This assumes a scalar link between two thermal nodes made of a material with a conductivity k (W/m/K), a cross section of A (m^2) and a length L (m). Figure 9.1 illustrates this scheme. For a two-dimensional problem, typically the conductors are computed with the scheme that is presented in Fig. 9.14. It can be noted that no diagonal couplings are considered.

In [59], the example that is also presented in Sect. 7.6 is used as benchmark to compare the performance of the PAT-based conductors with that of the classical scalar conductors.

The description of the model of the benchmark is found in Sect. 7.6. With the help of the finite element method, a reference temperature field has been produced. This reference temperature field is presented in Fig. 7.9.

In Fig. 9.15, the temperature fields produced with the two types of conductors are presented.

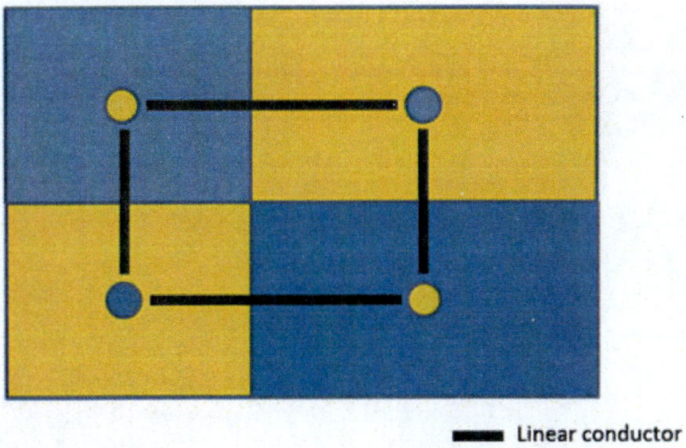

Fig. 9.14 Classical scheme to compute linear conductors for a 2-D structure [59]

Fig. 9.15 Temperature fields of the thermal model. On the left results obtained with classical conductors. On the right results with PAT-based conductors

Using the PAT method, these two temperature fields are mapped to the finite element model. The temperature mapping results are presented in Fig. 9.16.

Figure 9.16 shows the temperature fields generated in the FE model using the PAT method. The maximum temperature found using the PAT-based conductor calculation method is nearly identical to the peak temperature in the reference temperature field of shown in Fig. 7.9. The maximum temperature difference between the two models based on the two different conductor calculation methods indicates that the conductivity represented by the traditional conductors is lower than the one of the models with PAT-based conductors. It looks like that the conductivity of the model with the PAT-based conductors is very similar to the one of the reference models. A possible reason for the observed difference in modelled conductivity between the two conductor generation methods is the lack of diagonal couplings for the traditional conductor generation method.

In [59] are also the displacement and stress fields following from these two temperature fields compared (not shown here). The displacement and stress fields produced

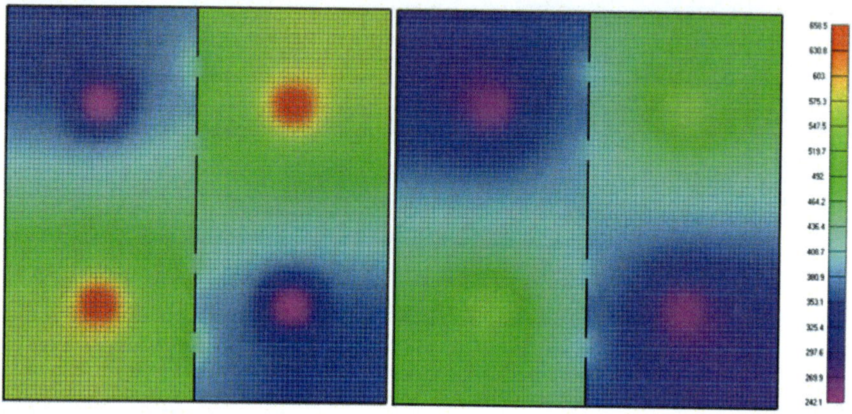

Fig. 9.16 Temperature fields mapped on the FE model. On the left results obtained with classical conductors. On the right results with PAT-based conductors

with the PAT-based temperature field reproduce the reference displacement and stress fields well. However, the temperature field produced with the classical conductors shows important deviations.

This example shows besides the ability to automate rather easily the generation of linear conductors for thermal lumped parameter models, it also provides an increase of accuracy of the conductive part of the thermal model.

Although the thermal mesh in this example is considered to have a reasonably high resolution, an exercise is done to increase the thermal mesh resolution with a factor of 4×4. This means that the number of thermal nodes is increased from 100 to 1600. For this regular shape, it was possible with the help of a script to generate the conductors with the traditional method between these nodes. The temperature field is then getting significantly closer to the reference temperature field; however, it has to be kept in mind that the large amount of thermal nodes in combination with a traditional conductor generation method has a significant impact on both thermal model preparation and run time.

9.3.3 Consideration for the Use of the PAT-Based Conductors

The set of conductors derived with the techniques discussed in Sects. 9.3.1 and 9.3.2 show some essential differences compared to a collection of one-dimensional conductors [11]:

- The conductors produced with the described methods are as good as the underlying finite element model. The finite element method is able to represent very well-conductive problems, and therefore, this method should be considered as a preferred method to generate conductors for continuous parts (see also previous example in which the PAT-based conductors out perform the classical conductors).

- The conductors cannot be considered independent from each other. These have to be used as a set of conductors that together are able to describe a 2-D or 3-D temperature field in the simulated part.
- The two techniques produce more conductors than just those that connect neighbouring thermal nodes. In other words, the methods also produce connectors between two thermal nodes that have no direct physical interface. The matrix $[\tilde{\Lambda}]$ from Eq. (9.46) and the matrix $[\tilde{C}]$ from Eq. (9.61) are in fact full matrices containing couplings from each thermal node to any other thermal node.
- Some conductors, especially those connecting non-neighbour thermal nodes, may be negative. Physical interpretation of these individual conductors may be suggested that heat is flowing from cold to warm. This is a built-in compensation mechanism that overall provides the correct physical representation of the conduction problem. Again, it is emphasised that the conductors must be used as a set of conductors. This set as a whole obeys the laws of physics.

Application of PAT-derived conductors

An Al-alloy strip, as shown in Fig. 9.17, is meshed with three thermal nodes 1, 2 and 3. The total length is $L = 1$ m, the width $b = 0.25$ m and the thickness $t = 10$ mm.

The strip is irradiated one-sided by a solar flux $Q_s = 1368$ W/m^2. The absorption coefficients α and emittance coefficients ε of the thermal node surfaces are given in Table 9.3.

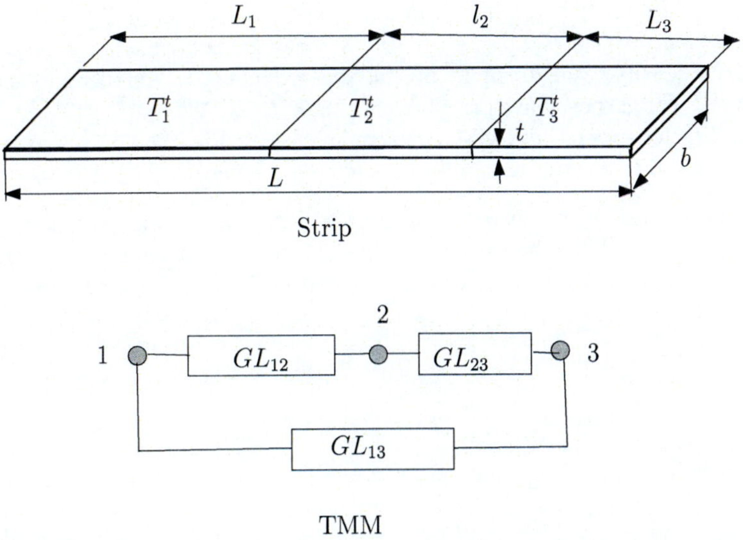

Fig. 9.17 Strip

Table 9.3 Physical properties strip

Thermal node	Length (m)	Area (m^2)	α	ε	Volume (m^3)	Q_{diss} (W)
1	$L_1 = 0.4$	$A_1 = L_1 b$	$\alpha_1 = 0.2$	$\varepsilon_1 = 0.2$	$V_1 = L_1 bt$	$Q_{diss,1} = 5$
2	$L_2 = 0.3$	$A_2 = L_2 b$	$\alpha_2 = 0.0$	$\varepsilon_2 = 0.8$	$V_2 = L_2 bt$	
3	$L_3 = 0.3$	$A_3 = L_3 b$	$\alpha_3 = 0.2$	$\varepsilon_3 = 0.1$	$V_3 = L_3 bt$	$Q_{diss,3} = 4$

Table 9.4 Radiators and capacitances

Thermal node	Radiation conductor ($W/m^2/K^4$)	Capacitance (J)
1	$GR_1 = A_1 * \varepsilon_1 \sigma (T_1 + T_{space})(T_1^2 + T_{space}^2)$	$C_1 = V_1 \rho c_p$
2	$GR_2 = A_2 * \varepsilon_2 \sigma (T_1 + T_{space})(T_1^2 + T_{space}^2)$	$C_2 = V_2 \rho c_p$
3	$GR_3 = A_3 * \varepsilon_3 \sigma (T_1 + T_{space})(T_1^2 + T_{space}^2)$	$C_3 = V_3 \rho c_p$

Table 9.5 Conductors

Thermal node	Length (m)	Area A (m^2)	GL (W/K)	GL (W/K) (PAT)
1–2	$L_{12} = 0.5(L_1 + L_2)$	$A_{12} = bt$	$GL_{12} = kA_{12}/L_{12}$	$GL_{12} = 2.9351$
2–3	$L_{23} = 0.5(L_2 + L_3)$	$A_{23} = bt$	$GL_{23} = kA_{23}/L_{23}$	$GL_{23} = 3.3296$
1–3				$GL_{13} = -0.59683$

The strip radiates one-sided heat into space with a temperature $T_{space} = 2.725$ K. The Boltzmann constant is $\sigma = 5.67 \times 10^{-8}$ $W/m^2/K^4$. The conductivity is $k = 205$ W/m/K, the density $\rho = 2700$ kg/m^3 and the specific heat $c_p = 900$ J/kg/K. Thermal nodes 1 and 3 are exposed to a dissipated heat flow Q_{diss} (W).

The characteristics of the TMM are given in Tables 9.3, 9.4 and 9.5.

Table 9.5 shows that two sets of linear conductors are considered as follows:

1. Linear conductors computed with the classical method with the expression kA/L. The conductor between node 1 and 3 is not considered ($GL_{13} = 0$).
2. Linear conductors computed with one of two PAT-based conductor generation methods.

The transient heat problem can be written as follows

$$[C]\left(\dot{T}(t)\right) + [K_c(t)]\left(T(t)\right) = \left(Q(t)\right), \tag{9.67}$$

or

$$\begin{bmatrix} C_1 & 0 & 0 \\ 0 & C_2 & 0 \\ 0 & 0 & C_3 \end{bmatrix} \begin{pmatrix} \dot{T}_1 \\ \dot{T}_2 \\ \dot{T}_3 \end{pmatrix} + \begin{bmatrix} GL_{12}+GL_{13}+GR_1 & -GL_{12} & -GL_{13} \\ -GL_{12} & GL_{12}+GL_{23}+GR_2 & -GL_{23} \\ -GL_{13} & -GL_{23} & GL_{23}+GL_{13}+GR_3 \end{bmatrix} \begin{pmatrix} T_1 \\ T_2 \\ T_3 \end{pmatrix} = \begin{pmatrix} A_1\alpha_1 Q_s + Q_{diss,1} \\ 0 \\ A_3\alpha_3 Q_s + Q_{diss,3} \end{pmatrix} + T_s \begin{pmatrix} GR_1 \\ GR_2 \\ GR_3 \end{pmatrix}. \tag{9.68}$$

Equation (9.67) will be numerically solved with the Culham and Varga scheme [95]

$$([C_n] + \Delta t[K_c])\,(T_{n+1}) = [C_n]\,(T_n) + \Delta t\,(Q_{n+1}), \tag{9.69}$$

where the matrices and vectors are evaluated at each time increment Δt. The initial temperatures are 300 K.

The transient problem is solved with the two models with the two different types of linear conductors. With both models, a total time of 200 h is simulated with the intention to obtain a steady-state solution after such a long time. For comparison, the steady-state solution of the problem is obtained with ESATAN-TMS for the model with PAT-based conductors. The results of these three analyses are presented in Table 9.6.

The effect of the different conductors can be observed for the results for this simple model. It must be noted that the approximation of the conductors based on the classical expression kA/L ignores the effect of the conduction of the two outer parts of the thermal nodes. The PAT method includes the full length of thermal nodes.

Problems

9.1 The conductor GL_A from the square surface to the interface $x/L,\ x = 0\ldots L$ of the thermal node A has to be calculated with the "Far Field" method. This problem is taken from [55]. The geometry and interface with adjacent thermal node are

Table 9.6 Evaluated thermal node temperatures

Thermal node	Solution 1 (°C)	Solution 2 (°C)	Solution 2 ESATAN (°C)
1	54.97	58.55	58.59
2	53.65	52.04	52.08
3	54.80	57.72	57.76

Fig. 9.18 Square with interface of length x/L

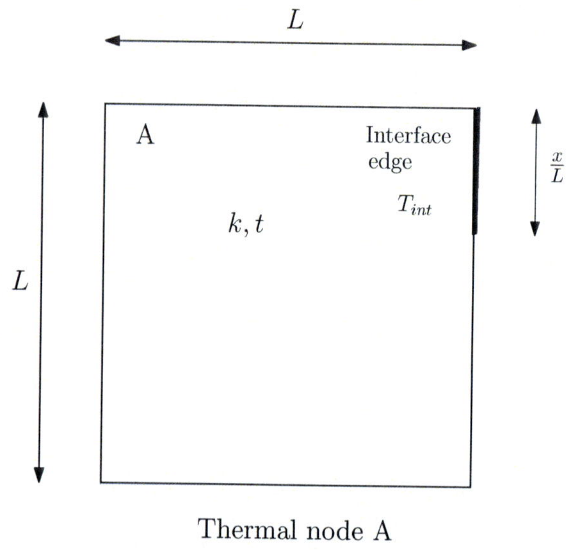

illustrated in Fig. 9.18. The material properties of the thermal node A are as follows: $L = 0.25$ m, the conductivity is $k = 200$ W/m°C, and the thickness is $t = 10$ mm. The interface temperature is set to $T_{int} = 20\,°C$. Perform the following assignments for $x/L = 0, 0.1, 0.2, \ldots 1$:

- Create a 2-D conduction FE model.
- Define "Far Field" points or edge.
- Calculate $GL_A(\frac{x}{L})$.
- Create a plot GL_A versus x/L.

9.2 Three volumes are coupled and shown in Fig. 9.19. The applied heat flows $Q_2 = Q_3 = 1$ W. The ratio $Ak/L = 1$. The temperature in volume 1 is $T_1 = 20\,°C$.

- Create a thermal model, and calculate the conductors GL_{12} and GL_{23}.
- Create a FE model, and calculate GL_{12} and GL_{23} using the method in Sect. 9.2.2.
- Generate the steady-state heat balance equation, and calculate the temperatures in the thermal nodes (volumes) 2 and 3.

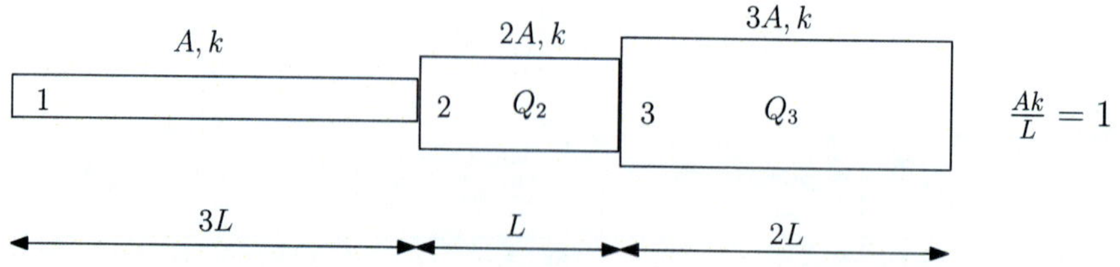

Fig. 9.19 Three coupled volumes 1, 2 and 3

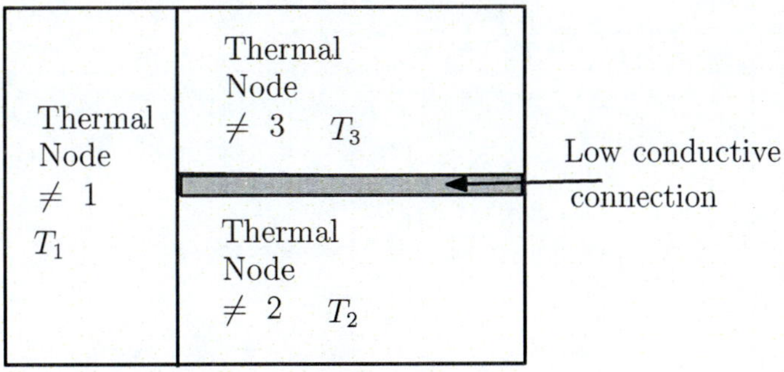

Fig. 9.20 Thermal mathematical model (TMM)

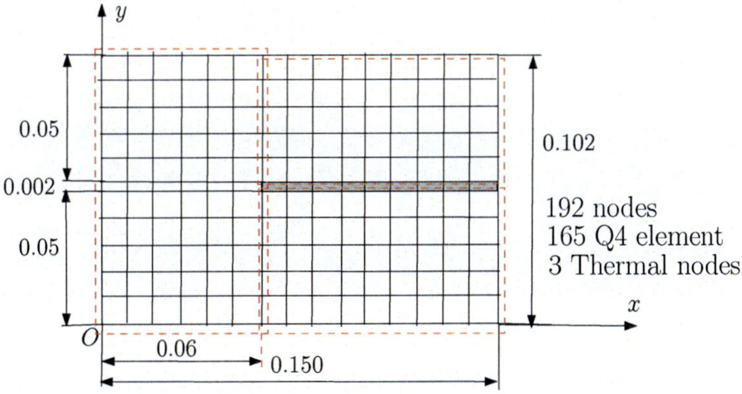

Fig. 9.21 FE conduction model

9.3 ESA/ESTEC presented on 7 February 2012 a simple PAT temperature interpolation problem, which will be numerically repeated here in this section. No dimensions and values of the conduction properties were provided, and therefore, these are assumed in order to perform the calculations. The thermal mathematical model (TMM) consists of three iso-thermal nodes 1, 2 and 3. The temperatures are T_1, T_2, T_3 °C. The TMM is shown in Fig. 9.20. This TMM completely covers the conduction finite element model, which is shown in Fig. 9.21. The TMM is shown with a red-dashed rectangular. The conduction finite element model consists of 165 iso-parametric Q4 elements and 192 nodes. The dimensions are given in meters. The thickness of the plate is $t = 1$ mm. The grey-coloured Q4 elements represent the low conduction gap of 2 mm. The iso-tropic conduction $k_{gap} = 1$ W/m/°C. The iso-tropic

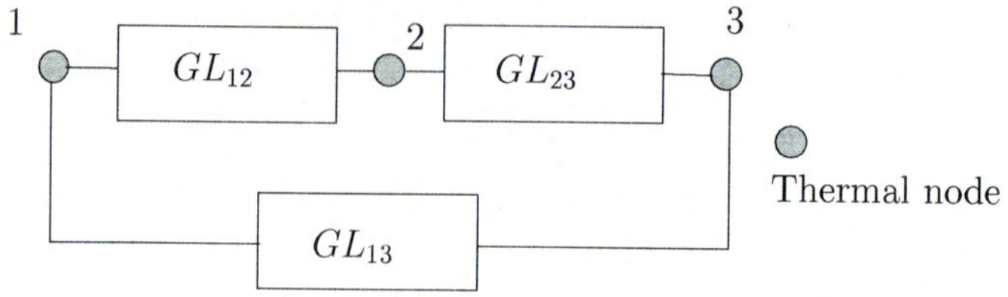

Fig. 9.22 Thermal network

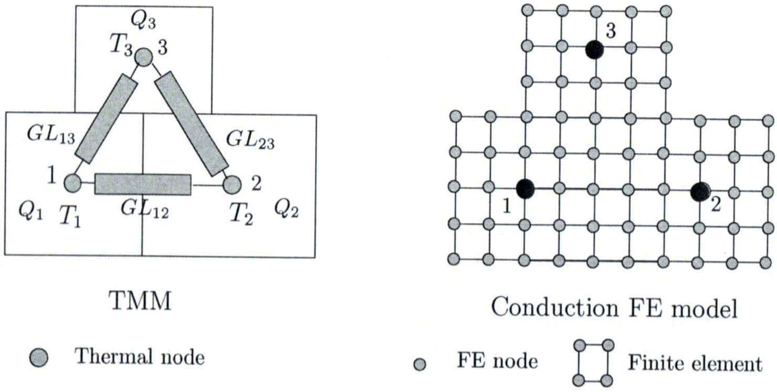

Fig. 9.23 TMM and conduction FE model

conduction for all other Q4 elements is $k = 237$ W/m/°C. Calculate the conductors GL between the thermal nodes 1, 2 and 3. The thermal network is shown in Fig. 9.22. The node numbers reflect the thermal nodes of the TMM.

9.4 Use the analysis results of Problem 8.1, and calculate the conductors GL between the thermal nodes.

9.5 Use the analysis results of Problem 8.2, calculate conductors GL between the thermal nodes 1, 2, 3, 4, 5 and 6, and draw thermal network.

9.6 In Fig. 9.23, a TMM with three thermal nodes and an associated conduction FE model are shown. The thermal nodes are connected with conductors GL_{12}, GL_{13} and GL_{23} (W/°C). The heat inputs at nodes 1, 2 and 3 are, respectively, Q_1, Q_2 and Q_3 (W). Perform following assignments:

- Set up the heat balance equations for the TMM.
- "Calculate" temperature T_1 applying $Q_1 = Q_0$ at FE node 1 and prescribe $T_2, T_3 = 0$. Q_2, Q_3 are then known too (why?). Calculate conductors GL_{12}, GL_{13} using the heat balance equations derived for the TMM.

Chapter 10
Estimating Uncertainties in the Thermoelastic Analysis Process

Abstract Instead of a single model for one discipline, thermoelastic analysis involves at least a thermal model and a structural model. In many cases, these two analysis steps are followed by a RF or optical simulation to determine the impact on the performance of the instrument of the thermoelastic responses. Each analysis step is accompanied with uncertainties. Besides the thermal and structural analyses, the transfer of temperature data introduces additional uncertainties. A method based on factors of safety, as is used in structural analysis to cover uncertainties, is discussed. Although the method can be pragmatic, this approach is often lacking a physical basis, may not be compatible with the requirements for specific cases, and the values used for these factors are to a large extent ambiguous. Stochastic approaches are known to be powerful for estimating ranges in the responses due to variation of parameters. Although the Monte Carlo method is well known and robust, it is computationally expensive. Promising alternative methods to quantify uncertainties of the responses, among many others, are the Latin hypercube sampling (LHS) method and the Rosenblueth $2k + 1$ Point Estimates Moments (PEM) method. The significance of the random design variables on the random output is detected through a sensitivity analysis, i.e. by a regression analysis method or as an intermediate result of the Rosenblueth PEM method. The section concludes with several worked out examples.

10.1 Uncertainties in the Thermoelastic Analysis Process

The thermoelastic analysis process is a multistep process of which the main steps are summarised in Fig. 4.1. Each of these steps introduces uncertainties and errors. In the following section, these errors and uncertainties are discussed.

S. Appel and J. Wijker, *Simulation of Thermoelastic Behaviour of Spacecraft Structures*, Springer Aerospace Technology, https://doi.org/10.1007/978-3-030-78999-2_10

10.1.1 Uncertainties from the Thermal Analysis

In Sect. 6 of the ECSS Thermal Analysis Handbook [35], a comprehensive overview of the possible sources of uncertainties for the results of a thermal analysis is given. Although the objectives of the thermal analysis for thermal control are different from those for thermoelastic, since both share the thermal analyses objective to compute temperatures, the principles and sources of uncertainties are therefore same. Typical sources of uncertainties are material and thermo-optical properties. Also, contact conductance values include a wide uncertainty range. Furthermore, variation of dimensions is one of the common uncertainties.

The thermal environment is also a source of uncertainties. This can for instance be caused by uncertainties in thermo-optical properties of objects around the instrument that is being verified. The properties of these other objects are under control of often other developers than those verifying the item under consideration. As a consequence, for instance, the predicted reflected solar and planetary heat radiation towards the instrument may have a certain level of uncertainty.

10.1.2 Uncertainties from the Temperature Mapping Process

The temperature mapping process is an other source of errors and uncertainties.

A first error source lays in the difference in the level of detail between the thermal and structural FE model. For instance is a bracket in the thermal model represented in such a way that that the thermal nodes can all be linked to the same parts in the structural model? Or the other way around: Is the thermal model able to provide temperature data for the complete bracket? Although good interaction between the thermal and structural engineers should prevent important differences in terms of level of detail and items to be modelled, it is not always easy to have a one to one correspondence between the models. In the space industry, the far from ideal case may occur in which the thermal analysis is subcontracted to a company different from the one doing the structural analysis. Incomplete correspondence could affect the thermoelastic response prediction. In all cases, close interaction between the engineers building the thermal and structural model is essential.

A second source comes from the fact that node boundaries in the thermal model and the element boundaries in the structural model do not coincide everywhere. Depending on the method, these situation may lead to ambiguous choices while defining the correspondence between thermal node and finite elements. Also, this problem can be minimised through good interaction between the thermal and structural engineers. The different temperature mapping methods (see also Chap. 7) deal in a different way with the non-congruent boundaries and have a different level of impact on the produced temperature field on the finite element model. An example of non-matching boundaries is shown in Fig. 10.1.

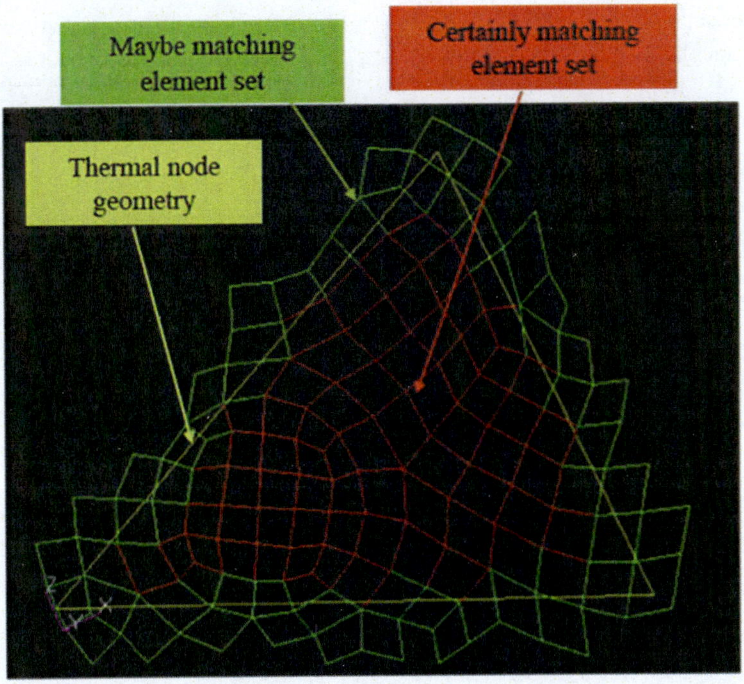

Fig. 10.1 Non-matching boundaries TMM and FE model

Besides the way of handling non-congruent mesh boundaries of the thermal and structural model, also the adequacy of the selected temperature mapping method plays an important role. For instance, the PAT method can produce, compared to the patch-wise temperature application method, with a lower thermal density the same quality mapped temperature field. As a consequence, thermal mesh convergence tests require less thermal nodes when the PAT method is used. In Chap. 7, different temperature mapping methods are compared for a limited number of cases. The differences in the temperature fields mapped on the structural finite element model can be significant. The selected temperature mapping method might therefore be an important contributor to the errors.

10.1.3 Uncertainties from the Thermoelastic Structural Response Analysis

In Chap. 6, various aspects of the FE modelling for thermoelastic analysis are described. With the suggestions in that chapter, the errors originating from the way of modelling can be reduced. On the other hand, there still remains the uncertainty of material properties and dimensions typically appearing as differences between the model and the as-build configuration.

10.1.4 Uncertainties from the Instrument Performance Impact Analysis

The final step of the full analysis chain is the evaluation of the impact of thermoelastic deformations on the performance of the instrument or spacecraft. This step is outside the scope of this book. Nevertheless, some simplified performance impact assessment is included in some of the examples.

Since this step is also based on models with their limitations and uncertainties. These models are relying on input data coming from the thermoelastic response analysis that have to be converted to the appropriate format and entities. This could be a source of errors and uncertainties on its own.

10.2 Use of Factors of Safety for Covering the Uncertainties

A way to deal with the above sources of uncertainties is to introduce factors similar to the ECSS "Structural factors of safety for spaceflight hardware" [34]. A factor of safety could be introduced for each uncertainty source.

Like the model factor K_m from the mentioned ECSS, the uncertainties in the thermal model can be covered with the thermal model factor K_{mt}. This factor will then act as a scaling factor for the thermoelastic responses, like stresses and displacements.

Similar to the thermal model factor, a thermal environment factor, K_{et}, may be introduced. This factor has some similarities with the project factor, K_p, from [34]. The structural project factor, K_p, is meant to cover the uncertainties in the mechanical loading. Similarly, the factor K_{et} could have this role in the thermal domain.

The uncertainty coming from the mapping process can be included likewise through the factor K_{map}. This factor is also applied to scale the thermoelastic responses (displacements and stresses) together with the other factors proposed in the this section.

In analogy with the factor K_m from the ECSS "Structural factors of safety for spaceflight hardware" [34], the uncertainty coming from the structural thermoelastic model can considered to be covered by the model thermoelastic structural model K_{ms}.

Combining the above factors, the uncertainty is then represented by the product of the factors

$$K_{total} = K_{mt} K_{et} K_{map} K_{ms} \qquad (10.1)$$

In this approach, the nominal displacements and nominal stresses, i.e. responses computed with the models based on the nominal model data, are then scaled with factor K_{total} to obtain the responses to be used for further evaluation and performance verification:

$$u = K_{total} u_{nominal} \tag{10.2}$$

$$\sigma = K_{total} \sigma_{nominal} \tag{10.3}$$

Depending on the type of performance, additional analysis might be required with for instance the thermoelastic deformation as input. This additional analysis may be an optical or RF simulation using the deformed RF or optical items of the instrument. This simulation on its own includes also uncertainties that might justify an additional uncertainty factor.

The ECSS [34] is suggesting for the factors K_m and K_p values that typically range from 1.1 to 1.3 depending on the maturity of the models and the phase of the project. Like with the factors K_p and K_m as proposed in the same ECSS [34], there is no physical basis for quantifying the previously introduced factors of safety and the selected values should then rely on experience, which is not always applicable to new types of instruments. Often the values of the factors K_p and K_m are the result of a compromise between customer and contractor. As a consequence, the values have a high level of ambiguity.

Since the quantified physical basis for the values of the factors is missing, it is also not certain if the intended conservatism that was foreseen to be introduced is sufficient to cover all the non-quantified uncertainties. So, despite that these factors are meant to introduce a level of conservatism in the results, they actually form an uncertainty on their own.

The concept of the factors of safety originates from the principle that computed responses (displacements, stresses, etc.) do not exceed a certain allowable maximum value. The application of the factors of safety assures that the computed responses is sufficiently below this allowable value. This concept does not work when the responses have to be higher than an allowable minimum value. Examples are lenses that must operate at low temperatures and are manufactured and assembled at room temperature. Expected thermoelastic deformation brings the lenses into a shape needed for the optical function. The factors of safety could be workable concept to assure that the thermoelastic deformation is not too high. The principles of factors of safety do not work to assure that thermoelastic lens deformation is sufficient, i.e. more than the minimum required, to have the right optical properties at the operational temperatures.

An other serious limitation of the factors of safety is that these simply scale the magnitude of a mode of response that is considered to be reliable. Thermoelastic problems have, however, shown that exactly this aspect, the way the structure was predicted to respond, showed the highest deviations from observations during test and flight.

Although the use factors of safety is a pragmatic approach, it clearly has some limitations and brings additional uncertainties. For those reasons, in the next section, stochastic methods are discussed to estimate the uncertainties in the analysis results as an alternative to the approach with factors of safety.

10.3 Uncertainty Assessment of Thermoelastic Analysis Using Probabilistic Analysis

Looking back at the modelling uncertainty factors K_{mt} and K_{ms} (see Sect. 10.2), it can be stated each of that these consist of two components. The first component is the uncertainty in the accuracy of the model in terms of discretisation and adequacy of level of detail of features of which the relevance is not known. The second component is the uncertainty on the impact on the performance of the instrument due to the variation of material and dimensional properties and manufacturing processes.

Let us assume that both the thermal model and structural model have successfully converged in terms of discretisation and all essential features are implemented in the models with sufficient detail. Also, the meshes of the thermal and structural model are nicely aligned such that no compromises are needed to be accepted for the temperature mapping process. Basically, with this state of the models, it can be assumed that the best possible models, in terms of discretisation, for the problem are available.

The modelling uncertainties that remain are the ones mentioned before, coming from the non-uniformity or variation in material and dimensional properties and applied manufacturing processes.

The question now is: What is the magnitude of the possible variation in performance of the instrument due do to all possible variation of the values of these properties? The uncertainty analysis aims to answer this question by determining the stochastic distribution in terms of mean and standard deviation of the output of analysis models.

In this chapter, three probabilistic methods are described that use the stochastic nature of the variation of the values of the physical properties. These methods are:

- Monte Carlo simulation (MCS).
- Modified MCS, Latin hypercube sampling (LHS).
- The Rosenblueth 2k + 1 Point Estimates Probability Moments Method.

For this purpose, the physical properties are identified as stochastic design variables. A stochastic design variable may be the conductivity of the material or the thickness of a plate. The plate is coming from a manufacturing process that is supposed to produce plates with a specific thickness within a certain tolerance. Assumed is that statistical analysis of several batches of these plates has learned that the actual thicknesses of these plates are rarely exactly the specified value, but the average thickness is very close to the specified value of the plates. The same statistical analysis provides the magnitude of variation around the specified nominal thickness in terms of standard deviation and type of distribution around the average.

So, for each stochastic design variable X, the following information is ideally known:

- Type of probability distribution $f_X(x)$, with $\int_{-\infty}^{\infty} f_X(x)dx = 1$.
- The mean value (average) μ_x, with the first probabilistic moment $\mu_x = \int_{-\infty}^{\infty} f_X(x)\, x dx$.
- The standard deviation σ_x, with $\sigma_x^2 = \int_{-\infty}^{\infty} f_X(x)(x - \mu_x)^2 dx = E(X^2) - \mu_x^2$, where $E(X^2) = \int_{-\infty}^{\infty} f_X(x)x^2 dx$ is the second probabilistic moment.
- The skewness $E(X - \mu_x)^3 = \int_{-\infty}^{\infty} f_X(x)(x - \mu_x)^3 dx$.

Stochastic design variables may have different distribution types, such as uniform, normal and log-normal.

In many cases, not all of information of the distribution of the design variables are known. A simplified way forward is to assume an uniform distribution $f_X(x) = \mathcal{U}(a, b)$ of the stochastic variable X with assumed or justified lower and upper bound values, where a is denoted to be the lower bound and b the upper bound of the stochastic variable. Giving:

- the mean value (average): $\mu_x = \frac{a+b}{2}$
- standard deviation: $\sigma_x = \sqrt{\frac{(b-a)^2}{12}}$
- the skewness $E(X - \mu_x)^3 = 0$.

When the stochastic variables are well defined, the stochastic analysis process can be applied. A generic visualisation of this analysis process is presented in Fig. 10.2.

Within the large block, the four analysis steps are joined into a single analysis chain. This chain of analysis steps is basically the same as was presented in Fig. 4.1.

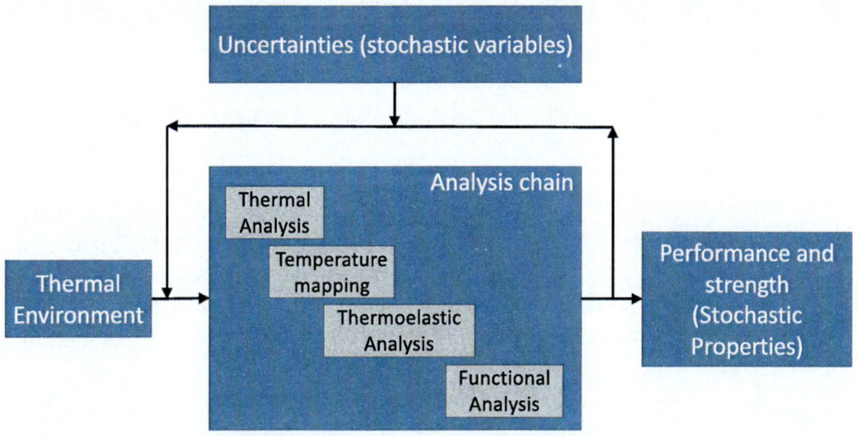

Fig. 10.2 General analysis process for stochastic analyses

The models that are the basis for this chain are the ones that have successfully been converged. The main inputs to the analysis chain are the uncertainties represented by stochastic variables and the thermal environment. In principle, the thermal environment also may include uncertainties. For simplicity reasons, the elements of the thermal environment that is also subject to uncertainty is considered to be included in additional stochastic variables representing the uncertainties in the thermal environment.

The loop back to the beginning of the analysis chain intends to indicate that the analysis chain has to be repeated. The number of times that this has to be repeated depends on the method that is used to produce the stochastic properties of the performance: the mean and standard deviation of each performance parameter.

At the end of this chapter, a number of examples are included showing the application of the stochastic methods.

10.4 Monte Carlo Simulation Method

To assess the uncertainty with the Monte Carlo simulation (MCS) method [78], a large number of simulation runs have to be done, varying the design variables stochastically. In accordance with [6], the following steps are performed:

1. Definition of the mathematical model(s).
2. Generation of memoryless random numbers for producing random values for the design variables.
3. Evaluation of stochastic responses with random values for the design variables.
4. Statistical analysis.

Steps 2 and 3 have to be repeated till convergence of the stochastic properties of the response has been reached.

Note that for the generation of random numbers, it is not verified whether a certain combination of design variable values has been used before. That is why the generation of random numbers is called memoryless.

The MCS process is illustrated in Fig. 10.3, where the stochastic variables x_i, $i = 1, 2 \ldots, n$ may have different probability distributions, but for simplicity, most times a uniform distribution will be assumed.

Assume that X represents a design variable like the Young's modulus E, the CTE α or the conductivity k. Let us now be assume as well that X has an uniform distribution with X_{min} and X_{max} being the minimum and maximum values of the design variable. With r_X being a random number between 0 and 1, the value for design variable X used in one response evaluation of with the mathematical model(s) can then be computed with:

$$X = X_{min} + r_X (X_{max} - X_{min}) \tag{10.4}$$

Many programming languages and numerical systems like MATLAB support the generation of random numbers based on a large variety of stochastic distribution functions. When the design variable is not uniformly distributed, then Eq. (10.4)

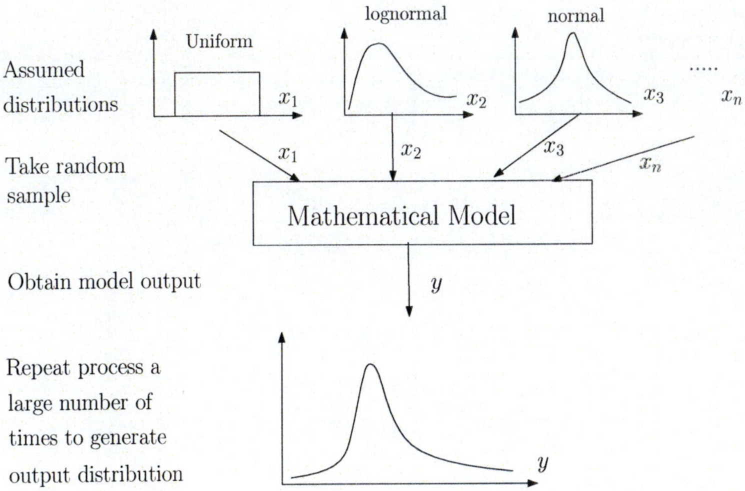

Fig. 10.3 Graphical depiction of the Monte Carlo simulation procedure

has to be replaced with the appropriate equivalent for the corresponding applicable distribution function. For each design variable, a random value has to be generated. All these random values of the design variables form the input to produce a single sample of the response variables. Every analysis loop produces a new sample for each response variable.

The mean, standard deviation, maximum and minimum values of the stochastic response variable(s) can be determined from the collection of samples.

The accuracy of the numerical solution obtained through MCS will improve with the number of response samples [6].

10.5 Modified MCS, Latin Hypercube Sampling Method

A possible alternative method is a variation on the MCS method, which is a type of stratified MCS known as Latin hypercube sampling (LHS) [47, 97].

LHS operates in the following manner to generate k samples from the n input design variables $X_1, X_2, \ldots, X_n$. The range of each random variable is divided into k non-overlapping intervals on the basis of equal probability $1/k$. One value of each interval is selected either at random or median or at mean with respect to the probability density in the interval. In the case a random value in the interval is chosen, the sample value $x_{i,j}$ in interval j for design variable X_i can be obtained by

$$x_{i,j} = F_{X_i}^{-1}\left(\frac{U + (j-1)}{k}\right), \quad j = 1, 2, \ldots, k, \tag{10.5}$$

Fig. 10.4 Selected values x_i from each interval

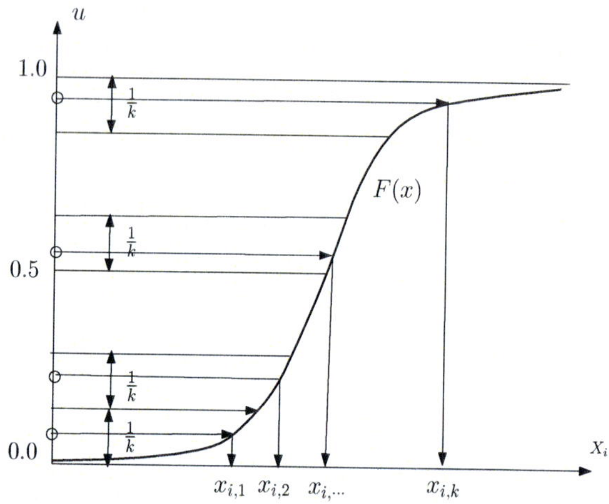

where U is a uniform distributed randomly generated number from the interval (0, 1), k is the number of intervals, and $F_{X_i}^{-1}$ is the inverse cumulative distribution function for the input variable X_i. This is illustrated in Fig. 10.4.

The k values obtained for X_1 are paired in a random manner with the k values of X_2. These k pairs are combined with the k values of X_3 to form triplets, and so on, until a set of k tuples with each n values is formed. This set of k tuples of length n is the Latin hypercube sample. The essence of the Latin hypercube is that in these combinations of the values of design variables each value $x_{i,j}$ of design variable i for interval j is only used once.

The stochastic properties of the response function are then determined from the k responses computed with the k tuples of n values of the design variables.

For given values of k and n, there exist in principle $(k!)^{n-1}$ possible independent interval combinations [47] for the design space that the Latin hypercube is spanning. With this number of combinations, the complete design space would be covered. The LHS method only evaluates k of these combinations.

An application is illustrated on a case with $n = 4$ random variables X_1, X_2, X_3, X_4, and the cumulative density function ($F(x)$ is divided in $k = 5$ intervals). Only 5 intervals (layers) are used here for simplicity. In this case, the output Y will be calculated for five combinations of four input variables, instead of $((k!)^{n-1} = 1.728 \times 10^6)$ combinations. X_1, X_2 is normal distributed, and the input values variables X_3, X_4 are uniformly distributed. Specific data of the stochastic inputs are shown in Table 10.1.

The randomly selected combinations of the 20 random inputs are shown in the left part of Table 10.2, and the right part of Table 10.2 shows the combinations (permutation and positions) of the input variables to calculate $Y = f(X_1, X_2, X_3, X_4)$. This pairing is showing the Latin hypercube method principle: every value $x_{i,j}$ of design variable i for interval j is only used once.

MCS versus LHS

The nonlinear coupled model

$$Y = 2X_1 + 4X_2 + 0.5X_1^2 + X_1X_2, \tag{10.6}$$

is taken from [68], where the variables $X_1,\ X_2$ are assumed to be independent and standard normally distributed ($\mathcal{N}(0, 1)$). The theoretical mean $y_{mean} = 0.5000$, and the standard deviation $\sigma_y = 4.6368$. These values were calculated with the wxMaxima computer algebra system. The nonlinear model will be numerically evaluated both MCS and with the LHS methods with the purpose to investigate the rate of convergence to the theoretical solutions. The random results of the calculation with both the MCS and LHS methods is provided in Table 10.3. The Latin hypercube samples are calculated with the MATLAB script "lhsgeneral.m".[1] Analysing the results in Table 10.3, it can be concluded that the LHS approach converges very quickly to the theoretical results.

It is of interest to see how the MCS and LHS random samples are distributed over the $X_1,\ X_2$ plane. Let us take 100 pairs $(X_1,\ X_2)$. This distribution of the pairs in the $X_1,\ X_2$ plane is shown in Fig. 10.5. The LHS method sampled pairs are better distributed over the $X_1,\ X_2$ plane compared to the MCS method sampled pairs. The better the distribution, the faster the convergence of the response variable Y to a more

Table 10.1 Distributions stochastic parameters

Parameter	Distribution	μ_x	σ_x	X_{min}	X_{max}	Notation
X_1	Normal	30	2			$\mathcal{N}(30, 2)$
X_2	Normal	35	1			$\mathcal{N}(35, 1)$
X_3	Uniform			25	30	$\mathcal{U}(25, 30)$
X_4	Uniform			45	50	$\mathcal{U}(45, 50)$

Table 10.2 Random inputs and permutations

	Input variables				Permutations				
$k = 1-5$	X_1	X_2	X_3	X_4	Pair	X_1	X_2	X_3	X_4
1	26.2754	33.1377	25.1564	46.8555	1	3	3	1	2
2	29.3421	34.6711	26.8555	47.6448	2	5	4	5	1
3	30.1453	35.0726	27.6448	47.6448	3	1	1	4	5
4	30.9089	35.4545	28.3763	48.3763	4	2	2	2	4
5	31.9740	35.9870	29.1909	49.1909	5	5	4	5	3

[1] "lhsgeneral.m" was developed by Iman Moazzen, 2060 Project, IESVic, University of Victoria, BC, Canada.

Table 10.3 Simulation random results MCS versus LHS

	MCS		LHS	
# Pairs (X_1, X_2)	μ_y	σ_y	μ_y	σ_y
10	−0.6899	4.7694	0.4399	3.4798
100	0.1028	4.3926	0.4937	4.8016
1000	0.5830	4.8375	0.4993	4.6517
10,000	0.5368	4.6193	0.4999	4.6262
100.000	0.4964	4.6215	0.5000	4.6380
1000.000	0.5000	4.6343	0.5000	4.6376
10,000,000	0.5004	4.6360	0.5000	4.6366

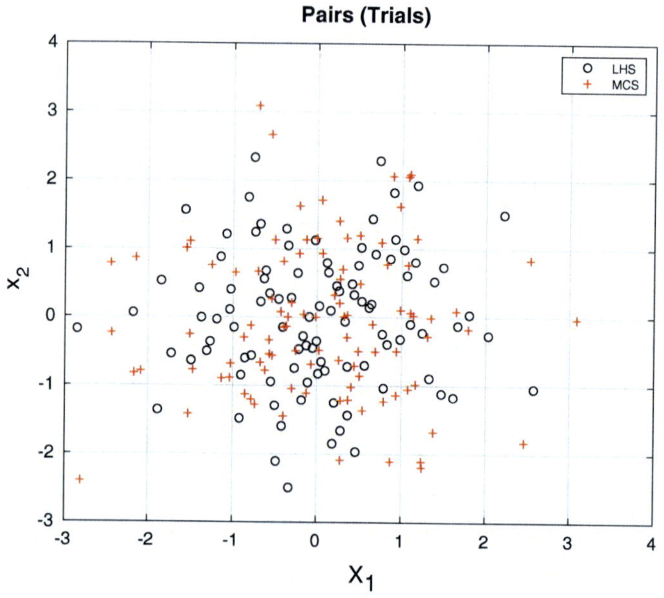

Fig. 10.5 Distribution 100 MCS LHS pairs $(X_1,\ X_2)$

accurate result. Thanks to the enforced distribution by splitting the design variable range into k non-overlapping intervals on the basis of equal probability $1/k$, the convergence is faster with less computational effort.

10.6 The Rosenblueth $2k + 1$ Point Estimates Probability Moment Method

With the use of the Rosenblueth point estimates for first- and second-order probability moments [72, 77], estimates of the mean and the variance of the stochastic response variable in combination with a mathematical model can be computed. The interesting feature of this method is that, when the number of stochastic design variables is k, the number of samples, i.e. results of analysis cases, to be produced for computation of the estimates is limited to $2k + 1$.

Considered is the response function $Y(X_1, X_2, \ldots, X_k)$ with X_i being stochastic design variables. Each design variable X_n has a mean value μ_n and a standard deviation σ_n.

These $2k + 1$ samples are the following:

- $Y_0 = Y(\mu_1, \mu_2, \ldots, \mu_k)$: The nominal or reference value of the response computed by using the mean values μ for all k stochastic design variables.
- $Y_{nm} = Y(\mu_1, \mu_2, \ldots, \mu_n - \sigma_n, \ldots, \mu_k)$: The response resulting from the analysis in which all values of the response variables are kept at their mean value μ , except the nth stochastic design variable that is set to the value $\mu_n - \sigma_n$ (m means minus).
- $Y_{np} = Y(\mu_1, \mu_2, \ldots, \mu_n + \sigma_n, \ldots, \mu_k)$: The response resulting from the analysis in which all values of the response variables are kept at their mean value μ, except the nth stochastic design variable that is set to the value $\mu_n + \sigma_n$ (p means plus).

The mean Y_n of two point estimates Y_{nm} and Y_{np} is given by

$$Y_n = \frac{Y_{np} + Y_{nm}}{2}, \quad n = 1, 2, \ldots, k, \tag{10.7}$$

and the coefficient of variation V_n can be obtained by

$$V_n = \frac{Y_{np} - Y_{nm}}{Y_{np} + Y_{nm}}, \quad n = 1, 2, \ldots, k \tag{10.8}$$

Assuming that all stochastic design variables are uncorrelated and have a non-skew distribution, the following approximation of the mean $\bar{Y} = \mu_Y$ and the coefficient of variation $V_Y = \sigma_Y / \mu_Y$ can be calculated with

$$\frac{\bar{Y}}{Y_0} = \prod_{n=1}^{k} \frac{Y_n}{Y_0} \tag{10.9}$$

and

$$1 + V_Y^2 = \prod_{n=1}^{k} (1 + V_n^2). \tag{10.10}$$

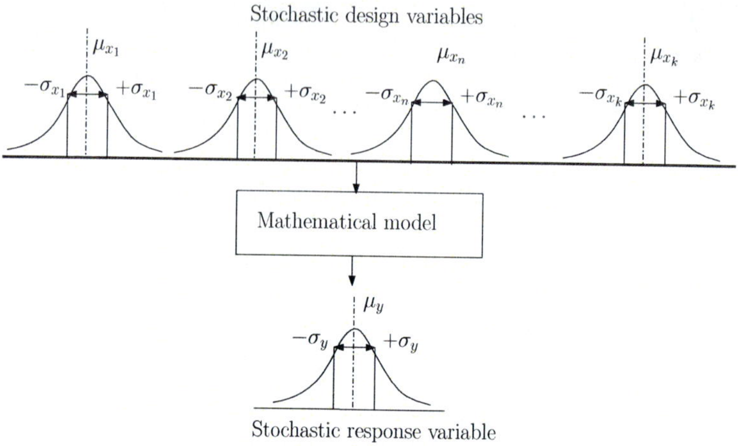

Fig. 10.6 Schematic illustration of the point estimates method

The values of V_n of Eq. (10.8) can be used as a sensitivity value of the design variables on the response variable(s). When all the coefficients of variation are normalised relative to the coefficient of variation for the response value, the individual sensitivity index S_n values can be obtained

$$S_n = \frac{V_n}{V_Y}, \quad n = 1, 2, \ldots, k \tag{10.11}$$

The sensitivity index values can provide useful information in identifying the parameters driving the uncertainty. The schematic of the Rosenblueth point estimate method is illustrated in Fig. 10.6. The probability density functions in Fig. 10.6 are displayed as if these have all normal distributions; however, other distributions like the uniform distribution may be applied as well. The mean and standard deviation of the stochastic design variables are the only input to the Rosenblueth $2k + 1$ PEM method.

It should be noted here that the method is based on a truncated Taylor series. For this reason, Rosenblueth stressed several times in his paper [77] that the dispersion of stochastic design variables should not be "too large". Rosenblueth is not specific about the quantification of "too large". This constraint may indicate that the accuracy has a sensitivity for the magnitude of the dispersion of the stochastic design variable.

In some of the following examples, the effect of the size of dispersion of the stochastic design variables is investigated.

Sandwich cantilevered beam

A cantilevered sandwich beam is shown in Fig. 10.7 as well as the design variables. The temperatures ΔT_1 and ΔT_2 are constant over the length of the beam and constant over the thickness of the face sheets. The core has a very low Young's modulus in

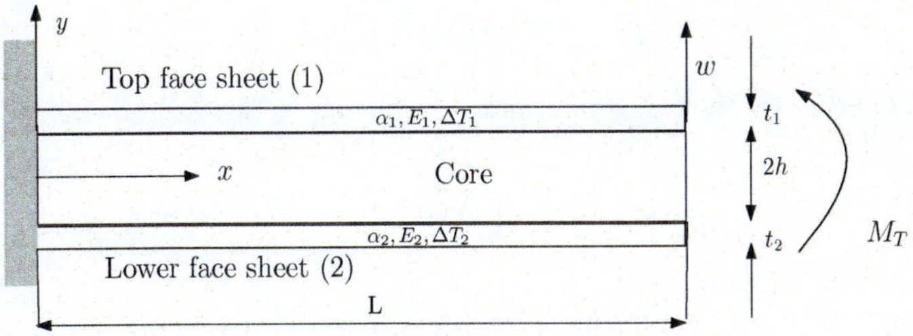

Fig. 10.7 Sandwich beam

Table 10.4 Stochastic design variables

Design variables	Dimension	Nominal value	Mean μ	Standard deviation σ
α_1	$m/m/°C$	$24 \times 10^{-6} \pm 5\%$	24×10^{-6}	6.9282×10^{-7}
α_2	$m/m/°C$	$24 \times 10^{-6} \pm 5\%$	24×10^{-6}	6.9282×10^{-7}
ΔT_1	°C	$50 \pm 2\%$	50	0.57735
ΔT_2	°C	$20 \pm 2\%$	20	0.23094
t_1	mm	$1 \pm 10\%$	1	0.057735
t_2	mm	$1.2 \pm 10\%$	1.2	0.069282
E_1	GPa	$70 \pm 1\%$	70	0.40415
E_2	GPa	$70 \pm 1\%$	70	0.40415
L	m	$0.5 \pm 1\%$	0.5	0.0028868
h	mm	$10 \pm 1\%$	10	0.057735

length direction, but this property is very large in the direction normal to the two face sheets. Shear deformation is ignored.

The numerical values of the considered stochastic design variables are listed in Table 10.4 together with their dispersion level (i.e. percentage level of the mean value). All design variables are assumed to have a uniform distribution. Note that for the purpose of this example also the temperature increase at the two face sheets ΔT_1 and ΔT_2 is included in the design variables. This could be considered as representing the variation in the temperature field produced by an imaginary thermal analysis.

The deflection w is calculated applying the classical beam bending theory with the Myosotis method [29]. The equivalent thermal bending moment is calculated using the following expression [19]

$$M_T = -\int_A \alpha E \Delta T y dy = -bh[\alpha_1 E_1 \Delta T_1 t_1 - \alpha_2 E_2 \Delta T_2 t_2], \tag{10.12}$$

assuming $t_1, t_2 \ll h$ and the width of the beam is b (m).

When ignoring the stiffness of the core, the equivalent bending stiffness EI_{eq} of the sandwich beam is

$$EI_{eq} = bh^2[E_1t_1 + E_2t_2]. \tag{10.13}$$

The deflection w at the tip of the sandwich beam can be calculated with

$$w = \frac{M_T L^2}{2EI_{eq}} = -\left(\frac{L^2}{2h}\right)\frac{\alpha_1 E_1 \Delta T_1 t_1 - \alpha_2 E_2 \Delta T_2 t_2}{E_1 t_1 + E_2 t_2}. \tag{10.14}$$

The deflection w is a stochastic response variable of which the mean μ and the standard deviation σ will be calculated with:

- The Rosenblueth $2k + 1$ PEM, $k = 10$ [72].
- The MCS method with 1,000,000 samples [78].

And, the results are reported in Table 10.5.

Both methods obtain very similar statistical values for the response variable w.

So far are in this example reasonably small dispersion levels used for the design variables. In the conclusions of the paper of Rosenblueth [77] is stated: "In this paper an approximate method is proposed, which is very simple and only sacrifices slightly provided that the dispersions of the variables are not too large". In the remaining of this example, the sensitivity of the stochastic properties of the response variable w with respect to the dispersion level of the design variables is investigated.

The design variables as listed in Table 10.4 are used again, but this time, the dispersion level will be for all stochastic design, variables the same and will vary simultaneously from 0% to ±25%. A percentage up to ±25% is of course very extreme.

The MCS analysis results are used again as a reference for the results with the Rosenblueth $2k + 1$ PEM method. The results obtained with both methods for the mean μ_w and standard deviation σ_w of the tip-displacement w are shown in Figs. 10.8 and 10.9, respectively.

The two figures show differences that are mostly noticeable beyond a dispersion level of 10% and especially for the standard deviation σ_w. Considering the reduced amount in computational effort for the Rosenblueth $2k + 1$ PEM method, the differences in results are quite limited.

Numerical example: PANELSAT

Table 10.5 Statistical values of deflection w

Method		Mean w_μ (m)	Standard deviation w_σ (m)
PEM	$k = 10$	−0.0035472	4.91910×10^{-4}
MCS	1,000,000 samples	−0.0035478	4.90326×10^{-4}

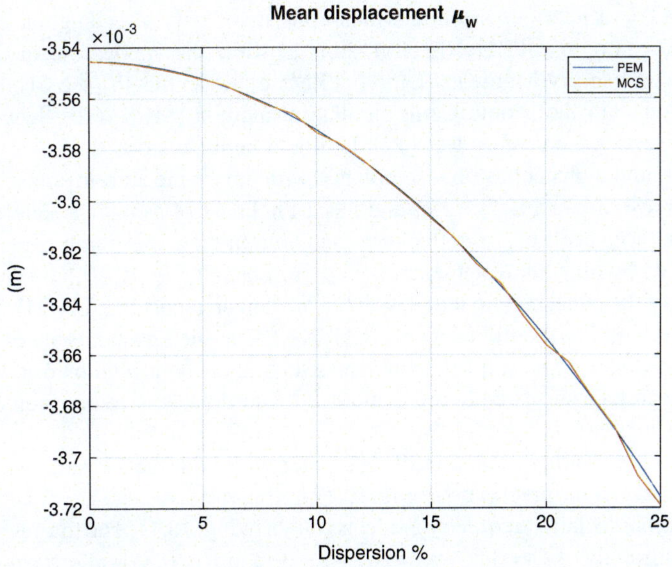

Fig. 10.8 Mean displacement w

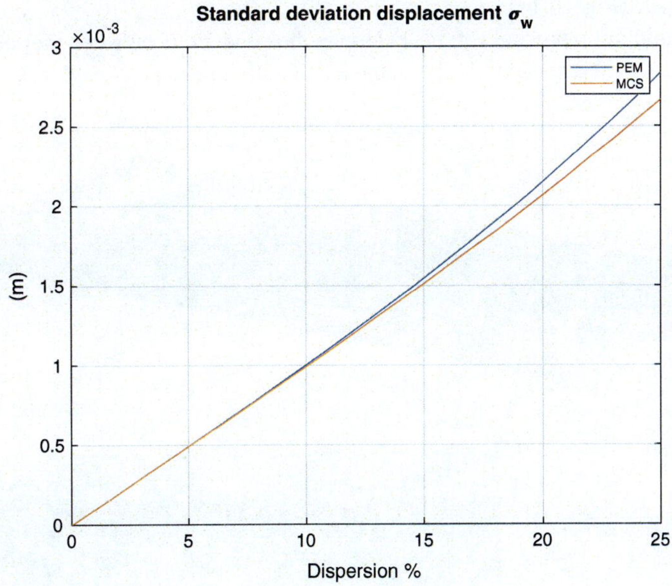

Fig. 10.9 Standard deviation displacement w

PANELSAT is a fictitious satellite which consists of only a rectangular sandwich panel. The example with PANELSAT aims to show the application of the Point Estimates Probability Moment Method (PEM) of Rosenblueth (see Sect. 10.6) in combination with the Monte Carlo simulation method (MCS) (see Sect. 10.4) of which the results are used as Ref. [12]. It should be understood that a single panel orbiting around a planet has no resemblance with any of the current spacecraft.

The sandwich panel is 1×2 m and has a thickness of 5 cm. The sandwich has aluminium face sheets of 1 mm thickness and aluminium honeycomb core material.

The MSC Nastran finite element model is presented in Fig. 10.10. The face sheets and edges of the panel are modelled with linear shell elements (CQUAD4). The core is modelled with linear solid elements (CHEXA). The shell elements are coinciding with the free faces of the solid elements representing the honeycomb core. A perfect joint between face sheets and core is assumed, and the adhesive between core and face sheets is omitted.

The panel is orbiting around Earth in a polar orbit with the +Y side of the panel pointing in the nadir direction (see Fig. 10.12).

The thermal radiative geometry model is shown in Fig. 10.11. The figures show the infrared emissivity of $\varepsilon = 0.5$ (−) for the +Y side and $\varepsilon = 0.8$ (−) for the remaining surfaces. The solar absorptivity for the +Y side is $\alpha = 0.5$ (−) and $\alpha = 0.2$ (−) for the remaining surfaces. The edges have the same thermo-optical properties as the zenith surface (Fig. 10.11).

In the simulation process for this example, the conductors between thermal nodes are computed with the method based on the PAT relations as explained in Sect. 9.3.2.

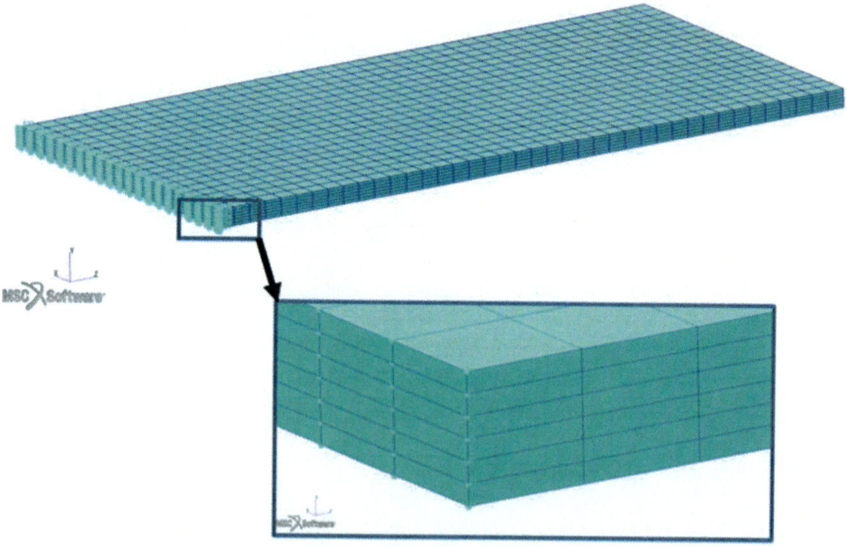

Fig. 10.10 PANELSAT FE model with boundary conditions indicated. Lower part: detail showing the high through thickness mesh resolution

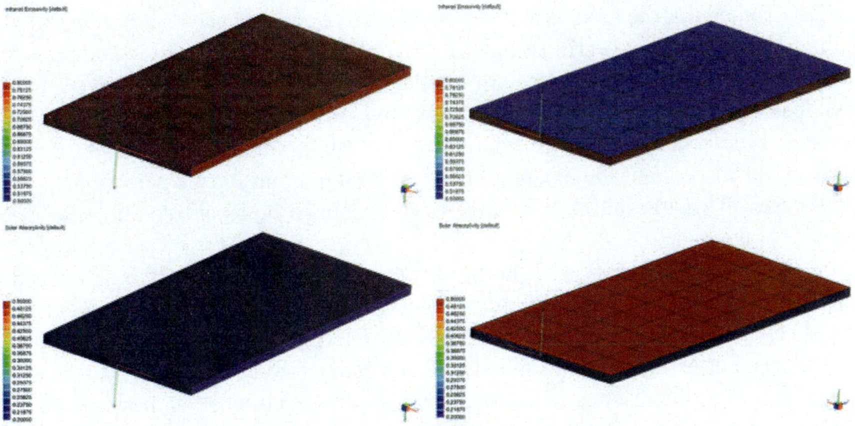

Fig. 10.11 PANELSAT radiative model with infrared emissivity and solar absorptivity indicated

Fig. 10.12 PANELSAT orbit around Earth with +Y axis nadir pointing

It is assumed that the models that are used for this example have gone successfully through the convergence verification (see Chap. 4).

As a result of the changing orientation to the sun in combination with entry and exit of eclipse, the temperature field is constantly changing with time with corresponding changes in deformation. The translation of the mid-point of the +Z edge of the panel is (for the sake of the example) considered important for the performance of PANELSAT. This translation is relative to the clamped boundary condition at the $-Z$ side of the panel.

In Fig. 10.13, the variation of the tip-displacement over two orbits is visualised together with the temperature evolution at the middle of the panel at the two opposite face sheets at corresponding moments in time. The figure shows strong variations in displacement as function of time. The displacements vary between −0.195 and −1.25 mm. The steep slopes in the displacement curves coincide with moments of increased temperature difference between top and bottom skin that occur during entry and exit of eclipse. In Fig. 10.13, the coinciding of the moments of strong variation of the tip-displacements and increased temperature difference between top (nadir) and bottom (zenith) face sheet is indicated with dotted lines.

The tip-displacement evolution shows extreme deep and sharp notches. Convergence checks on the time step size have been executed. Nevertheless, these notches are numerically not convenient.

The displacement field and the corresponding temperature field for one of the moments with highest deformation of the panel are presented in Figs. 10.14 and 10.15.

For many instruments, the peak-to-peak variation of the deformation is of interest, because it provides the range of deformation. In this example, the maximum and minimum tip-displacement are taken for that reason as performance results and the influence of stochastic design variables on the extreme displacement values is assessed. Therefore, in the following, the maximum and minimum values of the displacement over the transients are presented as responses of interest. Note that in this case the maximum value represents the smallest negative displacement and minimum the highest negative displacement. The minimum and maximum values are obtained by scanning over the transient data and searching for the highest and lowest value of the Y-displacement of the mid-point at the +Z edge of the panel.

In this example, the following model properties are taken as stochastic design variables:

- t_{nadir} and t_{zenith}: It is the face sheet thicknesses of the nadir and zenith side of the panel. Both model properties are considered as independent stochastic design variables. These parameters affect the stiffness of the panel as well as the face sheet conductance and heat capacity.
- k_{core}: It is the conductivity of the core material, which affects both the through thickness and the in-plane conductance.
- ε_{nadir} and ε_{zenith}: It is the infra-red emissivity of the two sides of the panel. Also, these parameters are considered independent stochastic design variables. The parameters affect the infra-red radiative heat exchange with the planet and deep

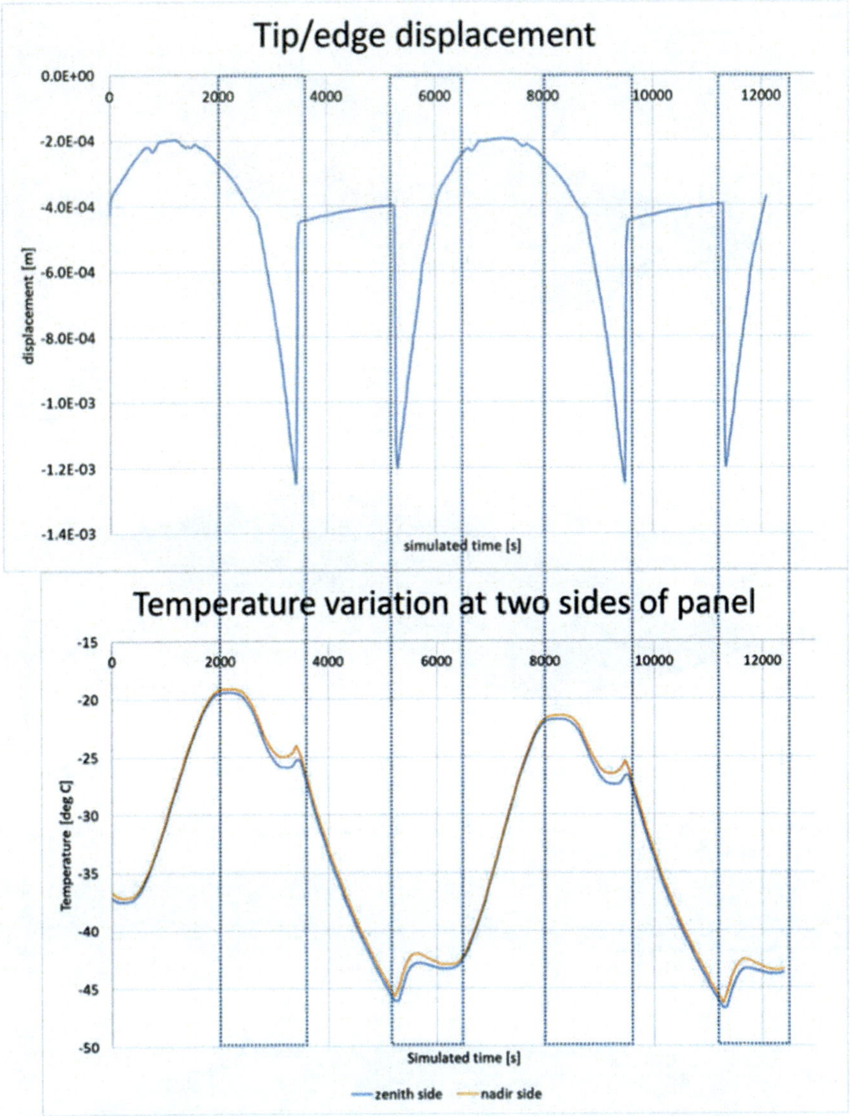

Fig. 10.13 As a function of time the PANELSAT tip edge displacement together with the corresponding varying temperature field at the middle of the panel at opposite sides

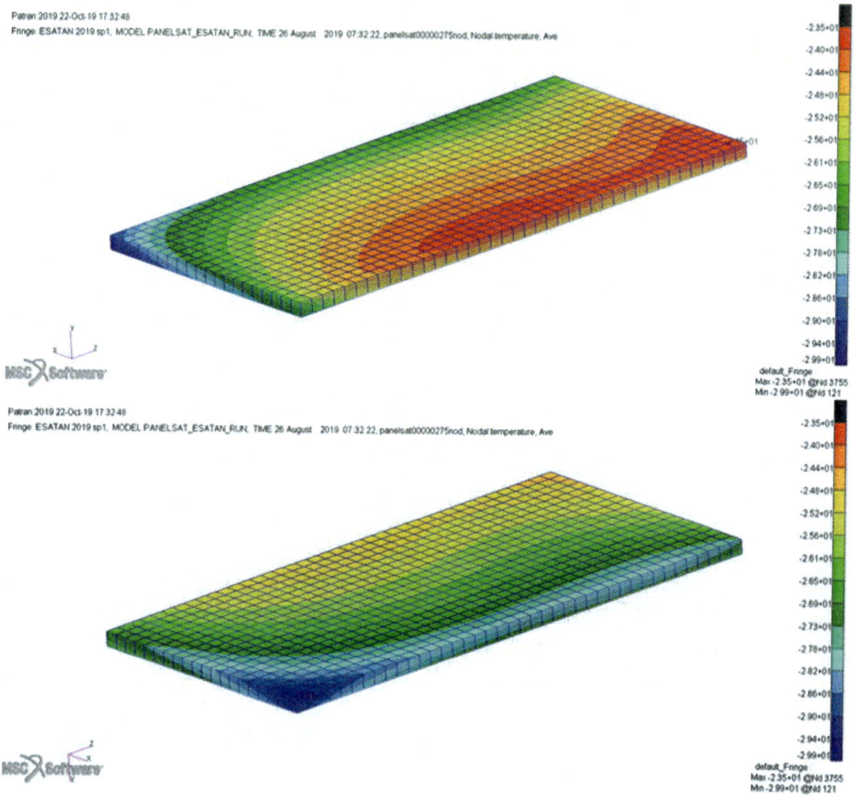

Fig. 10.14 Temperature field at nadir side (top picture) and zenith side (bottom picture) of panel at the moment of the highest deformation

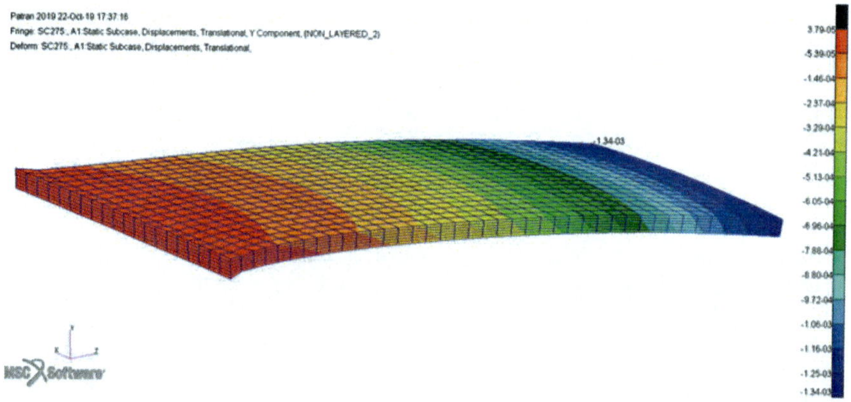

Fig. 10.15 PANELSAT thermally induced displacements

space. It must be noted that when these parameters change, the corresponding infra-red reflectivity has to be changed accordingly.

- α_{nadir} and α_{zenith}: It is the solar absorptivity of the two sides of the panel. Also, these parameters are considered independent stochastic design variables. The parameters affect the absorption of direct solar and albedo radiation. Also, here the corresponding solar reflectivity has to be changed when these parameters change.
- CTE_{nadir} and CTE_{zenith}: It is the coefficients of thermal expansion of the nadir and zenith face sheet material. These are considered to be two independent design variables. One could consider this situation to be relevant when the material of the two face sheets is coming from two different manufacturing batches or when the material appears to be not homogeneous.

The fact that the CTE_{nadir} and CTE_{zenith} for the two face sheets are considered as independent variables implies that both variables can be in their opposite extremes (one at the smallest and one at its highest value in the variation range). This might lead to high ranges in deformation responses.

The above list of design variables means that in total, nine stochastic design variables are used. All the other material and geometrical properties are kept at their nominal level.

In Table 10.6, nominal or mean values of the stochastic design variables, that are used for the PANELSAT example, are presented.

To find out what the impact is on the minimum and maximum tip-displacement due to random variation of the above-listed stochastic design variables, both Monte Carlo is used and the $2k+1$ PEM method.

As is explained in Sect. 10.6 care must be taken with selection of the size of the dispersion of the stochastic design variables when the $2k+1$ PEM method is used. For this reason, different values of the dispersion of the stochastic design variables have been investigated.

All the stochastic design variables are assumed to have a uniform probability distribution around the mean value that is considered to be equal to the nominal

Table 10.6 Nominal values of stochastic design variables

Variable	Nominal/mean value
t_{nadir}	0.001 m
t_{zenith}	0.001 m
k_{core}	8.5 W/m K
ε_{nadir}	0.5
ε_{zenith}	0.8
α_{nadir}	0.5
α_{zenith}	0.2
CTE_{nadir}	2.32E−5 m/m K
CTE_{zenith}	2.32E−5 m/m K

value. In this example, the dispersion is expressed as a percentage of the mean value. So, when a dispersion of 10% is indicated, it means that the lower bound is at 90% of the mean values and upper bound is at 110% of the mean value.

The percentages of dispersion that are used are 10, 5, 2.5 and 1.25% and are applied to all design variables simultaneously.

For each of these dispersion values, the $2k + 1$ PEM method is applied. The Monte Carlo simulation was run in parallel to act as a reference value for the estimates produced by the $2k + 1$ PEM method of the computed stochastic properties of the two responses: the minimum and maximum edge displacement. For the Monte Carlo simulations for each magnitude of dispersion 1800 analyses were executed. For each analysis all nine stochastic variables were given a random value within the dispersion range. With both the Monte Carlo and the $2k + 1$ PEM runs, the mean and standard deviation of the minimum and maximum tip-displacement are estimated and presented for the different dispersion levels in Figs. 10.16 and 10.17.

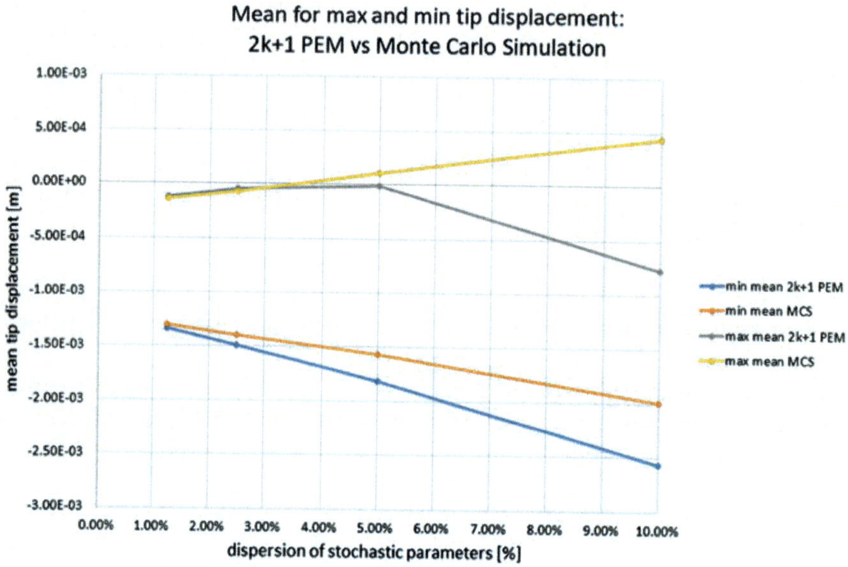

Fig. 10.16 PANELSAT correlation of the mean for maximum and minimum tip-displacement (with among others two independent CTE design variables)

Table 10.7 Mean and standard deviation for minimum and maximum tip-displacement for dispersion of 10% with CTE parameters comparing the 2k + 1 PEM method and Monte Carlo (MCS)

	min $2k + 1$	min MCS	max $2k + 1$	max MCS
μ	−2.565E−03	−1.987E−03	−7.721E−04	4.454E−04
σ	9.191E−03	4.297E−03	4.112E−02	4.112E−03

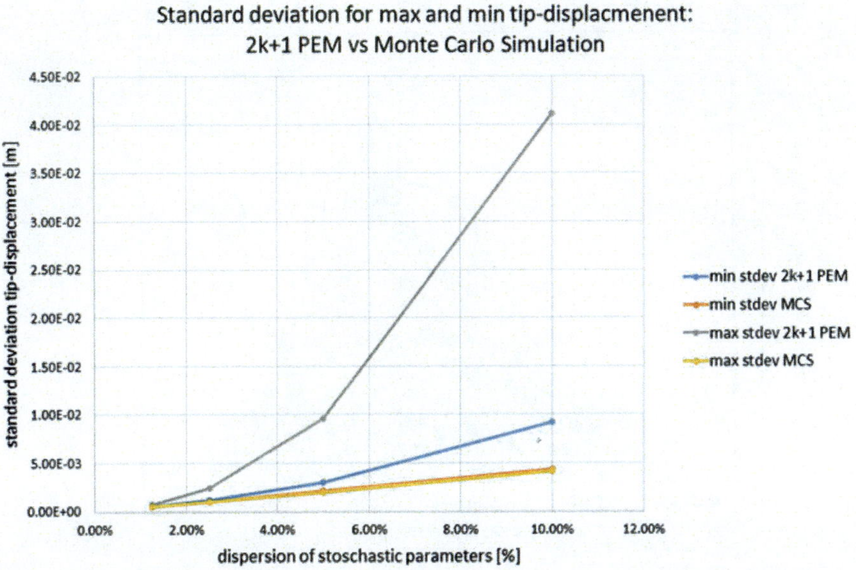

Fig. 10.17 PANELSAT correlation of the standard deviation for maximum and minimum (with among others two independent CTE design variables)

From Fig. 10.17, it can be observed that the curves for the standard deviations produced with the Monte Carlo runs appear to be linear functions of the dispersion. This should be the case because the standard deviation is proportional to the dispersion of the stochastic design variables. This gives confidence that an adequate amount of analyses were done with the Monte Carlo approach to produce this property and that therefore the results of these analyses can be used as reference to judge the quality of the estimates produced with the $2k + 1$ PEM method.

What can be observed clearly from Fig. 10.17 is that the curves of the standard deviation for the $2k + 1$ PEM method do not have this linear behaviour and are even estimating for a dispersion of 10% a standard deviation of more than 4 cm for the maximum value of the tip-displacement. This is not a realistic value, and also for a dispersion of 5%, the produced standard deviation 9.5 mm is a non-physical value. For the dispersion levels of 5 and 10%, it may well be that for this example for some of the design variables the dispersion level is too high.

From Fig. 10.16, it can be observed as well that the mean values are also not well correlating. For the 10% dispersion of the design variables, the mean and standard deviation for the tip-displacements are numerically compared in Table 10.7.

An additional interesting observation is that the mean and standard deviation of the MCS results do not match with the one of a fitted normal distribution (compare the mean and standard deviation in the column "min MCS of Table 10.7 with the values in Fig. 10.18").

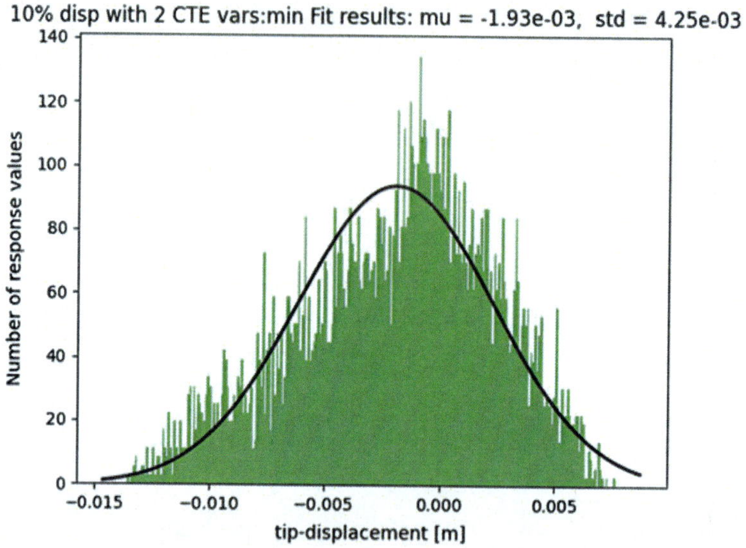

Fig. 10.18 Fitted normal distributions to minimum tip-displacements produced with the MCS method (with among others two independent CTE design variables)

Table 10.8 Coefficient of variation $|V_n|$ of the minimum and maximum tip-displacement for dispersion of 10% using the $2k + 1$ PEM method

Variable #	Name	$\|V_n\|$ for minimum	$\|V_n\|$ for maximum
1	t_{nadir}	2.40E−03	4.26E−02
2	t_{zenith}	1.29E−03	3.51E−02
3	k_{core}	5.57E−02	5.23E−02
4	ε_{nadir}	1.90E−03	2.12E−03
5	α_{nadir}	4.14E−02	1.20E−01
6	ε_{zenith}	1.68E−02	9.87E−02
7	α_{zenith}	3.32E−03	1.45E−01
8	CTE_{nadir}	1.63E+00	7.15E+00
9	CTE_{zenith}	1.67E+00	7.12E+00

Also, a level of skewness in the distribution of the response values can be observed in Fig. 10.7.

The question now is which design variables are causing these unrealistic values?

To answer this question, use can be made of the coefficient of variation values V_n of the response due to variation of each individual parameter that have been calculated for the computation the standard deviation of the responses (see Eq. (10.8)). In Table 10.8, these values are presented for the response of minimum and maximum tip-displacement for the case in which design variables have a 10% dispersion.

Table 10.9 Mean and standard deviation for minimum and maximum tip-displacement for dispersion of 10% without CTE parameters comparing the $2k+1$ PEM method and Monte Carlo (MCS)

	min $2k+1$	min MCS	max $2k+1$	max MCS
μ	−1.25202E−03	−1.25683E−03	−1.93120E−04	−1.94725E−04
σ	8.97010E−05	9.15466E−05	4.40009E−05	4.42642E−05

The results of Table 10.8 show that the CTE design variables have by far the highest values for the coefficients of variation and are therefore expected to be the driving parameters behind the excessively high standard deviations for the response of the tip-displacement. There is therefore a high likelihood that the dispersion of these design variables are causing the $2k+1$ PEM estimate to be off. To have this confirmed, a case with again a dispersion of 10% is run, but this time leaving out the design variables CTE_{nadir} and CTE_{zenith}. Since for the Rosenblueth method all the Y_{nm} and Y_{np} values are already available, it is quite straightforward to compute the estimates for the mean and standard deviation without the influence of the CTE parameters. It is, however, needed to rerun all the cases for the MCS analysis for the set of parameters without the CTE parameters.

In Table 10.9, the estimates for the mean and standard deviation are presented without the influence of the CTE parameters, but still with a dispersion level of 10%. The results show that without the design variables CTE_{nadir} and CTE_{zenith}, but maintaining the high dispersion of 10% for the remaining design variables, the $2k+1$ PEM method estimate is reproducing quite well the Monte Carlo results.

Trying to fit again the data from the Monte Carlo simulation with a normal distribution, it can be seen that the fitted normal distribution matches also well the stochastic properties of the Monte Carlo results (compare the mean and standard deviation in the column "min MCS of Table 10.9 with the values in Fig. 10.19"). Also, no obvious sign of skewness in the response of the minimum tip-displacement is visible.

So, there is definitely something going on that is caused by the two CTE design variables.

So far, two independent design variables were considered for the CTE for the zenith and nadir face sheet. With a dispersion of 10%, it is possible to have sample combinations of both design variables that differ up to 20%. This is extremely high. It would imply a not so well-controlled manufacturing process of the face sheets or the two face sheets are coming from two different material suppliers, which are quite rare.

The choice of having two independent design variables for the CTE design variables also implies that bending of the panel can occur even when the panel has a uniform temperature. With a dispersion level of 10% for each of the CTE design variables, this is most likely too much for the $2k+1$ PEM method for the current example.

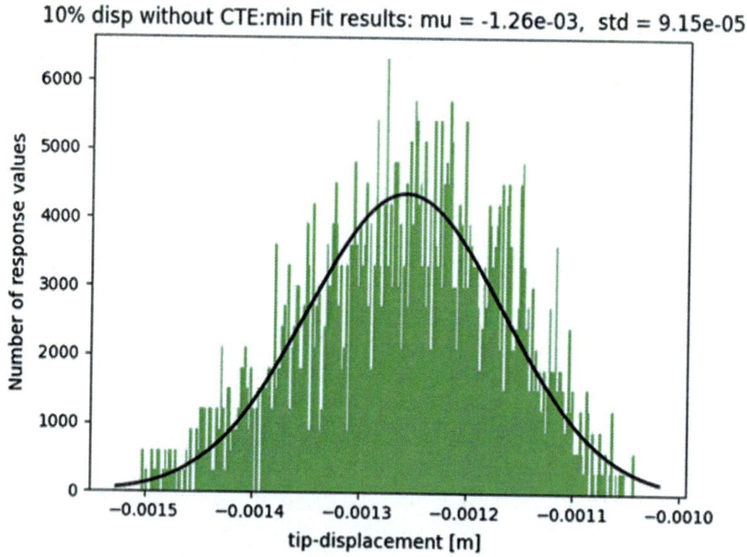

Fig. 10.19 Fitted normal distributions to minimum tip-displacements produced with the MCS method (without CTE design variables)

From Figs. 10.16 and 10.17, it can be learned that when the dispersion levels are reduced in magnitude, the deviation between the MCS and 2k + 1 PEM method is also significantly reduced to a level that could be considered acceptable for a dispersion of 2.5%.

It must be noted that for this example case the structure is subjected to more than 60 ° of temperature variation relative to the reference temperature. The high temperature excursion is likely to amplify the effect of CTE difference in top and bottom face sheet.

The above consideration has led to the hypothesis that with a single CTE design variable for both the zenith and nadir face sheet, above observed problems with the results would be reduced. To verify this hypothesis, a dedicated analysis with a 10% dispersion level for all design variables is again run. This time CTE_{nadir} and CTE_{zenith} is one-to-one correlated and has in each analysis the same value. This implies that there are now eight instead of nine independent design variables. For the $2k + 1$ PEM method, this means that only 17 analysis runs have to be performed. The MCS is also repeated with 1800 runs, sampling values for the 8 design variables.

In Table 10.10, the mean and standard deviation are compared with for both the MCS and $2k + 1$ PEM methods. In this case, there is almost perfect correspondence between both the methods.

Trying to fit again the data from the Monte Carlo simulation with a normal distribution, it can be seen that the fitted normal distribution matches well the stochastic properties of the Monte Carlo results (compare the mean and standard deviation in

Table 10.10 Mean and standard deviation comparing the 2k + 1 PEM method and Monte Carlo (MCS) for minimum and maximum tip-displacement for dispersion of 10% with a single CTE design variable associated to top and bottom face sheet

	min $2k + 1$	min MCS	max $2k + 1$	max MCS
μ	−1.252E−03	−1.256E−03	−1.925E−04	−1.930E−04
σ	1.115E−04	1.224E−04	4.489E−05	6.745E−05

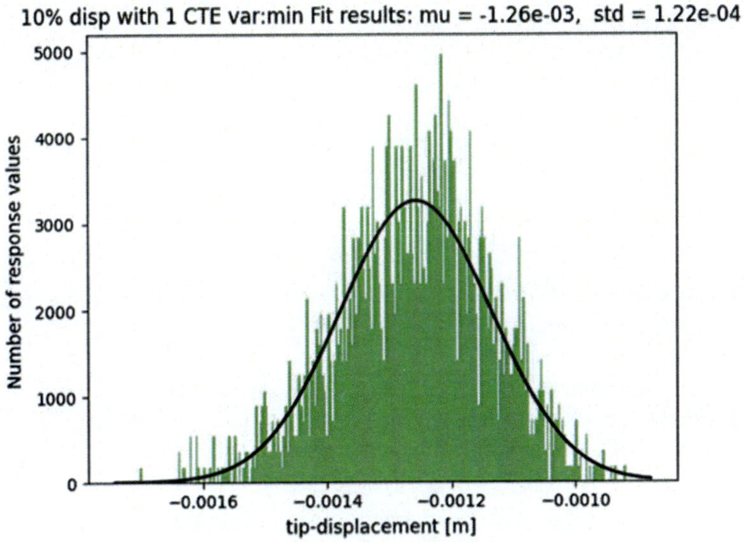

Fig. 10.20 Fitted normal distributions to minimum tip-displacements produced with the MCS method with a single CTE design variable

the column "min MCS of Table 10.10 with the values in Fig. 10.20"). A slight trace skewness may be observed. This may also be caused by the "just" 1800 result values.

Concluding remarks on the example This example shows the strength of the $2k + 1$ PEM method compared to the Monte Carlo simulation method. Instead of the 1800 simulation runs for the Monte Carlo simulation, only 17 cases for the 8 design variables are sufficient to accurately reproduce the results. Even when the 17 cases have to be repeated, the number of analysis cases is still manageable with a little effort. The example showed that limitations of the method can be reached and how this can be detected.

In this example, it was quite obvious that the dispersion level of two of the design parameters was affecting the quality of the estimates for the mean and the standard deviation of the responses. It might not always be that obvious. In this case, the Monte Carlo simulation results were available for reference and showed a proportional relation between the standard deviation of the responses and the standard deviation of the input variables.

When no reference MCS results are available, the proportionality verification, as a way to verify the validity of the dispersion value, has then to be performed by generating additional results with the $2k + 1$ PEM method. These results are obtained through an extra set of runs in which the dispersion levels of the design variables are scaled by a factor 2 or 0.5. The corresponding estimates of the standard deviation of the responses should then show the same scaling factor. If this proportionality is not demonstrated, then the values of the coefficients of variation for the different design variables (for instance, a list similar to Table 10.8) may help to identify which of the design variables may be causing the problem.

The experience with the $2k + 1$ PEM method for thermoelastic problems in the space industry is still quite limited. Further application of the method will extend the experience and will provide more guidelines on the use of the method.

Probabilistic PAT temperature interpolation, $2k + 1$ PEM approach

This following example is based on the same model presented in the example "Example: Effect of changing conductivity" in Sect. 8.4.

Here, the objective is to investigate the influence of stochastic plate thicknesses and conductive values on the PAT interpolation (mapping) process.

The description of the model is briefly repeated. The thermal mathematical model (TMM) consists of three lumped parameter nodes 1, 2 and 3. The temperatures are $T_1, T_2, T_3 = 50.0, 100.0, 0.00\,°\text{C}$. The TMM is shown in Fig. 10.21.

The TMM mesh of Fig. 10.21 completely covers the corresponding FE model that is shown in Fig. 10.22 to which thermal node temperatures are foreseen to be mapped. The boundaries of the thermal nodes of the TMM are indicated with red dashed lines. The FE model consists of 165 iso-parametric, 4 node elements and 192 nodes. The dimensions are given in metres.

For the mapping with the PAT method, the structural FE model has to be converted into a conductive FE model. The element thickness values and the material conductivity properties determine the conduction matrix that is used for the PAT method. The elements that correspond with thermal node 1 have thickness t_1 and conductivity

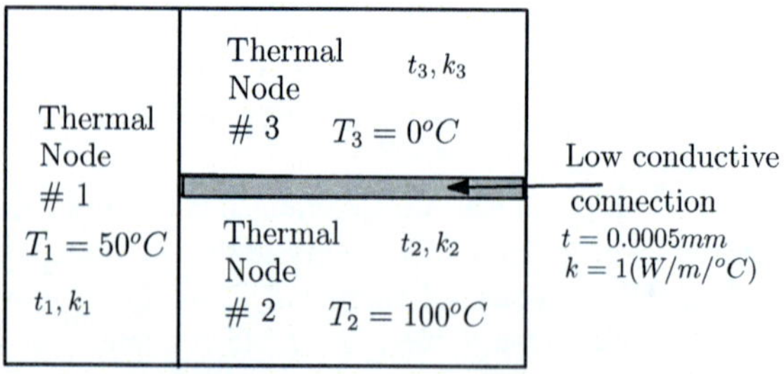

Fig. 10.21 Thermal mathematical model (TMM)

Table 10.11 Strip stochastic variables

Parameter	Range	Mean μ	Standard deviation σ
t_1 (mm)	$2 \pm 5\%$	2	0.0577
t_2 (mm)	$3 \pm 5\%$	3	0.0866
t_3 (mm)	$4 \pm 5\%$	4	0.1155
k_1 (W/m °C)	$100 \pm 10\%$	100	5.7735
k_2 (W/m °C)	$150 \pm 10\%$	150	8.6603
k_3 (W/m °C)	$200 \pm 10\%$	200	11.5470

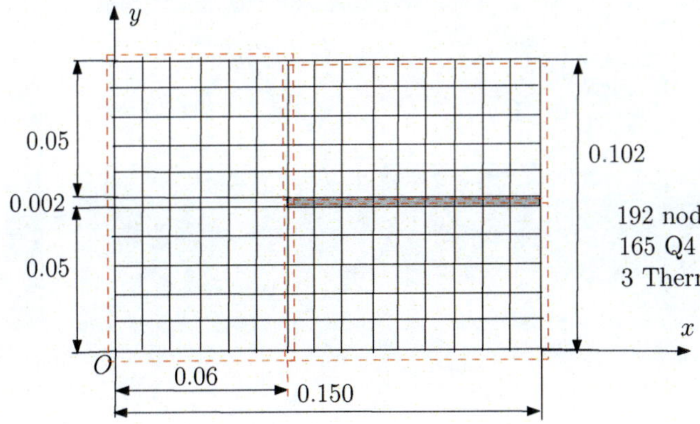

Fig. 10.22 Finite element conduction model

k_1 assigned. In a similar way are the thickness values t_2 and t_3 and conductivity k_2 and k_3 properties for the elements that correspond to thermal nodes 2 and 3.

The thicknesses t_1, t_2, t_3 and the conductivity k_1, k_2, k_3 are chosen as stochastic design variables and are considered uniformly distributed, and mean values μ and standard deviation σ are given in Table 10.11. The Rosenbluth's $2k + 1$ point estimate method (PEM) is applied [72].

The grey-coloured 4 node elements represent the low conductive gap with width $b = 2$ mm, and the thickness is $t = 0.5$ mm. The isotropic conduction is taken to be $k = 1$ W/m/°C.

The total conduction matrix $[K_c]$ is a 192×192 matrix, and the A-matrix has 3 rows and 192 columns. The thermal node temperatures are in this example considered not affected by the parameters of the conductive FE model used for the temperature mapping. With thermal nodal temperatures $\left(T^t\right)$, the temperature at the FE model nodes can be evaluated using Eq. (8.31). The resulting interpolated nodal temperatures, with nominal values, are presented as a contour plot shown in Fig. 10.23. The maximum temperature is $T_{max} = 111.08$ °C, and the minimum temperature is $T_{max} = -7.56$ °C. The contour plot was generated with a temperature increment of $\Delta T = 5$ °C.

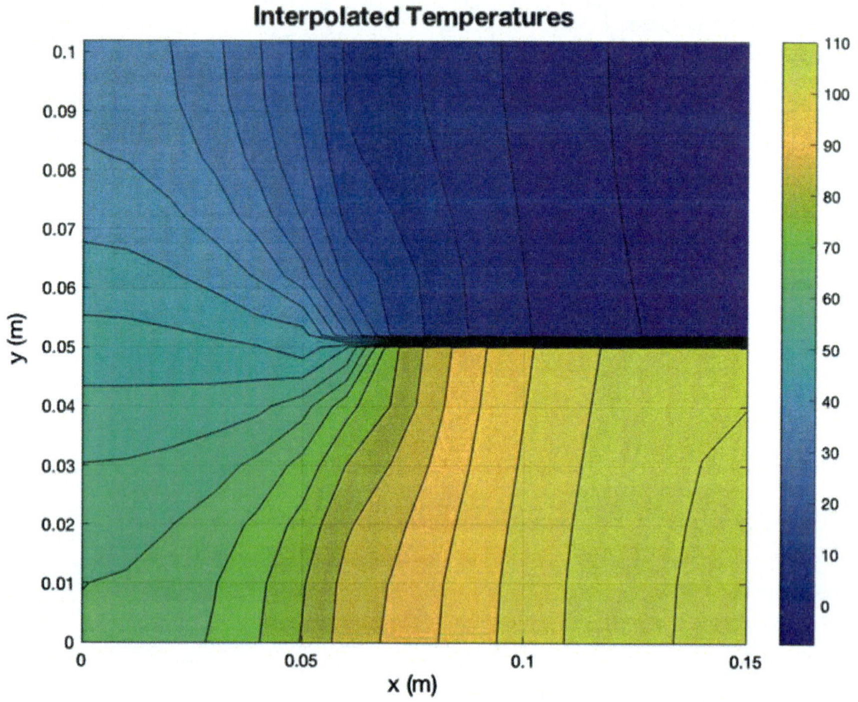

Fig. 10.23 Interpolated temperatures

Table 10.12 Interpolated nodal temperatures

Parameter (nodal temp.)	Nominal	Method	Mean μ	Standard deviation σ
T_1 (°C)	65.5552	PEM	65.5481	0.3564
T_{16} (°C)	111.0785	PEM	111.0817	0.4677
T_{177} (°C)	38.0171	PEM	38.0183	0.3889
T_{192} (°C)	−7.5575	PEM	−7.5651	0.4128

The mean value μ and standard deviation σ of the PAT interpolated temperatures in the four corners (nodes: 1, 16, 177 and 192) are presented in Table 10.12. The mean value of the thicknesses and the conduction values are used to calculate the nominal values of the mapped FE node temperatures.

The uncertainties in the plate thicknesses and conductivity values of, respectively, 5% and 10% have no significant influence of the PAT mapping process. The prescribed average temperature constraints imposed by the PAT method appear to be more dominant.

It must be noted that in this example only the effect on the temperature mapping due to the change of design variables is investigated. A change in thermal node temperatures due to the change of conductance between the thermal nodes, as a result of changes in the selected design variables, is not taken into account.

10.7 Sensitivity Analysis

Once the mathematical models have gone through the convergence tests and uncertainties in the modelling is minimised, then the starting point is reached to find out which physical thermal and structural parameters are having the highest influence on the performance of the instrument or spacecraft. In other words a sensitivity analysis has to be carried out, which means a study of the relative importance of different input factors on the model output. Saltelli in [79, 80] gave the following definition of sensitive analysis: "*The study of how uncertainty in the output of a model (numerical or otherwise) can be apportioned to different sources of uncertainty in the model input*".

Several sensitivity methods exist. Global sensitivity analysis methods are one category of sensitivity analysis methods that can be divided in two general sub-classes of methods. The first sub-class contains the variance-based methods [85]. A thorough discussion of these types of methods would go beyond the scope of this book. However, an aspect of this type of methods has been briefly discussed in Sect. 10.6 about the Rosenblueth 2k + 1 point estimates method [77]. The combination of Eqs. (10.8) and (10.10) provides the ratio $V_n/V_Y, n = 1, 2 \ldots$ which is in fact the sensitivity index. This sensitivity index can be considered, with the use of the words of Saltelli that are quoted above, as a measure of the relative importance of the input variables.

The variance-based sensitivity analysis methods are a more enhanced and sophisticated version of the so-called regression-based methods that try to fit a linear relation through the data obtained from stochastically selected samples. The description of the regression method summarised below is to a large extent based on the instructions in [80].

Imagine a model, for instance consisting of a thermal model and a structural thermoelastic model, that can be characterised by means of the following general equation

$$Y = f(X_1, X_2, \ldots, X_n) \tag{10.15}$$

where $X_i,\ i = 1, 2, \ldots, n$ are the stochastic design parameters (variables) and Y the uncertain output. The uncertain input parameters can be physical parameters such as CTE or Young's modulus, described with an associated probability density function (PDF), e.g. uniform, normal, log normal or other.

Considered is now an experiment in which samples of $X_i,\ i = 1, 2, \ldots, n$ are drawn with for instance with the Monte Carlo method, Latin hypercube method or otherwise. This leads then to a sample matrix $[M]$ of (most times independent) inputs

$$[M] = \begin{bmatrix} x_1^{(1)} & x_2^{(1)} & \ldots & x_n^{(1)} \\ x_1^{(2)} & x_2^{(2)} & \ldots & x_n^{(2)} \\ \vdots & \vdots & \vdots & \vdots \\ x_1^{(N)} & x_2^{(N)} & \ldots & x_n^{(N)} \end{bmatrix}. \tag{10.16}$$

Each row in $[M]$ is a trial set to evaluate $y^{(k)}$ that is determined by computing Y for each row of $[M]$ resulting in the response vector

$$(y) = \begin{pmatrix} y^{(1)} \\ y^{(2)} \\ \vdots \\ y^{(N)} \end{pmatrix} \tag{10.17}$$

Now are μ_{x_j} and σ_{x_j} indicated as the mean and standard deviation of the design variable X_j, determined from the values in column j of matrix $[M]$. In a similar way, the mean and standard deviation of the response of model Y can be indicated as μ_y and σ_y. This makes it possible to normalise the sample values of x_j and the responses y_j, relative to their corresponding standard deviation, to so-called standardised variables.

$$\begin{aligned} \tilde{x}_j^{(k)} &= \frac{x_j^{(k)} - \mu_{x_j}}{\sigma_{x_j}} \\ \tilde{y}^{(k)} &= \frac{y^{(k)} - \mu_y}{\sigma_y} \end{aligned} \tag{10.18}$$

The values of $\tilde{y}^{(k)}$ can be approximated by $\tilde{y}'^{(k)}$ with a linear regression model in which

$$\tilde{y}'^{(k)} = \frac{\hat{y}^{(k)} - \mu_y}{\sigma_y} = \sum_{j=1}^{n} \beta_j \tilde{x}_j^{(k)} \tag{10.19}$$

where $\hat{y}^{(k)}$ is an approximation of $y^{(k)}$ and the values of β_j are determined via a least squares method in which

$$\sum_{k=1}^{N} \left(\hat{y}^{(k)} - y^{(k)}\right)^2 \tag{10.20}$$

is minimised.

The values β_j are called standardised regression coefficients (SRC) and can be considered as a simple measure for the sensitivity of the model response Y with respect to its input design variables X_j.

An indicator on how good the β_j values achieve the regression is the factor R_y^2 which is defined as

$$R_y^2 = \frac{\sum_{k=1}^{N}\left(\hat{y}^{(k)} - \mu_y\right)^2}{\sum_{k=1}^{N}\left(y^{(k)} - \mu_y\right)^2} \tag{10.21}$$

The value of R_y^2 is between 0 and 1. A large value, for instance greater than 0.7, means that the regression model is able to represent a large part of the variation of Y. It means as well that the regression has worked and can form a basis for sensitivity assessment. When the β_j coefficients are sorted on their absolute values, it provides an indication of the importance of the corresponding design variable.

A large value of R_y^2 also means that the model has a high degree of linearity. For linear models, this is when $R_y^2 = 1$, implying no scatter of samples. In such a case, the following holds [79]

$$\sum_{j=1}^{n} \beta_j^2 = 1 \tag{10.22}$$

Sensitivity analysis on a thermoelastic model

The Al-alloy strip is idealised with a FE model consisting of 20 rod elements and 21 nodes. The strip and the FE model are shown in Fig. 10.24. The strip is fixed in node 1. The length of the strip is $L = 1$ m, the thickness is $t = 10$ mm, and the width $b = 25$ cm. The conductivity is $k = 205$ W/m/°C. Young's modulus is $E = 70$

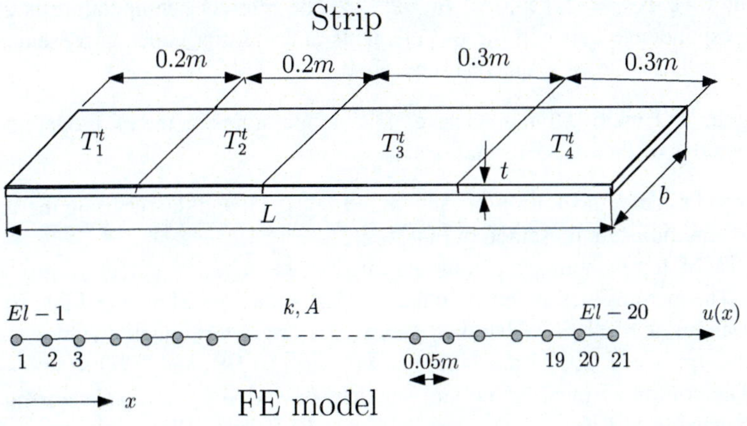

Fig. 10.24 Al-alloy strip and FE model

Table 10.13 Thermal nodes, overlaid elements, nodes

Thermal node	Temperature (°C)	FE model rod elements	FE model nodes
1	40	1–4	1–5
2	60	5–8	5–9
3	30	9–14	9–15
4	50	15–20	15–21

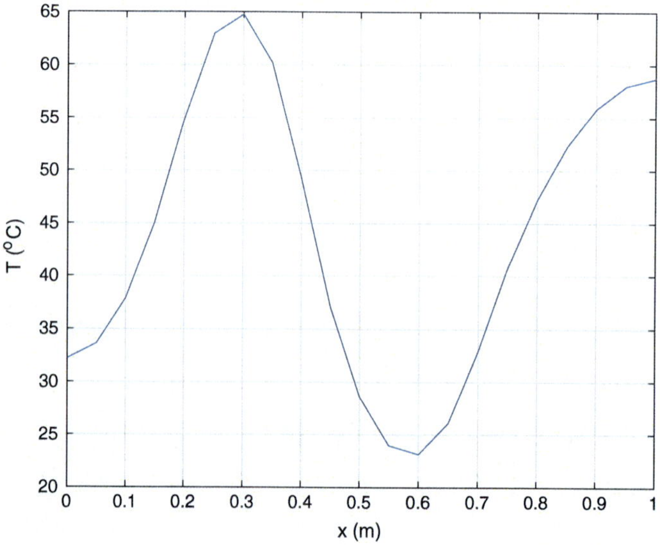

Fig. 10.25 PAT interpolated temperatures

GPa. The CTE is $\alpha = 2.4 \times 10^{-5}$ (m/m/°C). The reference temperature is zero. The correspondence between the rod elements and thermal nodes is presented in Table 10.13. The following analysis steps are done:

- Using the PAT method to map thermal node temperatures on the FE model nodes.
- Calculations of thermal deformations.

The mapped nodal temperatures are shown in Fig. 10.25, and the resulting thermoelastic deformations are presented in Fig. 10.26.

The TMM temperatures are now assumed to be random and uniformly distributed. The mean values are equal to the nominal values and shown in Table 10.13. The variations are ±10%. All other parameters are deterministic. The sensitivity values $\beta^2_{x_i}$, $i = 1, 2, 3, 4$ are obtained with Eq. (10.19), and 1000 samples are taken. The computed mean value and standard deviation of tip-displacement u_{tip} are, respectively, 1.0192×10^{-3} and 3.1638×10^{-5} (m). The sensitivity values β^2_i, $i = 1, 2, 3, 4$ are presented in Table 10.14. Since thermal nodes 2 and 4 have the highest absolute temperature variation range, it could be expected beforehand

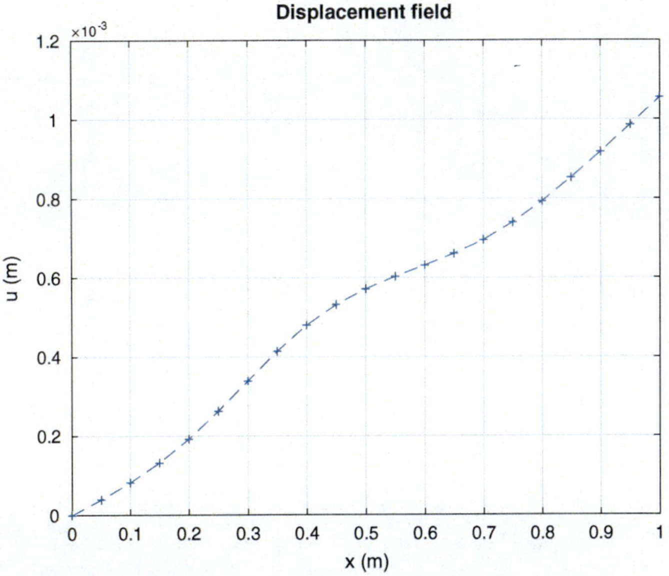

Fig. 10.26 Thermoelastic displacements of the tip of the strip

Table 10.14 Sensitivities of temperature on output

Thermal node T^t	μ_T (°C)	$\mu_T \times \pm 10\%$ (°C)	β_i^2
1	40	±4	0.1234
2	60	±6	0.2849
3	30	±3	0.1579
4	50	±5	0.4252
$\sum \beta_i^2$			0.9914

that the corresponding thermal node temperature T_2^t and T_4^t would show high sensitivities on the output u_{tip}.

Four scatter plots are shown in Fig. 10.27. Corresponding with the higher sensitivity for thermal nodes 2 and 4, a higher linearity can be observed in the scatter plots (b), for thermal node 2, and (d), for thermal node 4.

Probabilistic thermoelastic analysis thin strip

A cantilevered strip (thin plate) is shown in Fig. 10.28. The length of the strip is $L = 0.5$ m, the width is $b = 20$ cm, and the thickness is $h = 2$ mm. The strip is made of Al-alloy with a CTE $\alpha = 24 \times 10^{-6}$ m/m/°C. The reference temperature

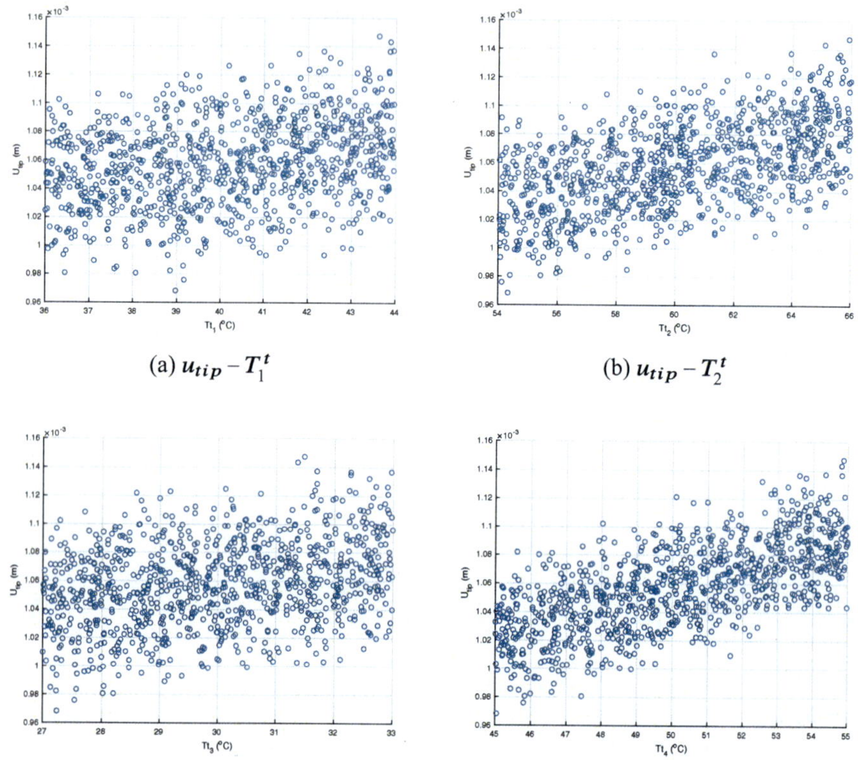

(a) $u_{tip} - T_1^t$ (b) $u_{tip} - T_2^t$

(c) $u_{tip} - T_3^t$ (d) $u_{tip} - T_4^t$

Fig. 10.27 Scatter plots $u_{tip} - T_i^t,\ i = 1, 2, 3, 4$

$T_{ref} = 0\,°\text{C}$. The angular rotation $\phi(L)$ and deflection $w(L)$ at the tip (m) can be calculated with [44]

$$\begin{aligned} \phi(x) &= -\alpha\frac{(T_1 - T_2)}{h}x, \\ w(x) &= -\alpha\frac{(T_1 - T_2)}{2h}x^2 \end{aligned} \qquad (10.23)$$

where T_1 is the temperature of the top surface and T_2 the temperature of the lower surface. In case $T_1 = 20\,°\text{C}$ and $T_2 = 22\,°\text{C}$, the angular rotation $\phi(L) = 0.012$ rad and the deflection $w(L) = 0.003$ m.

The probabilistic analysis will be done using two different approaches [72]

- Monte Carlo simulation (MCS). The equations in (10.23) are 1,000,000 times evaluated. The five variables involved are assumed uniformly distributed. The bounds are shown in Table 10.15.

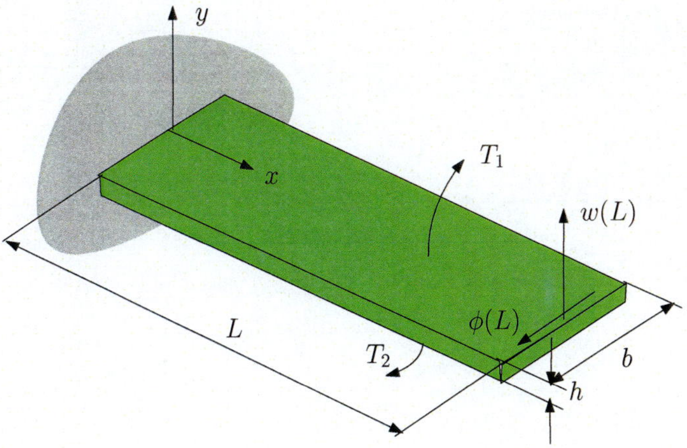

Fig. 10.28 Cantilevered strip

Table 10.15 Strip stochastic design variables

Parameter	Range	Mean μ	Standard deviation σ
L (m)	0.49–0.51	0.5	5.7735×10^{-3}
h (mm)	1.9–2.1	2.0	5.7735×10^{-2}
T_1 (°C)	19–21	20	5.7735×10^{-1}
T_2 (°C)	21–23	22	5.7735×10^{-1}
$\alpha \times 10^6$ (m/m/°C)	22–26	24	1.1547

Table 10.16 Strip tip-displacement and rotation

Parameter	Method	Mean μ	Standard deviation σ
$w(L)$L (m)	MCS	3.0022×10^{-3}	1.2423×10^{-3}
$\phi(L)$L (rad)	MCS	1.2008×10^{-2}	4.9619×10^{-3}
$w(L)$L (m)	PEM	3.0029×10^{-3}	1.2667×10^{-3}
$\phi(L)$L (rad)	PEM	1.2010×10^{-2}	5.0594×10^{-3}

- Rosenbluth 2k + 1 point estimate method (PEM). Again uniform probability density function of then $k = 5$ uncorrelated stochastic variables is assumed. The mean values and the standard deviations are given in Table 10.15.

With both probabilistic approaches, the mean μ and standard deviation σ of the displacement $w(L)$ and $\phi(L)$ are calculated with, respectively, 1,000,000 (MCS) and $2 \times 5 + 1 = 11$ (PEM) samples. The results of the probabilistic calculations are shown in Table 10.16.

The MCS and PEM analyses give the same mean values and standard deviations; however, the number of samples when the Rosenbluth 2k + 1 PEM is applied is much

Table 10.17 Sensitivity analysis

	β_i^2			
	100,000 samples		V_n/V_Y	
Parameter	$w(L)$	$\phi(L)$	$w(L)$	$\phi(L)$
L (m)	0.0004	0.0001	0.0547	0.0274
h (mm)	0.0070	0.0059	0.0684	0.0685
T_1 (°C)	0.5060	0.4973	0.6843	0.6852
T_2 (°C)	0.4657	0.4778	0.6843	0.6852
$\alpha \times 10^6$ (m/m/°C)	0.0116	0.0126	0.1140	0.1142
$\sum \beta_{x_i}^2$	0.9909	0.9938	0.9922	0.9941

lower when applying the MCS. Applying the Rosenbluth 2k + 1 PEM, the probability density function for $\phi(x)$ is not determined by the probabilistic calculations.

Afterwards, a sensitivity analysis is made:

- Regression method (MCS), $\beta_i^2,\ i = 1, 2, \ldots, k$
- Rosenblueth $2k + 1$ method, $V_i/V_Y,\ i = 1, 2, \ldots, k$

The outcomes of the sensitivity analyses are presented in Table 10.17.

The sensitivity analysis for both with the regression method and with the Rosenbluth $2k + 1$ PEM shows that the temperature $T_1,\ T_2$ are most important on the outputs $w(L),\ \phi(L)$. In addition, it may be concluded that the Rosenbluth $2k + 1$ PEM performs very well. $L,\ h$ may be assumed deterministic.

Problems

10.1 The probability value $p = \Phi(z)$, $(0 \leq p \leq 1)$ of the standard normal variable $z = (x - \mu)/\sigma_x$ is given by

$$p = \Phi(z) = \frac{1}{\sqrt{2\pi}} \int_{-\infty}^{z} e^{\frac{-t^2}{2}}\, dt$$

The inverse problem is defined as $z = \Phi^{-1}(p)$, where $\Phi^{-1}(p),\ 0 \leq p \leq 0.5$ is given by [4]

$$z_p = \Phi^{-1}(p) = -1 + \frac{2.30753 + 0.27061u}{0.99229u + 0.04481u^2},\quad u = \sqrt{-\ln p^2},$$

and when $p \geq 0.5$, $p^* = 1 - p$, then $z_p = \Phi^{-1}(p) = -\Phi^{-1}(p^*)$.

Draw 10 random values of $p = \mathcal{U}(0, 1)$ and calculate $X_p = \mu_x + z_p\sigma_x$, $\mu_x = 30$ and $\sigma_x = 1$.

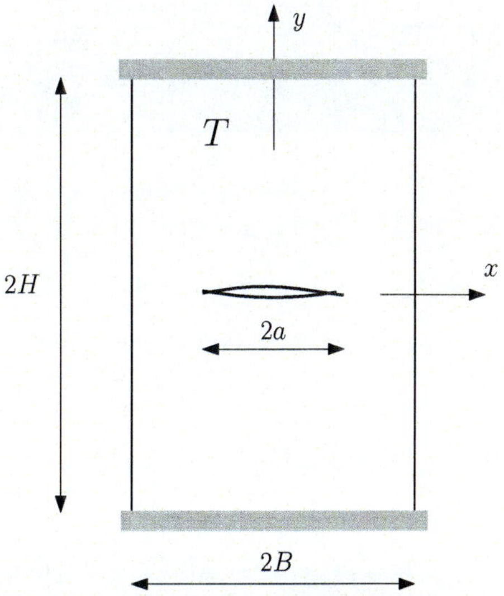

Fig. 10.29 Cracked plate under thermal loading

10.2 Cracked plate under thermal loading is shown in Fig. 10.29. The analytical expression for the thermally induced stress intensity factor of the rectangular plate with a crack size a is given as follows

$$K_{ic} = -\alpha E(T - T_0)\sqrt{\frac{\pi a}{\cos(\pi a/2B)}}\left[1 - 0.025\left(\frac{a}{4B}\right)^2 + 0.06\left(\frac{a}{2B}\right)^4\right] \text{ (MPamm}^{1/2}\text{)}. \tag{10.24}$$

The definition of the design variables and respective values are given in Table 10.18. Lognormal distribution is indicated by $LN(\lambda, \zeta)$, in which $\lambda = \ln(\mu_x) - 1/2\zeta^2$ and $\zeta^2 = \ln\left(1 + (\sigma_x/\mu_x)^2\right)$, [6]. Perform the following assignments:

Table 10.18 Mean and COV of different variables

Variable	Description	Distribution	μ_x	$\frac{\sigma_x}{\mu_x}$
T_0	Initial hot temperature	Lognormal	100 °C	0.1
T	Reference temperature	Lognormal	20 °C	0.1
a	Crack size	Lognormal	10 mm	0.1
B	Width of plate	Lognormal	200 mm	0.1
E	Young's modulus	Lognormal	210 MPa	0.1
α	CTE	Deterministic	12.5×10^{-6} mm/mm/°C	–

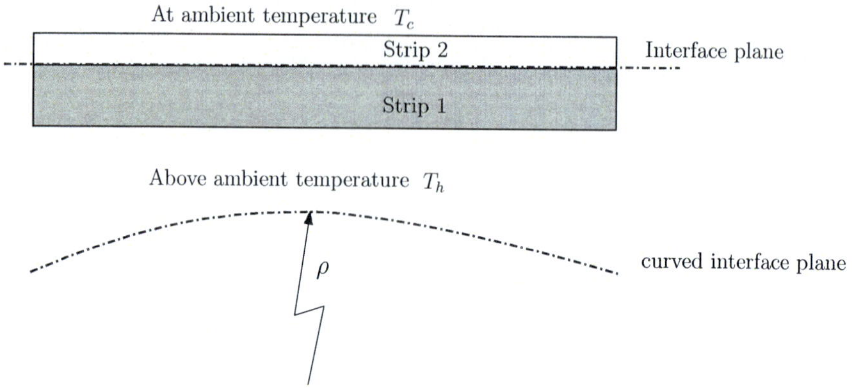

Fig. 10.30 Bimetallic strip

- Write a script in MATLAB/Octave/Python.
- Calculate with the MCS method, the mean and standard deviation of the thermal stress intensity factor K_{ic}. Apply 1000 loops with each six samples.
- Generate a plot of the density function of K_{ic}.
- Create 6 scatter plots; on the y-axis K_{ic} and on the x-axis the design variables T_0, T, a, B, E and α, respectively. What can you conclude from the scatter plots?
- Perform the Rosenblueth $2k + 1$ PEM analysis.
- Write a script in MATLAB/Octave/Python3.
- Calculate with the $2k + 1$ PEM the mean and standard deviation of the thermal stress intensity factor K_{ic}.
- Calculate with Eq. (10.8) the coefficients of variation V_n for all input design variables T_0, T, a, B, E and α, respectively, and draw your conclusions about the relative importance of the input parameters. Compare these conclusions with the conclusion drawn from the scatter plots.

10.3 Thermostatic bimetal strip

When a bimetallic strip is uniformly heated along its entire length, it will bend or deform into an arc of a circle with a certain radius of curvature. The value of that radius of curvature ρ is dependent on the geometry and metal components making the strip. This problem is based on [7]. The bimetallic strip is shown in Fig. 10.30. The radius of curvature of the metallic strip is given by

$$\rho = \frac{t\left[3(1+m)^2 + (1+mn)\left(m^2 + \frac{1}{mn}\right)\right]}{6(\alpha_2 - \alpha_1)(T_h - T_c)(1+m)^2}, \tag{10.25}$$

where

- $t = t_1 + t_2$ the total thickness
- t_1, t_2 are the individual strip thicknesses
- $m = \frac{t_1}{t_2}$ is the ratio of thicknesses

Table 10.19 Material properties, dimensions and characteristics

Strip #	Material	Thickness (mm)	CTE (m/m/°C)	Young's modulus (nominal) (GPa)
1	Invar 36	$t_1 = 0.8$	$\alpha_1 = 1.45 \times 10^{-6}$	$E_1 = 141$
2	Nickel silver	$t_2 = 0.4$	$\alpha_2 = 13 \times 10^{-6}$	$E_2 = 119$

Table 10.20 Uncertainty ranges in material properties and dimensions

Strip #	Material	Thickness (mm)	Young's modulus (GPa)
1	Invar 36	$t_1 = 0.8, \pm 0.5$ %	$E_1 = 137–145$
2	Nickel silver	$t_2 = 0.4, \pm 0.8$ %	$E_2 = 118–120$

- $n = \frac{E_1}{E_2}$ is the ratio of Young's moduli
- T_h, T_c are the hot and cold temperatures
- E_1, E_2 are the individual strip linear Young's moduli
- α_1, α_2 are the CTEs for the two metals.

The ambient temperature is $T_c = 20$ °C, and the elevated temperature is $T_h = 80 \pm 0.5$ °C. The material properties, dimensions and characteristics are provided in Table 10.19.

Assess the uncertainty of the radius of curvature Eq. (10.25) by computer simulation both with the MCS (1,000,000 samples) and Rosenbluth $2k + 1$ PEM methods (define k), uncertainty ranges given in Table 10.20. Uniform distributions shall be assumed.

10.4 Please derive and reproduce the expressions below.

The stresses for a thick-walled pipe occurs under internal pressure P (Pa) and thermal gradient $\Delta T = T_i - T_o$ ($T_{ref} = 0$) (°C). The pipe and dimensions are shown Fig. 10.31. This example is partly based on [18]. Expressions for the plain strain radial, circumference and axial stress can be found in [75] for the mechanical part and in [44] for the thermals stress part. The radial stress σ_r (Pa) is given by

$$\sigma_r = \sigma_r^P + \sigma_r^T = \frac{P}{a^2 - 1}\left[1 - \left(\frac{r_o}{r}\right)^2\right] + \frac{\alpha E \Delta T}{2(1-\nu)\ln a}\left[-2\ln\left(\frac{r_o}{r}\right) + \frac{\left(\frac{r_o}{r}\right)^2 - 1}{a^2 - 1}\ln a\right], \tag{10.26}$$

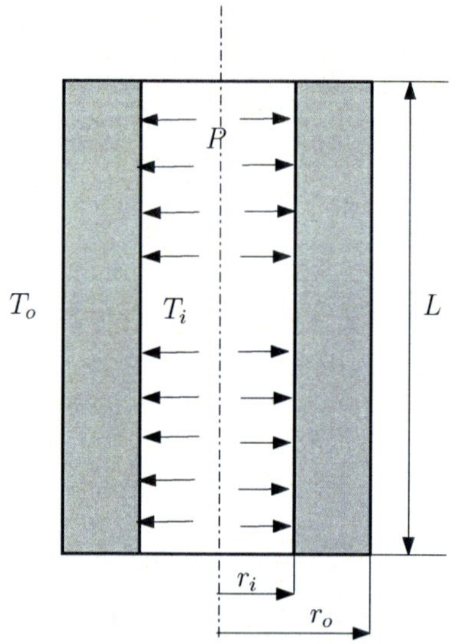

Fig. 10.31 Schematic diagram of axisymmetric pipe section

the circumferential stress σ_θ (Pa) is given by

$$\sigma_\theta = \sigma_\theta^P + \sigma_\theta^T = \frac{P}{a^2-1}\left[1+\left(\frac{r_o}{r}\right)^2\right] + \frac{\alpha E \Delta T}{2(1-\nu)\ln a}\left[1-\ln\left(\frac{r_o}{r}\right)+\frac{\left(\frac{r_o}{r}\right)^2+1}{a^2-1}\ln a\right], \tag{10.27}$$

and the axial stress σ_z (Pa) is

$$\sigma_z = \sigma_z^P + \sigma_z^T = \frac{P}{a^2-1}2\nu + \frac{\alpha E \Delta T}{2(1-\nu)\ln a}\left[1-2\ln\left(\frac{r_o}{r}\right)-\frac{2\ln a}{a^2-1}\right]. \tag{10.28}$$

The von Mises stress σ_{vM} (Pa) can be obtained by [76]

$$\sigma_{vM} = \frac{1}{\sqrt{2}}\sqrt{(\sigma_r-\sigma_\theta)^2+(\sigma_r-\sigma_z)^2+(\sigma_\theta-\sigma_z)^2}, \tag{10.29}$$

where $a = r_o/r_i$. The pipe made up of HK40 material is considered with a pipe length $L = 100$ mm, inner radius $r_i = 25$ mm and outer radius $r_o = 50$ mm. The material properties are Young's modulus $E = 138$ GPa, Poisson's ratio $\nu = 0.313$, CTE $\alpha = 15 \times 10^{-6}$ /m/m/°C. The internal pressure $P = 40$ MPa, the inside temperature $T_i = 500$ °C and the outside temperature $T_o = 420$ °C, the thermal gradient $\Delta T = T_1 -$

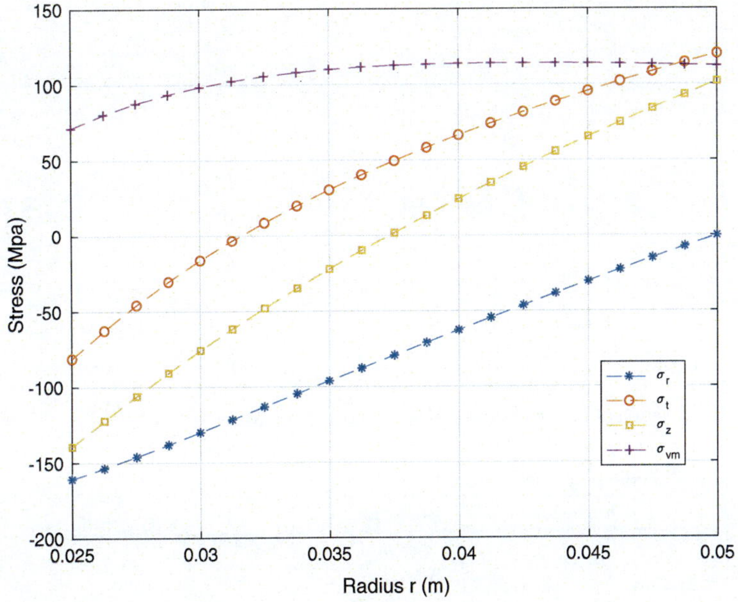

Fig. 10.32 Variation of radial, circumferential, axial and von Mises stresses

$T_o = 80\,°\text{C}$. The reference temperature $T_{ref} = 0\,°\text{C}$. The allowable yield stress $\sigma_y =$ 241 MPa.

Translate Eqs. (10.26), (10.27), (10.28) and (10.29) into a MATLAB script and plot stresses. The stresses σ_r, σ_θ, σ_z and σ_{vm} are shown in Fig. 10.32. The maximum values of $\sigma_{vm} = 114.11$ MPa at $r = 42.5$ mm. Let us assume a factor of safety $FoS = 1.5$. The margin of safety can obtained by [94]

$$MoS = \frac{\sigma_y}{FoS \times \sigma_{vm}} - 1 = \frac{241}{1.25 \times 114.11} - 1 = 0.41. \tag{10.30}$$

Create a FE model of the pipe with load and boundary conditions and perform the thermoelastic analysis with your available FE package.

Calculate the stresses when $\Delta T = 0$, and subsequently, $P = 0$. Plots of the stresses are shown in Fig. 10.33.

Further calculation will have a probabilistic character, and the Young's modulus E, the temperature gradients ΔT and the internal pressure P are considered to be random. The random nature of these material properties are presented in Table 10.21. The parameters λ, ζ can be calculated by $\lambda = \ln \mu + 0.5\zeta^2$ and $\zeta = \ln(1 + (\sigma/\mu)^2)$ [6].

Create 10, 100, 1000 sets of samples with the LHS method (MATLAB script `lhsgeneral.m`) and calculate the mean μ_{vm} and the standard deviation σ_{vm} of the von Mises stress σ_{vM} at radius $r = 42.5$ mm. The results are shown in Table 10.22.

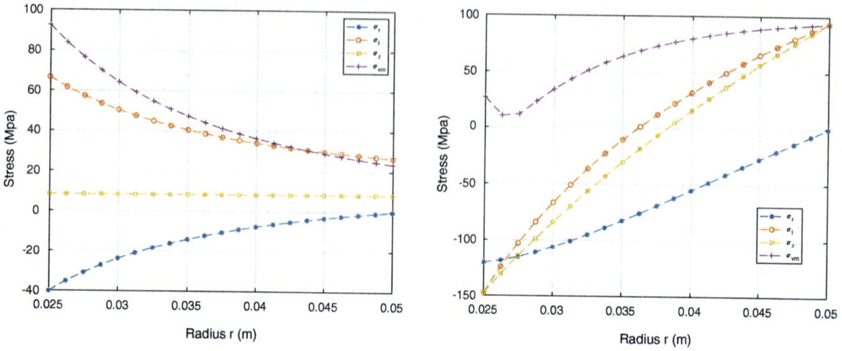

(a) $\Delta T = 0,\ P = 40$ MPa (b) $P = 0,\ \Delta T = 80\,^{o}$C

Fig. 10.33 Variation of radial, circumferential, axial and von Mises stresses

Table 10.21 Random properties materials

Material	Dimension	Distribution	μ	σ/μ
E	GPa	*Lognormal*, $\mathcal{LN}(\lambda, \zeta)$	138	0.2
ΔT	°C	*Normal*, $\mathcal{N}(\mu, \sigma)$	80	0.05
¶	MPa	*Normal*, $\mathcal{N}(\mu, \sigma)$	40	0.1

Table 10.22 Probabilistics of von Mises stress σ_{vM} at $r = 42.5$ mm

$\neq$ LH samples	Mean μ_{vm} (MPa)	Standard deviation σ_{vm} (MPa)
10	114.1102	6.0496
100	114.1193	6.1390
1000	114.1198	6.1279

In spacecraft structures design, it is usual to verify the strength capability against $\mu_{vm} + 3\sigma_{vm}$. Considering a factor of safety, $FOS = 1.5$ and the margin of safety MoS can be obtained by

$$Mos = \frac{\sigma_{yield}}{Fos \times (\mu_{vm} + 3\sigma_{vm})} = 0.21.$$

It is questionable if the FOS is still applicable. The probability of failure can be calculated with

$$P_f(\sigma_{vM} \leq \sigma_{yield}/FoS) = 1 - \int_{-\infty}^{\sigma_{yield}/FoS} \mathcal{N}(\mu_{vm}, \sigma_{vm}) x dx \approx 0.$$

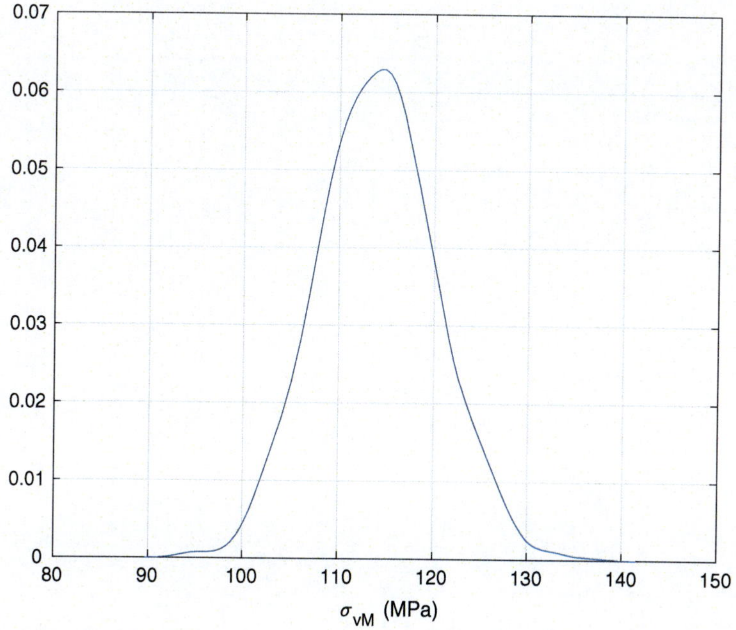

Fig. 10.34 Density function of von Mises stress at $r = 42.5$ mm

A normal distribution of the von Mises stress σ_{vM} is assumed. A plot of the density function of the von Mises stress σ_{vM} is illustrated in Fig. 10.34.

At last a sensitivity analysis with the Rosenbluth $2k + 1$ PEM is done using Eqs. (10.8) and (10.10). It turns out Young's modulus E is most important followed by the thermal gradient ΔT.

Appendix A
Detailed Description of "Box on Plate" Experiment

Abstract

Every instrument on a spacecraft has to be structurally mounted on a panel. More and more instruments require high thermal stability of the platform these are mounted on, or at least a high accuracy of the prediction of the thermoelastic distortion. In this experiment, the question that is tried to be answered is: What is the effect of heat dissipating boxes on the orientation of instruments, all mounted on the same panel of the spacecraft?

A.1 Background and Research Question

It might be worthwhile to put attention to the level of detail with which various features are modelled. The features discussed in this experiment are electronic boxes mounted on a sandwich panel. Electronic boxes are everywhere on a spacecraft. Many of them support the functioning of scientific instruments. For that reason, these boxes are located near the instrument itself and in many cases on the same panel.

In general, the design of the mechanical support of the instrument is getting rightfully quite some attention. Especially for instruments requiring high stability under time varying thermal environments, the support of those instruments has to assure that deformation of the under laying structure leads to no or very small deformations in the instrument. The base plate of the instrument, for optical instruments often referred to as optical bench, should absorb only deformations up to a level that is not affecting the accuracy of the instrument. For that reason, the materials selected for these base plates have a high stiffness and low CTE.

Support concepts such as iso-static mounting systems with blades and bi-pods are then implemented to mechanically disconnect the instrument from the spacecraft panel on which the instrument is mounted. These systems in general allow the interface plane to rotate causing no or very small interface loads in the instrument. This stress-free, and therefore deformation free, aspect of the stress-free rotation is with-

S. Appel and J. Wijker, *Simulation of Thermoelastic Behaviour of Spacecraft Structures*, Springer Aerospace Technology, https://doi.org/10.1007/978-3-030-78999-2

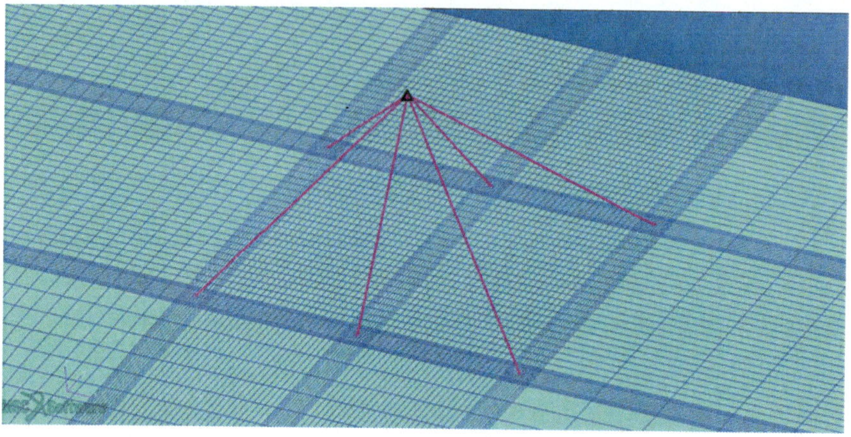

Fig. A.1 Simplified modelling of electronic box with a rigid element and a lumped mass, typical for dynamic simulations

out any doubt important for the performance of the instrument, but the same rotation will likely cause a tilting of the orientation of the instrument. This may be however essential for the performance of the instrument itself as well as the interaction with other instruments on the spacecraft.

The panel on which instruments are mounted is commonly chosen to have a high stiffness. One of the reasons is to support the often high mass of the instrument under dynamic environments. Sandwich panels with a large core thickness are chosen as a solution. It is also believed that this high stiffness of the sandwich is a sort of guarantee for thermal stability of the panel itself.

In system-level models, equipments, like electronic boxes, are often not modelled in detail and are simplified to, in many cases, a lumped mass connected via a rigid element to the sandwich panel (see Fig. A.1). The question now is, considering the stringent stability requirements of the instrument mounted on the same panel as the heat dissipating equipment, is the simplified modelling of these boxes adequate? If a more detailed modelling is required, is it then always needed? Or are there circumstances that would allow a less sophisticated representation of the boxes?

This appendix tries to answer these questions with the help of a numerical experiment.

A.2 Description of Numerical Experiment

In this numerical experiment, an electronics box mounted on a sandwich panel. The objective is to examine:

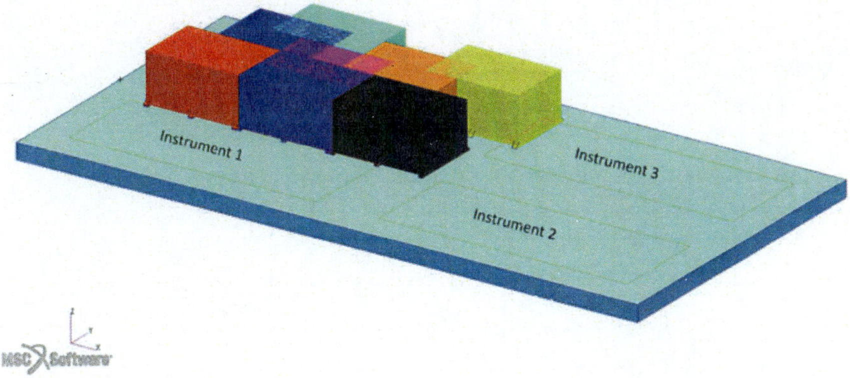

Fig. A.2 Some of the 32 positions of a box on a plate in the upper left quadrant

- The effect on the tilting of three imaginary instruments due to the thermoelastic deformation induced by an electronics box on the same panel.
- How this tilting changes with a change of position of the box.
- Whether the height of the box has an influence on the deformation of the panel and hence on the tilting of neighbouring instruments.
- The effect on the tilting of the imaginary instruments of a combined change of size and heat dissipation of the electronic box.

Considered is a sandwich panel with a width of 1 m and a length of 2 m. The face sheets are 0.5 mm thick aluminium. For the honeycomb core, a 50 mm thickness has been chosen to assure high bending stiffness for the panel.

Three large box configurations are considered. All have have a footprint of 20 x 30 cm but have three different heights: 30, 15 and 7.5 cm. Also, one small box configuration is included in the assessment with dimensions of 15 × 10 × 15 cm. Imagined is that the panel is split into four quadrants. The box is positioned at one of the 32 locations in the upper left quadrant of the panel. Investigated is how three imaginary instruments, located in the other three quadrants, are tilted due to the deformation induced by a box positioned in the upper left quadrant. In Fig. A.2, a 3-D impression of the different box positions is presented. In Fig. A.3, a top view of the different positions is displayed together with markers indicating for each box position the origin of the box coordinate system. Along the edges of the panel are the X and Y coordinates of the origin locations shown. This picture will be referenced when the results are discussed.

A.3 Model Description

The model represents a box on a sandwich plate. The box is made of aluminium plate material. Most of the outer surface has a plate thickness of 1.6 mm. The base plate and the lower edge of the side panels have a higher thickness of 4 mm. The box is

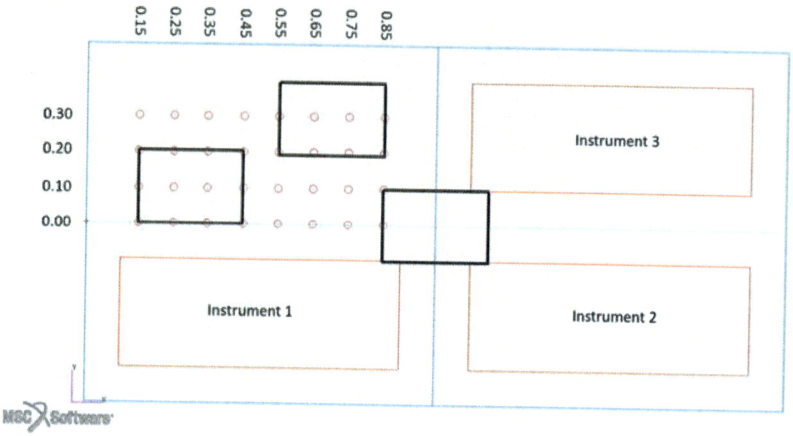

Fig. A.3 32 box positions (red circles) with coordinates and box footprint

Fig. A.4 Thickness distribution of outer surfaces of the electronics box

considered to be bolted through six feet to the sandwich plate. The plate material of the feet also has a thickness of 4 mm. The thickness distribution of the box material is visualised in Fig. A.4.

Although the mesh is quite detailed at the outside, no effort has been made to make up an interior design of the box and is therefore left empty. The representation of the external housing is considered adequate for the purpose of this experiment, since no important mechanical function should be assigned to the PCBs and the internal frame to hold these. With this approach, the thermomechanical function of a generic electronics box is considered to be well captured.

The box is mounted on a sandwich plate of 1 m wide and 2 m long. The sandwich has 0.5 mm thick aluminium face sheets and aluminium core material 1/8-5056.0015

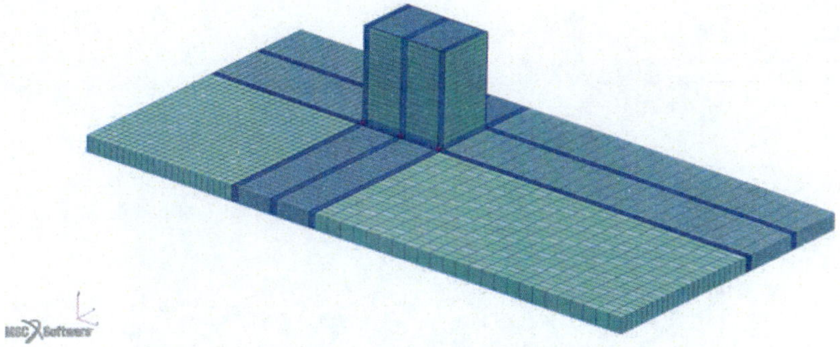

Fig. A.5 Detailed finite element representation of an electronics box on a sandwich plate

of 50 mm high. The sandwich is equipped with block inserts that extend to one-third of the thickness of the sandwich. The honeycomb material is the same over the full size of the panel. So, no heavier honeycomb is introduced in the vicinity of the block inserts.

Both the honeycomb core material, as well as the block inserts, are modelled with solid elements. The face sheets are modelled with shell elements and are sharing the GRIDs with the outer faces of the solid elements for the core and the block inserts. The finite element mesh of the box on the plate is presented for one position in Fig. A.5.

The displacements of the edges of the top and bottom face sheets are constraint. This implies a clamped boundary condition at the edges of the plate.

There is also a corresponding very detailed thermal model as presented in Fig. A.6. The structural details are also implemented in the thermal model. For instance, the solid aluminium block inserts are also implemented in the thermal model. The thermal mesh has formed the basis for the structural mesh. This means that the geometry of the thermal node is meshed with finite elements of the structural model. This allows for accurate definition of correspondence between the entities of the two models.

In the thermal model, the core is represented by three layers of solid thermal nodes. The thermal nodes representing the block inserts are part of the top layer of thermal nodes.

Each of the solid thermal nodes is meshed with 27 solid elements. The surface thermal nodes are meshed with nine shell elements. As indicated above, these elements form the mesh of the structural model. It also forms the mesh that is used for the computation of the linear thermal conductors. The generation technique based on the PAT method (see Sect. 9.3.2) has been applied to compute linear conductors (GLs) within the box and the panel.

The thermal environment consists of a heat dissipation by the electronics that is uniformly applied to the surfaces of all the six walls of the box. This is a simplified assumptions, since in practice, the heat dissipation depends on the distribution of the electronic components over the interior of the box. In many cases, the most dissipating components are located near or on the base plate that is using the sandwich panel as heat sink. This aspect is not considered, and therefore, this simplified heat distribution

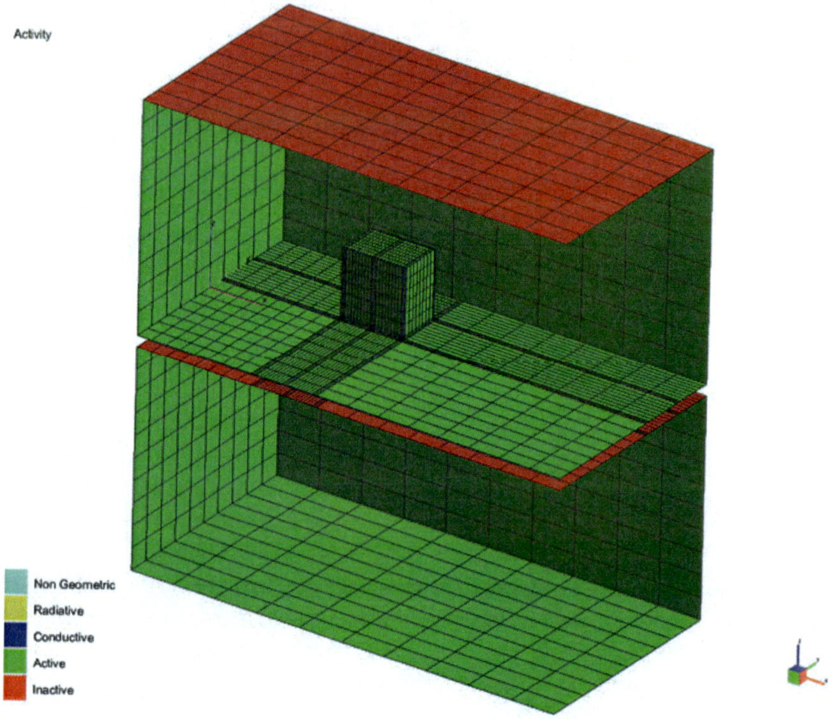

Fig. A.6 Detailed thermal model for representation of an electronics box on a sandwich plate

is applied. It might however be that in some cases, the effects that are observed in this exercise are less significant due to a different distribution of the heat dissipation.

Both the top and bottom side of the panel are having a radiative boundary condition of 20 °C. One could imagine the panel under consideration as if it is inside a spacecraft surrounded by panels with a temperature of 20 °C. This implies that in this example, the box is the only source of thermal gradients and therefore the only source that can cause thermoelastic distortion. In real-life applications, there are many of these sources on a single panel and its interfacing panels. The surrounding radiative boundaries are modelled with artificial covers at the top and bottom side. In Fig. A.6, these covers are presented with two panels of each of the covers removed to allow for internal view. For all surfaces, the infrared emissivity is set to 0.5.

The box and the plate are connected conductively via a contact feature implemented through the ESATAN-TMS DEFINE_CONTACT_ZONE method. The contact conductance is considered uniform, while in reality, it will have a lower value in the middle of the base plate and higher at the edges corresponding to the distribution of the contact pressure imposed by the fasteners at the edge of the base plate. Heat transfer trough the bolts is ignored. Although the contact conductance distribution and value can be important for the temperature field in the box, the sensitivity of this parameter is not investigated here.

A.4 Analysis Approach

As a start, one single configuration is examined. For this purpose, a steady-state thermal analysis is ran and a subsequent temperature mapping to the FE model is applied using the PAT method. The resulting temperature field for one box location is presented in Fig. A.7. From the picture, it can be seen that a strong spatial thermal gradient exists in the box. Another observation is that the effect of this rather big box on the temperature field of the panel seems quite local. The temperature increase relative to the 20 °C reference temperature is relatively mild.

Looking at the temperature distribution in a cross section of the sandwich panel under the box as presented in Fig. A.8, one can notice a quite light through thickness thermal gradient. The largest temperature difference between top and bottom face sheet is limited to about 1 °C.

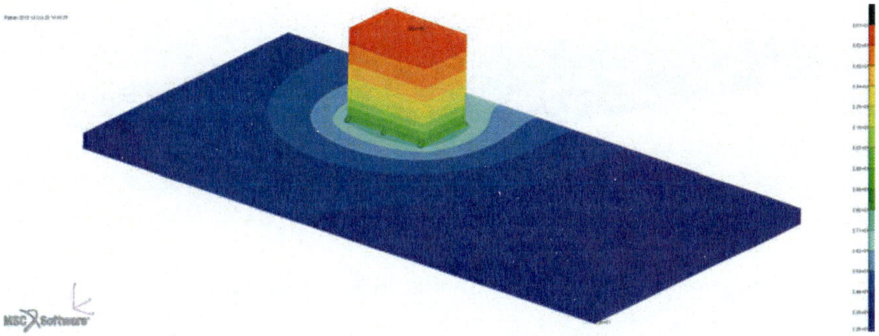

Fig. A.7 Temperature field induced by the dissipation of the electronics box

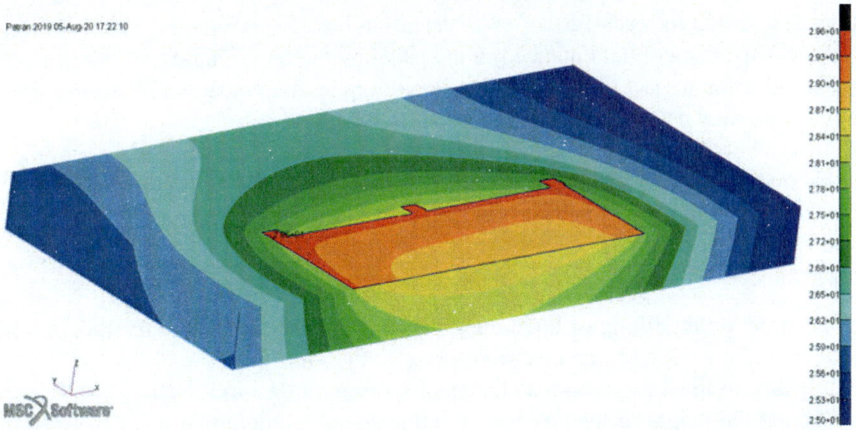

Fig. A.8 Temperature field induced by the dissipation of the electronics box through the cross section of the sandwich panel

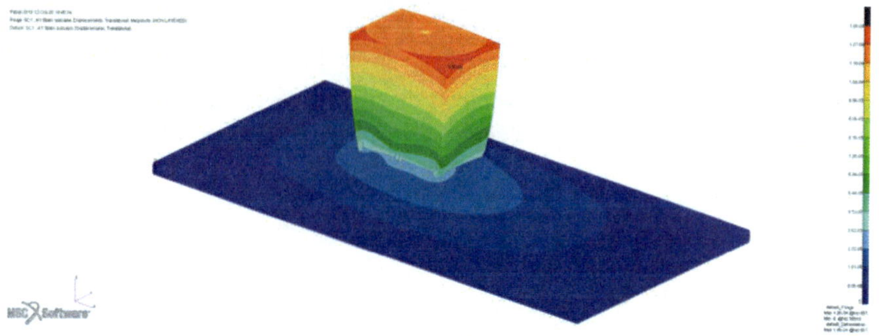

Fig. A.9 Displacement induced by the temperature field

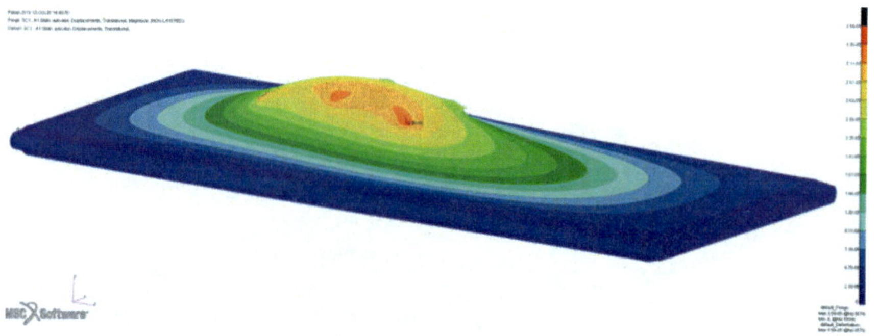

Fig. A.10 Displacement (out of plane bending) on the sandwich panel induced by the temperature field

The corresponding displacement field as presented in Figs. A.9 and A.10 has an effect that seems to reach further away from the box. Especially, Fig. A.10 shows well the displacement field induced in the panel by the box. This shows that if one would judge the impact only on the temperature field, then incorrect or incomplete conclusions may be drawn for the deformation field.

A local effect that is not visible from Fig. A.10 is the increase of thickness of the honeycomb. The net effect is that the lower face sheet is still moving upward. This is a phenomena that only thanks to detailed modelling can be observed. This aspect is here not further investigated.

The objective of the exercise explained above is recalled and is to examine:

- The effect on the tilting of three imaginary instruments due to the thermoelastic deformation induced by an electronics box on the same panel.
- How this tilting changes with a change of position of the box.
- Whether the height of the box has an influence on the deformation of the panel and hence on the tilting of neighbouring instruments.
- The effect on the tilting of the imaginary instruments of a combined change of size and heat dissipation of the electronic box.

For this purpose, the model is constructed such that all dimensions can be altered to generate easily different configurations of the panel with the box mounted on it. The parameters that are varied for this specific investigation are:

- Height of the box
- Width of the box
- Length of the box
- Position of the box on the plate (two coordinates)
- Heat dissipation of the box

For a single box configuration, 32 analysis cycles are run for each of the box positions in the upper left quadrant of the plate (see Fig. A.3). For each of the 32 positions of the box, the steady-state thermal analysis has been run, temperatures have been mapped on the structural finite element model, and the corresponding displacement field has been computed.

The displacements of the corner points of the footprints of the virtual instruments (see Figs. A.2 and A.3) have been recovered to determine the vectors corresponding to the two diagonals of these tilted and deformed footprints. The vector product of these two diagonal vectors per instrument provides the normal of the tilted interface plane of the instrument. After normalisation, the X-component of this normal vector equals to the Y-rotation. Likewise, the Y-component of the normal vector provides the X-rotation of the interface plane. All these rotations are in radians, but these are converted to arc-seconds for the presentation of the results.

Due to the fact that the four corner points of the imaginary instruments are relatively far away from each other, the tilting angles produced by the method described above can be considered as average rotation of the panel surface covered by the instrument foot print. This means that locally, the rotations may be higher or lower. This may be important for instruments with a smaller foot print such as a star tracker.

For the investigation, four geometrical configurations were included. Three of them have a base of 30 × 20 cm and are used with three different heights: 30, 15 and 7.5 cm. Because these three configurations all have a rather large foot print, also a box of half the size of the largest box, i.e.15 × 10 × 15 cm, is included in the investigation.

Except for the largest box, all boxes are analysed with two sets of heat dissipation levels. In the first set, every box is subjected to 30 W dissipation. In the second set, the heat dissipation per unit volume is kept constant, such that the heat dissipations per box size become as presented in Table A.1. In the latter set, the dissipation per box is proportional to its size.

For this investigation, in total, 224 box and plate configurations are constructed, analysed and post-processed.

Table A.1 Heat dissipation per box with constant power per unit volume

Box dimensions (h × w × d) [cm]	Heat dissipation (Q) [W]
30 × 20 × 30	30
15 × 20 × 30	15
7.5 × 20 × 30	7.5
15 × 10 × 15	3.75

A.5 Detailed Results and Evaluation

A.5.1 Structure of this Section

In this section, the results of the 224 runs are presented through filled contour plots. These contour plots are found in Figs. A.11, A.12, A.13, A.14, A.15, A.16, A.20, A.21, A.22, A.23, A.24, A.25 and A.26, A.27, A.28, A.29, A.30, A.31. Along the axis of the plots, the coordinates of the box positions are indicated. These coordinates are the ones explained in Figs. A.2 and A.3.

The first part of the evaluation of the results is focussed on global observations of the responses of the imaginary instruments.

The second and largest part of this section is dedicated to obtaining understanding of the differences in the levels of deformation induced by the box configurations considered in combination with different heat dissipation levels.

None of these parts are claimed to be conclusive but have the intention to focus on the most interesting and important aspects.

A.5.2 Global Observations

The analyses use camped boundary conditions at the edges of the sandwich panel. In normal applications, the edges are attached to other panels that provide a softer interface. Clamped boundary conditions cause lower levels of the deformation of the sandwich panel. Although not reported, the cases in this experiment are all run as well with free-free boundary conditions using inertia relief. This confirmed that clamped boundary conditions provide a lower bound for the deformation of the sandwich panel and also must be kept in mind that for larger spacecraft, the panels may be bigger than the one used for this numerical experiment. A larger panel will have a lower stiffness and the will cause the deformation responses to be higher. On the other hand, the larger panels have often attachments to supporting panels such as shear webs.

The results presented are all due to the effect of a single heat dissipating box on the panel. It is likely that when multiple boxes are mounted to the same side of the

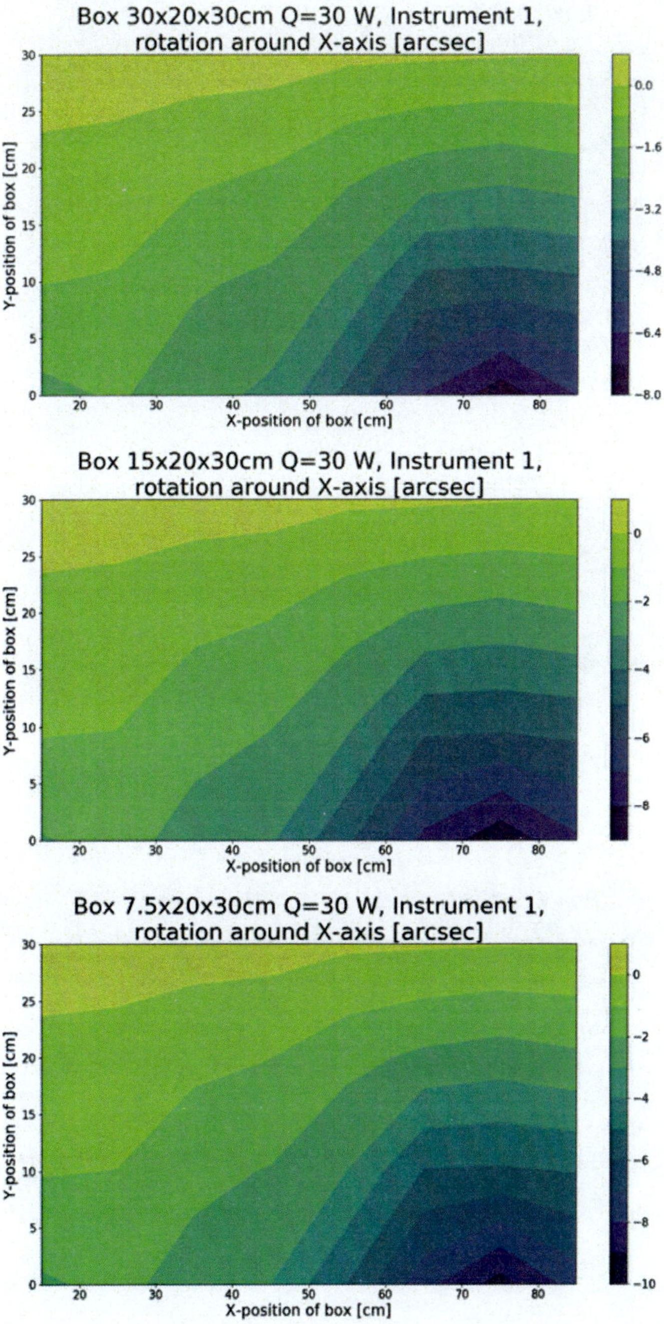

Fig. A.11 Influence of box size on the rotation around the X-axis for imaginary instrument 1 with equal dissipation of 30 W for each box

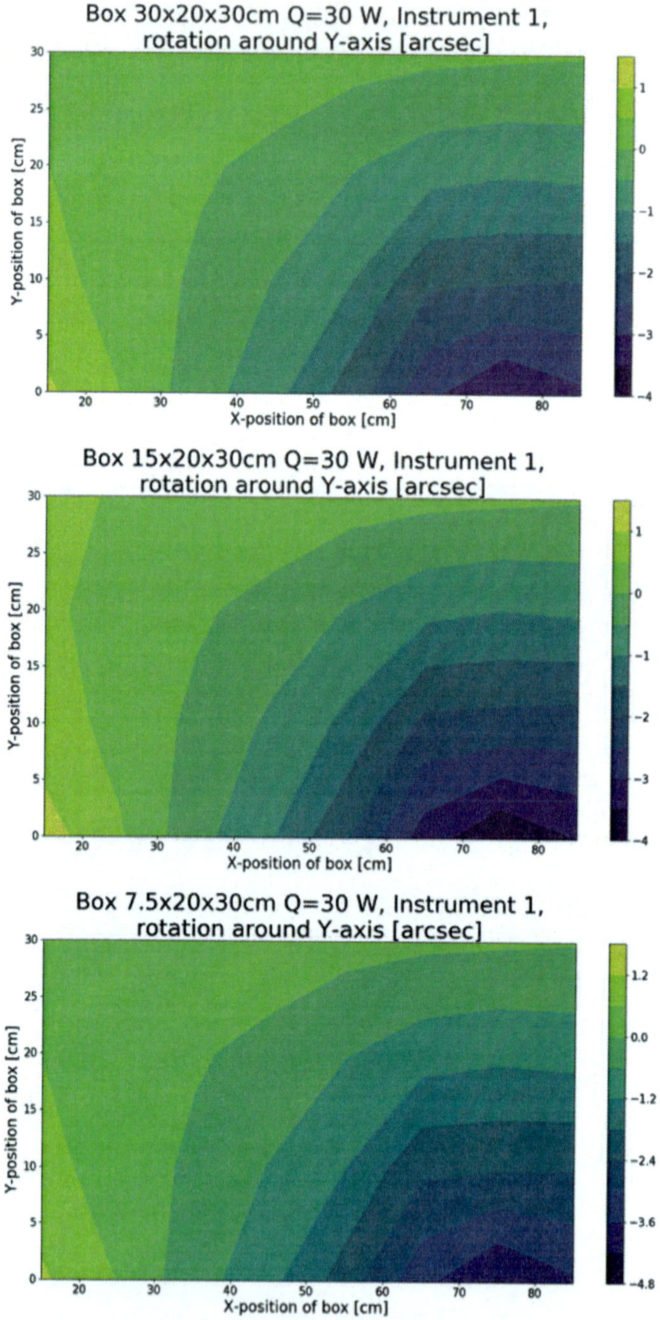

Fig. A.12 Influence of box size on the rotation around the Y-axis for imaginary instrument 1 with equal dissipation of 30 W for each box

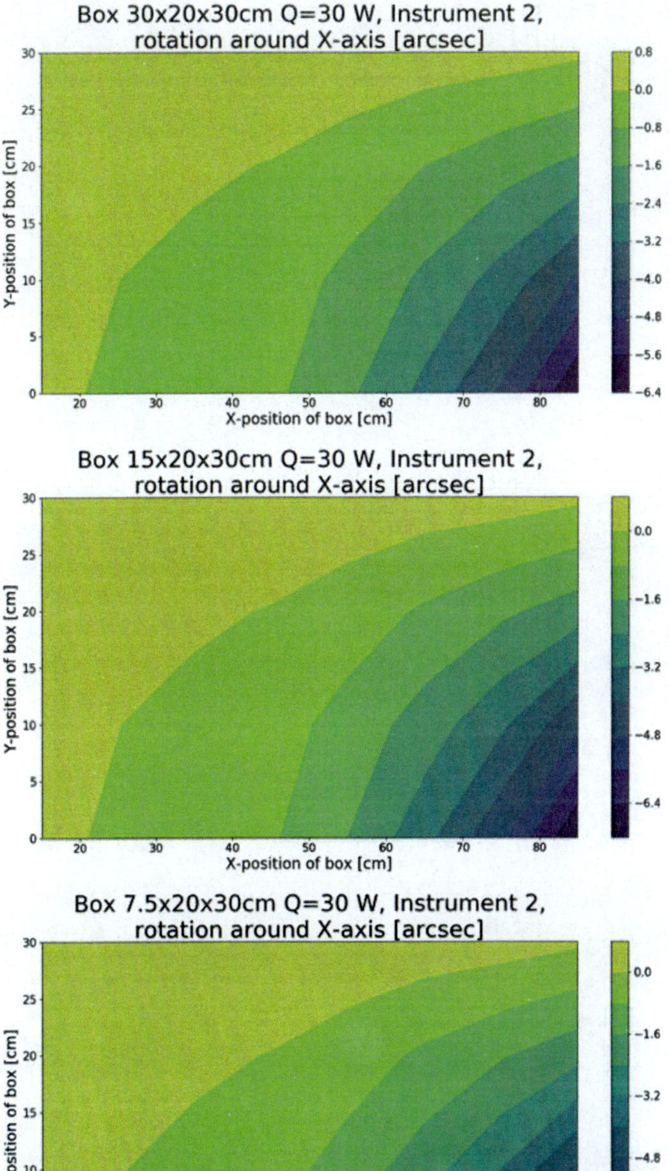

Fig. A.13 Influence of box size on the rotation around the X-axis for imaginary instrument 2 with equal dissipation of 30 W for each box

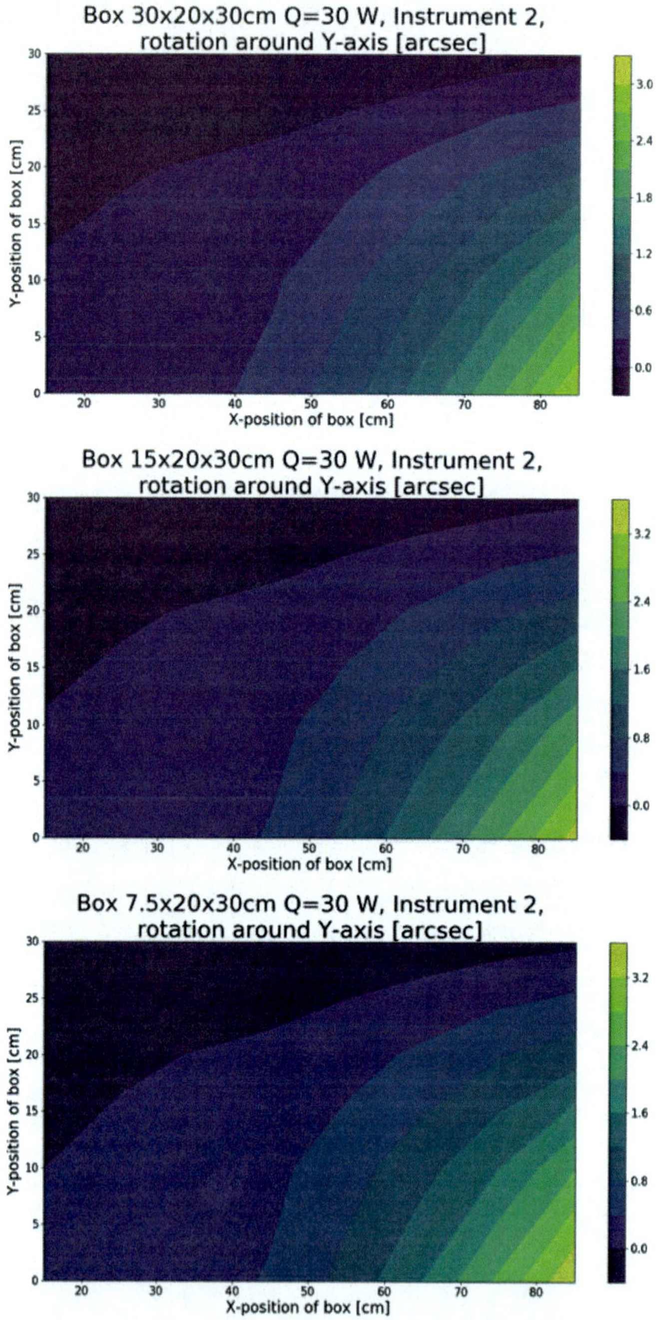

Fig. A.14 Influence of box size on the rotation around the Y-axis for imaginary instrument 2 with equal dissipation of 30 W for each box

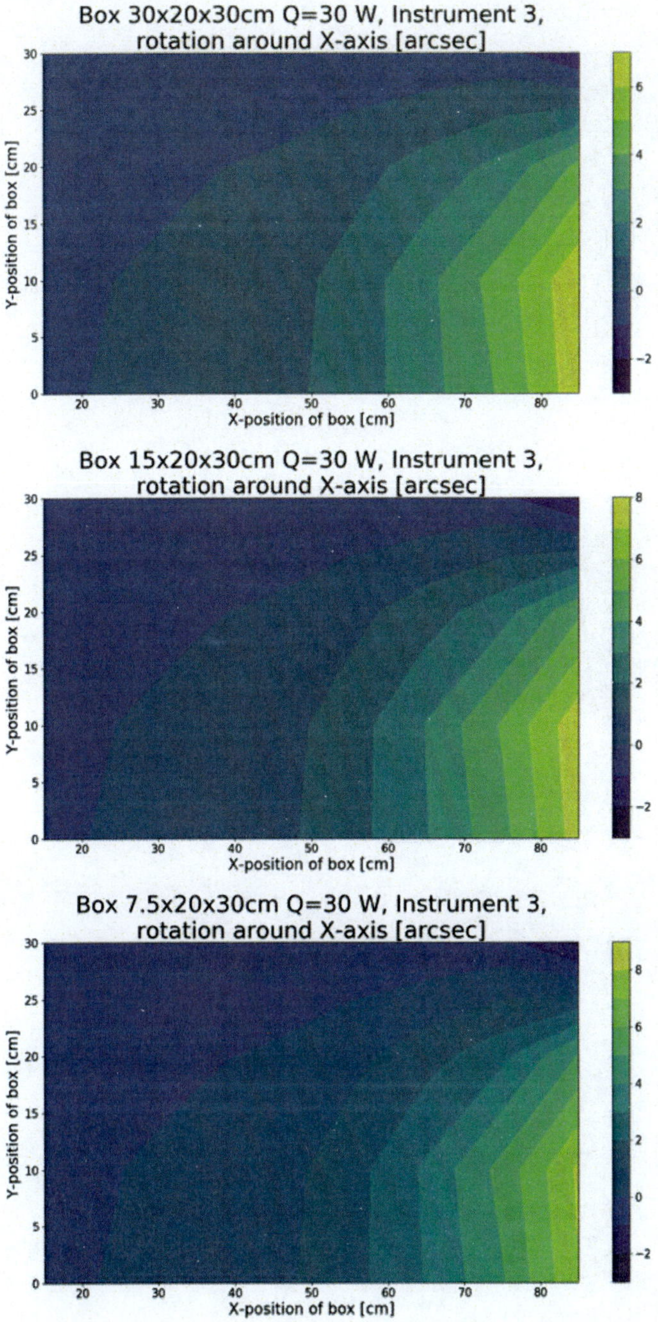

Fig. A.15 Influence of box size on the rotation around the X-axis for imaginary instrument 3 with equal dissipation of 30 W for each box

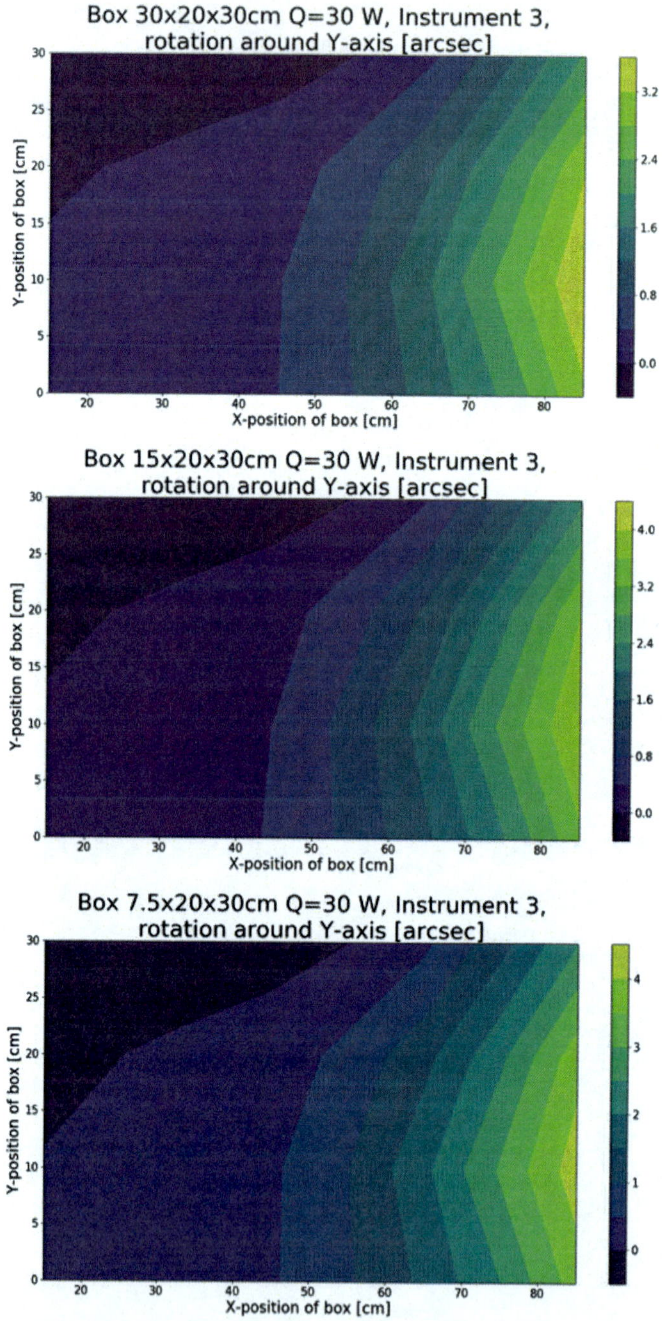

Fig. A.16 Influence of box size on the rotation around the Y-axis for imaginary instrument 3 with equal dissipation of 30 W for each box

panel, the combined effect of these individual boxes can lead to a superposition of the deformation levels.

Scanning through the various contour plots in the next sub-section, one can find tilting angles for the imaginary instruments due to the effect of a single electronics box up to 10 arc s. This is well approaching the levels that were considered critical for the instruments on several recently launched spacecraft and for instruments on spacecraft under development. In many cases, the rotation of two instruments relative to each other is also of importance. For the rotations around the X-axis can be seen that imaginary instruments 2 and 3 have opposite rotations and therefore important relative rotations. The same effect can be observed for the rotations around the Y-axis for instruments 1 and 2.

As one would expect, the box positions near the edges of the panel introduce the lowest tilting of the imaginary instruments. This is along the lines X = 15 cm and Y = 30 cm. The edges of the panel constrain the deformation that the box tries to induce in the panel. These positions might then be the best places to locate the electronic boxes in case one wants to limit the effect on instruments on the same panel.

Since the effect of the boxes located near the edges of the panel is quite low, one could consider to allow a more simplified modelling of the boxes at these locations. However, it is recommended to obtain first the confirmation that the effect is really small. The influence of a box on the rest of the panel may depend on a number of circumstances. As can be seen in the next section, the thermal gradient in the box appears an important parameter.

The constraining effect of the edges becomes less strong when the box is positioned further away from the edges. In general, the highest deformation is induced in the panel when the box is positioned near the centre of the sandwich panel. This is consistently shown by the contour plots.

In Fig. A.3, it can be seen that along the line Y = 0 cm and X = 85 cm for the bigger boxes considered, there is some interference between the box footprint and the footprints of the imaginary instrument. This may have an effect on the results. Nevertheless, the results follow a trend that physically makes sense.

A.5.3 Understanding the Effect of Different Box Configurations

The first results discussed are those for the three boxes with the footprint of 30 × 20 cm and all with simulated internal heat dissipation of 30 W. The results are presented in sets of three plots, all three presenting the same quantity being the rotation around X or Y, for the same instrument, but for the three different heights of the boxes. The combined presentation of the results related to the different box heights intends to ease the identification of the effect of the box height on the instrument tilting. These results are presented in Figs. A.11, A.12, A.13, A.14, A.15 and A.16.

The tilting of the virtual instruments is in general higher with the reduction of the height of the box. Does this mean that a box with a lower height is inducing a higher bending of the sandwich panel? What is causing this effect? A first possible cause could be that a smaller box with the same internal heat dissipation is getting higher temperatures. To get the confirmation of this expectation, the temperature fields of the three boxes need to be investigated. The location with the highest temperatures of each box is at the top panel of the box.

The temperature field of the top panel has a maximum temperature approximately in the middle. In Fig. A.17, is shown the temperature field in the top panel of a box of 30 × 20 × 30 cm at the position X = 35 cm and Y = 30 cm on the panel for a heat dissipation of 30 W.

The maximum temperatures of the top panel of the boxes with three different heights are compared in Table A.2 for the same position on the sandwich panel and the same internal heat dissipation of 30 W.

The temperatures in Table A.2 for the different box height do not support the expectation that the box with the smaller height reaches a higher temperature and therefore would then cause a higher deformation of the sandwich panels. In fact, the opposite temperature trend is shown. The smaller box is slightly less warm than the

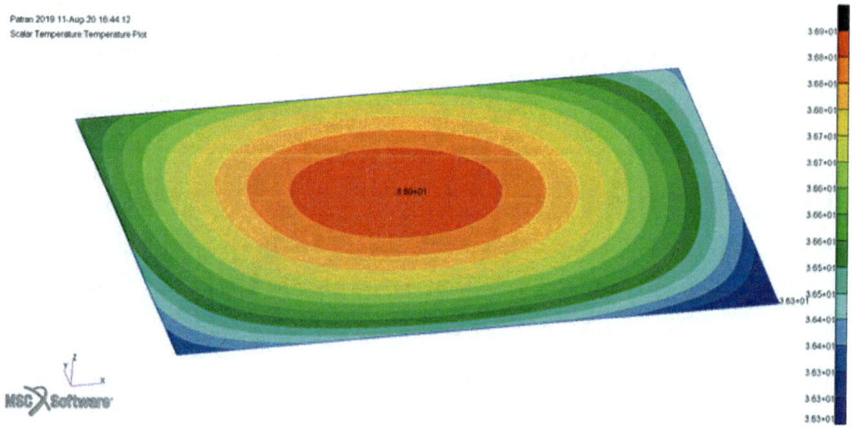

Fig. A.17 Temperature distribution of the top panel of the box with dimensions 30 × 20 × 30 cm for position X = 35 cm and Y = 30 cm. Q = 30 W

Table A.2 Maxiumum temperature at top panel for three box configurations at position X = 35 cm and Y = 30 cm. Q = 30 W

Box dimensions (h × w × d) [cm]	Top panel Max temperatures [°C]
30 × 20 × 30	36.9
15 × 20 × 30	36.1
7.5 × 20 × 30	35.2

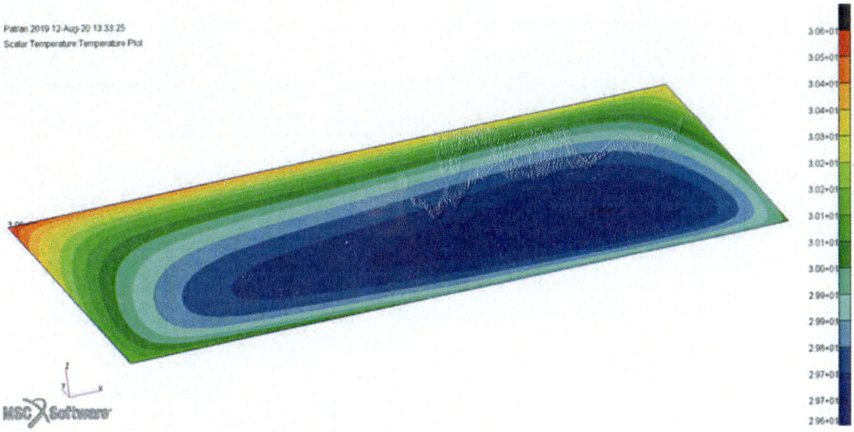

Fig. A.18 Temperature distribution of the base plate of the box with dimensions 30 × 20 × 30 cm for position X = 35 cm and Y = 30 cm. Q = 30 W. (feet removed from plot)

two bigger boxes. A possible explanation for this effect is that for the smaller box, the heat source, i.e. the uniform heat applied to the box walls, is closer to the heat sink, which is the sandwich panel. Because of the shorter heat flow path, a lower conductive resistance is present allowing the heat to flow easier from the box to the panel, resulting in a lower temperature.

So the top panel temperature is not a basis to explain that the smaller box is causing a higher deformation. Could it then be that the base plate temperature provides a better indication of the cause of the difference in deformation? In Fig. A.18 is the temperature field in the base plate of the box shown for a box of 30 × 20 × 30 cm at the same position and heat dissipation as was used for Fig. A.2, i.e. X = 35 cm and Y = 30 cm for a heat dissipation of 30 W.

The minimum temperatures of the base plate of the boxes with the three different heights are compared in Table A.3 for the same position on the panel and the same internal heat dissipation of 30 W.

The temperatures of the base plate do increase with reducing height of the boxes. The increase of the temperature of the base plate relative to the reference temperature of 20 °C for the largest box is 9.6 °C for the smallest box this is 11.2 °C. This may be the explanation for the higher deformation of the sandwich panel but the relative difference in temperature increase is quite small and is not so convincing.

What else could then explain the observed higher deformation of the sandwich for the smallest box? Is the temperature difference between top and bottom panels of the box then a determining factor? In Table A.4, the temperature differences between top and bottom panel of the boxes are listed. Also, these values do not support an explanation for the differences in deformation between the different box sizes. The smallest box shows the smallest difference in temperature between top and base panel, what is opposite to what is needed to support the explanation.

Table A.3 Minimum temperature at base plate for three box configurations at position X = 35 cm and Y = 30 cm. Q = 30 W

Box dimensions ($h \times w \times d$) [cm]	Base panel Min temperatures [°C]
30 × 20 × 30	29.6
15 × 20 × 30	30.7
7.5 × 20 × 30	31.2

Table A.4 Delta temperature between top and base panel for three box configurations at position X = 35 cm and Y = 30 cm, Q = 30 W

Box dimensions (h × w × d) [cm]	Top panel Max temperatures [°C]	Base panel Min temperatures [°C]	ΔT [°C]
30 × 20 × 30	36.9	29.6	7.3
15 × 20 × 30	36.1	30.7	5.4
7.5 × 20 × 30	35.2	31.2	4.0

The observations so far were limited to temperature levels and difference in temperatures. What is not yet checked is the temperature difference per unit box height, in other words the temperature gradient through the height of the box. With the delta temperature values from Table A.4, the average thermal gradient values in Table A.5 are computed.

The average thermal gradient values in Table A.5 show important relative differences between the different box sizes that could well explain the difference in sandwich panel deformation for the different box configurations. One can imagine a box to be a sort of sandwich with the top and bottom panels being the face sheets of the “box sandwich”. The boxes with their relatively high wall thicknesses, acting as face sheets of the “box sandwich”, compared to the sandwich face sheet, have a relatively high stiffness compared to the panel. A thermal gradient through the thickness of the “box sandwich” will induce a curvature in this “box sandwich”. The deformation of the box with its high stiffness is therefore well able to be imposed as deformation to the less stiff sandwich panel. The curvature of the box due to the thermal gradient is proportional to the thermal gradient.

Figure A.19 shows nicely the deformed shape of the 7.5 cm high electronics box. This deformed shape illustrates well the mechanical principle of the curvature induced by the box into the panel.

This looks like a consistent explanation for the three of the seven box and dissipation configurations analysed. Can the thermal gradient over the height of the box also explain the relative differences for the other configurations? To verify this, first the contour plots are presented for the same three box configurations with footprint of 20 × 30 cm and heights of 30, 15 and 7.5 cm, but now with the internal heat dis-

Table A.5 Average thermal gradients over the height of the for three box configurations at position X=35 cm and Y=30 cm. Q=30 W

Box dimensions (h × w × d) [cm]	$\Delta T/\Delta h$ [°C/cm]
30 × 20 × 30	0.24
15 × 20 × 30	0.36
7.5 × 20 × 30	0.53

Fig. A.19 Curvature in the shape of the box due to the thermal gradient along the height of the box

sipation proportional to the volume of the box. This implies keeping the dissipation density the same for the three boxes.

The results for these three box configurations are presented in the same way as before per imaginary instrument and per tilting angle axis in Figs. A.20, A.21, A.22, A.23, A.24 and A.25.

For this set of results, the opposite effect of what is found for the previous set is observed: The tilting of the imaginary instruments reduces with the height of the boxes. In this case also, the total heat dissipation per box reduces with the height of the box. Also now, the thermal gradients are calculated to confirm that this quantity is also for this set of cases the driving factor for the deformation of the sandwich panel. Again the results are taken for the box position on the panel of X = 35 cm and Y = 30 cm. The gradients are presented in Table A.6

The results in Table A.6 support well the earlier identified relation between the thermal gradient and the imposed deformation of the sandwich panel. Figures A.20, A.21, A.22, A.23, A.24 and A.25 show that the box with the lowest height this time is producing the smallest deformation in the sandwich panel. This is now confirmed to be caused by the fact that this box has the lowest thermal gradient over the height of the box.

The boxes discussed so far had a rather large footprint. For that reason, two sets of data were produced for a box half the size of the largest box discussed so far. This means a box with dimensions (h × w × d) of 15 × 10 × 15 cm. This box configuration has been analysed with the 30 W internal heat dissipation and with 3.75 W (1/8 portion of 30 W). The latter corresponds to heat dissipation density of

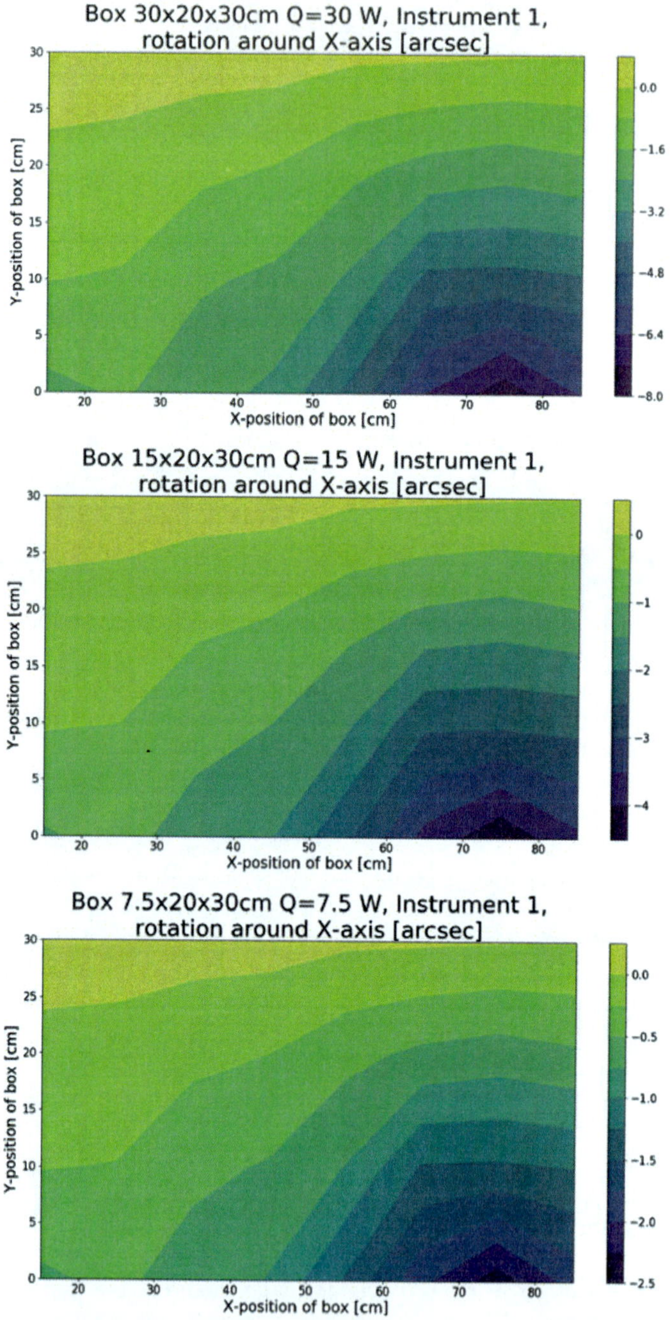

Fig. A.20 Influence of box size on the rotation around the X-axis for imaginary instrument 1 with equal dissipation density for each box

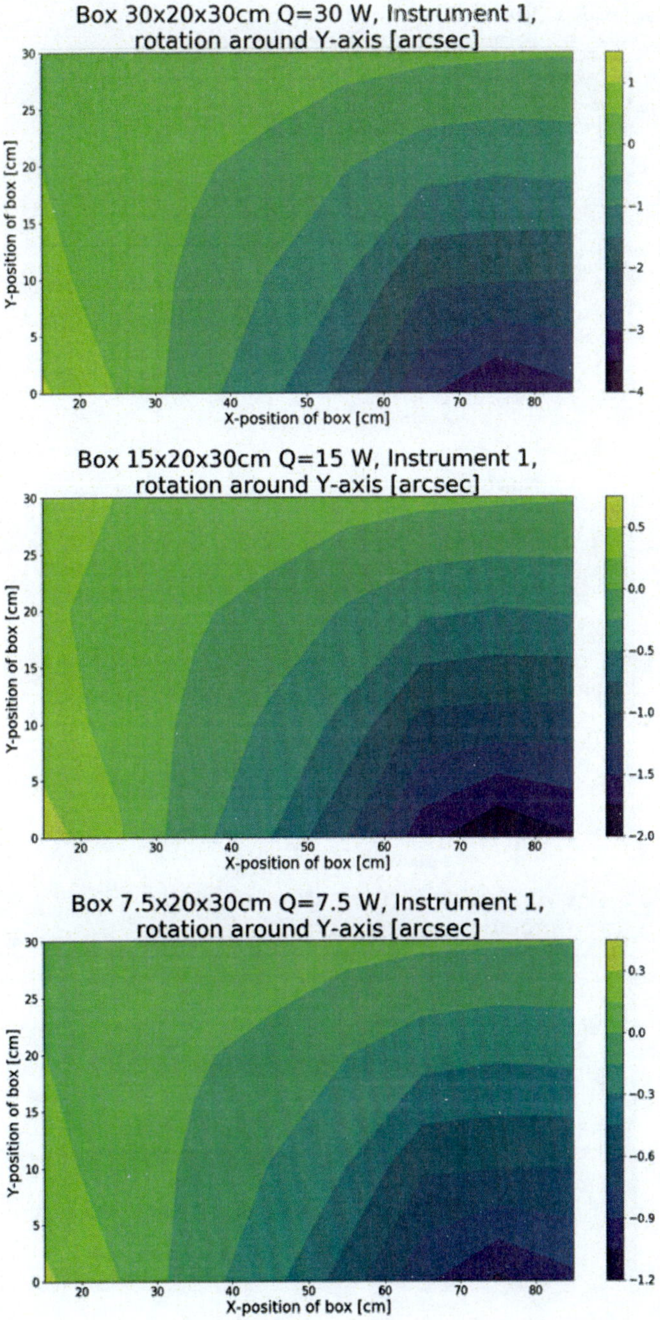

Fig. A.21 Influence of box size on the rotation around the Y-axis for imaginary instrument 1 with equal dissipation density for each box

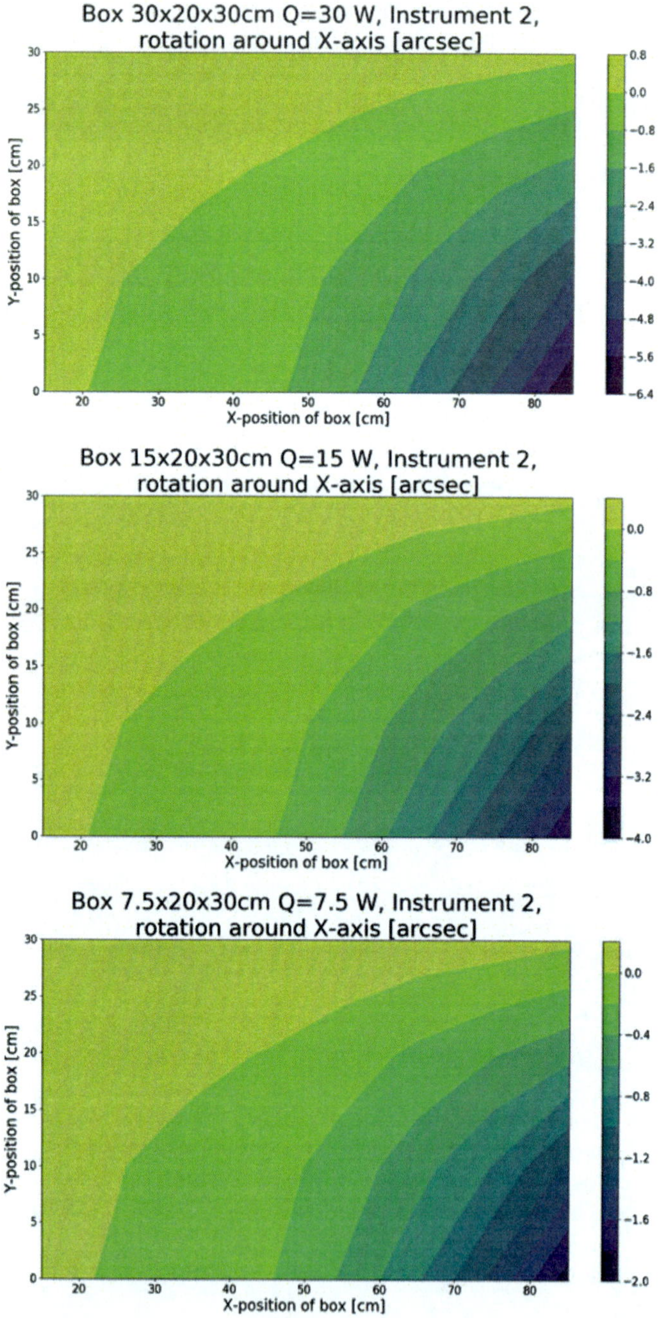

Fig. A.22 Influence of box size on the rotation around the X-axis for imaginary instrument 2 with equal dissipation density for each box

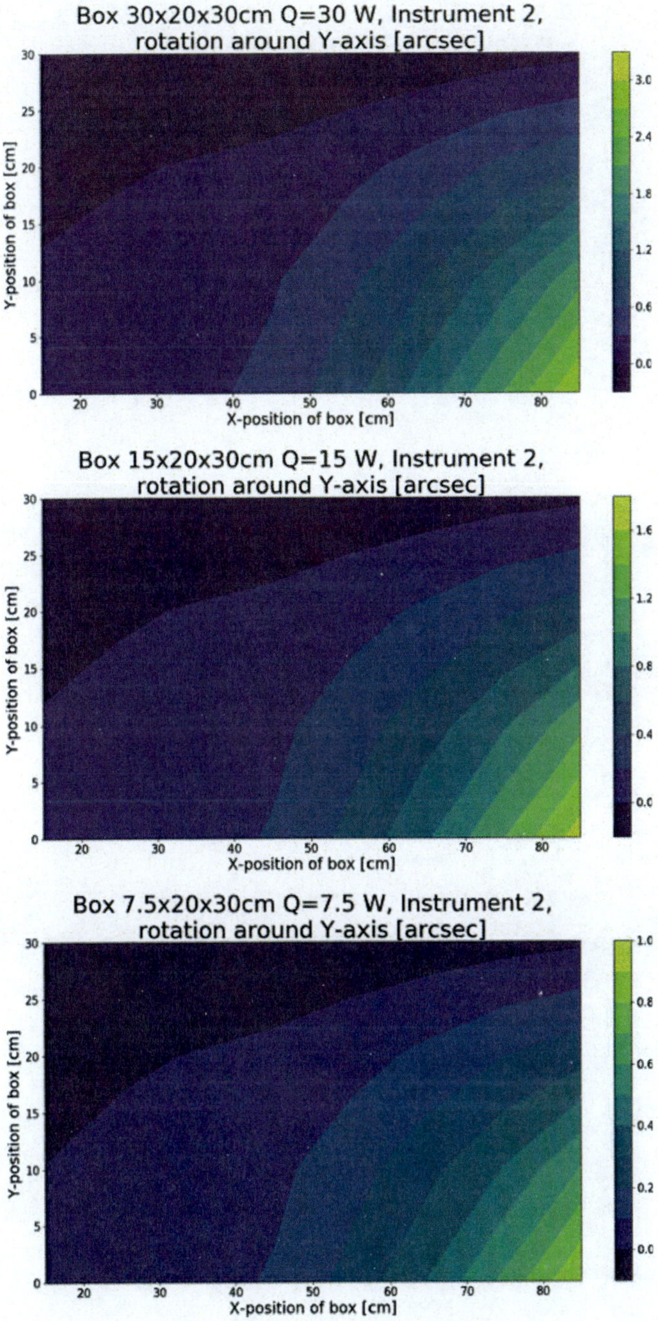

Fig. A.23 Influence of box size on the rotation around the Y-axis for imaginary instrument 2 with equal dissipation density for each box

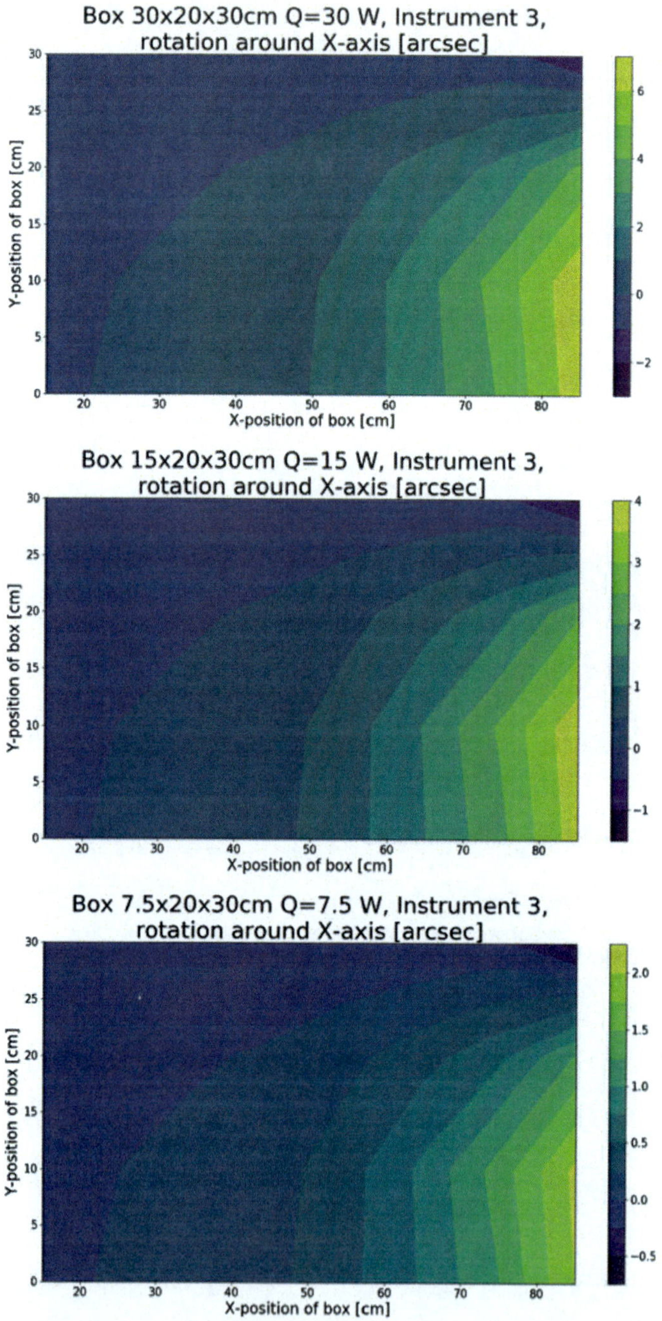

Fig. A.24 Influence of box size on the rotation around the X-axis for imaginary instrument 3 with equal dissipation density for each box

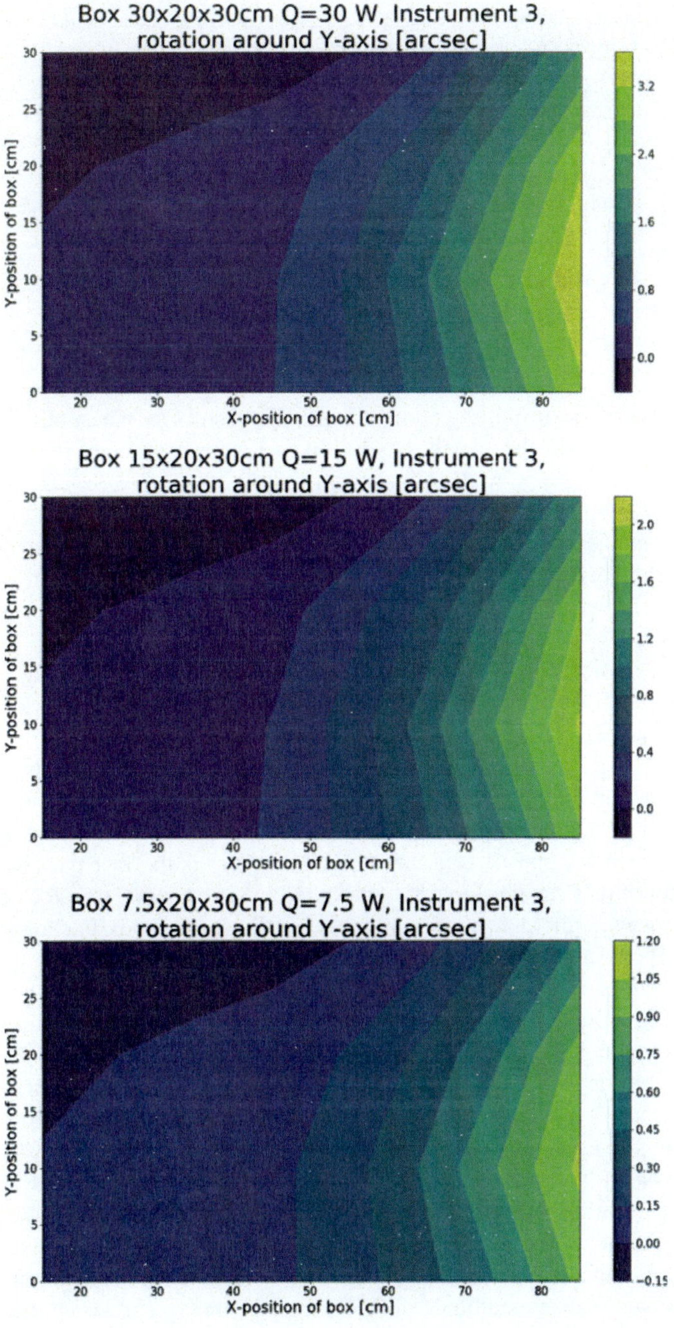

Fig. A.25 Influence of box size on the rotation around the Y-axis for imaginary instrument 3 with equal dissipation density for each box

Table A.6 Thermal gradients over the height of three box configurations at position X = 35 cm and Y = 30 cm with constant heat dissipation density per box

Box dimensions (h × w × d) [cm]	Q total [W]	Top panel max temperatures [°C]	Base panel min temperatures [°C]	ΔT [°C]	$\Delta T/\Delta h$ [°C/cm]
30 × 20 × 30	30	36.9	29.6	7.3	0.24
15 × 20 × 30	15	28.1	25.4	2.7	0.18
7.5 × 20 × 30	7.5	23.9	22.9	1.0	0.13

the largest box with a heat dissipation of 30 W. The results for the smaller boxes are presented together with those of the largest box to provide a reference for comparison in Figs. A.26, A.27, A.28, A.29, A.30 and A.31.

As was done with the previous two sets of data, the gradients along the height of the boxes are computed and presented for comparison to check the relation between the relative levels of deformation and the relative levels of thermal gradients. In Table A.7, the gradients in two small box configurations are compared with the gradient in the large box.

From the gradient levels presented in Table A.7, one could expect that the small box with a heat dissipation of 30 W would produce the highest distortion of the sandwich panel of the three boxes listed in the table. It turns out that this is not the case. The larg box with 30 W heat dissipation is causing a higher distortion of the sandwich than the smaller box with 30 W heat dissipation, despite the latter has the highest thermal gradient. Apparently, also another effect plays a more dominant role. It is considered most likely that this other effect is the fact that the smaller box has a four times smaller footprint (measured in surface area) than the large box. As a consequence, the deformation induced by the smaller box has a more local character. At the same time, the larger box is influencing directly a larger part of the panel. This makes it likely that the small box, despite its higher thermal gradient, is not producing more deformation in the panel than the larger box. This implies that besides the thermal gradient over the height of the box, also the size of the box an important factor is for the magnitude of the deformation induced by a box into the sandwich panel.

Linked to size aspect is the thermal expansion of the stiff base plate. A larger box has a larger base plate and has therefore the ability to impose the same thermal strain to a larger part of the top skin of the sandwich panel compared to the base plate of the smaller box. This imposed stretching (or shrinking) of one of the face sheets will cause bending deformation in the sandwich panel. The numerical experiments do not indicate the relative magnitude of this skin stretching or shrinking effect. Clear is however that besides the level of thermal gradients, also the size of the foot print is an important parameter for the level of distortion that is transferred from the box to the panel.

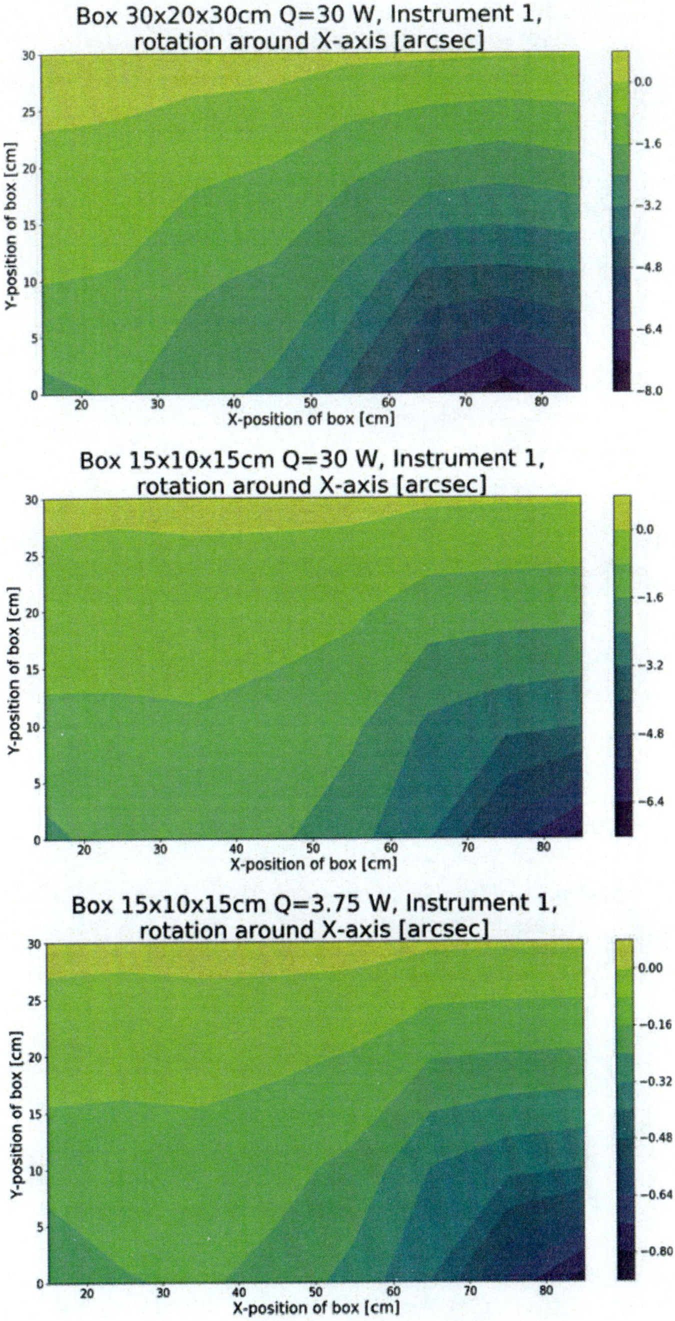

Fig. A.26 Comparing the rotations around the X-axis for imaginary instrument 1 for half size box with large box

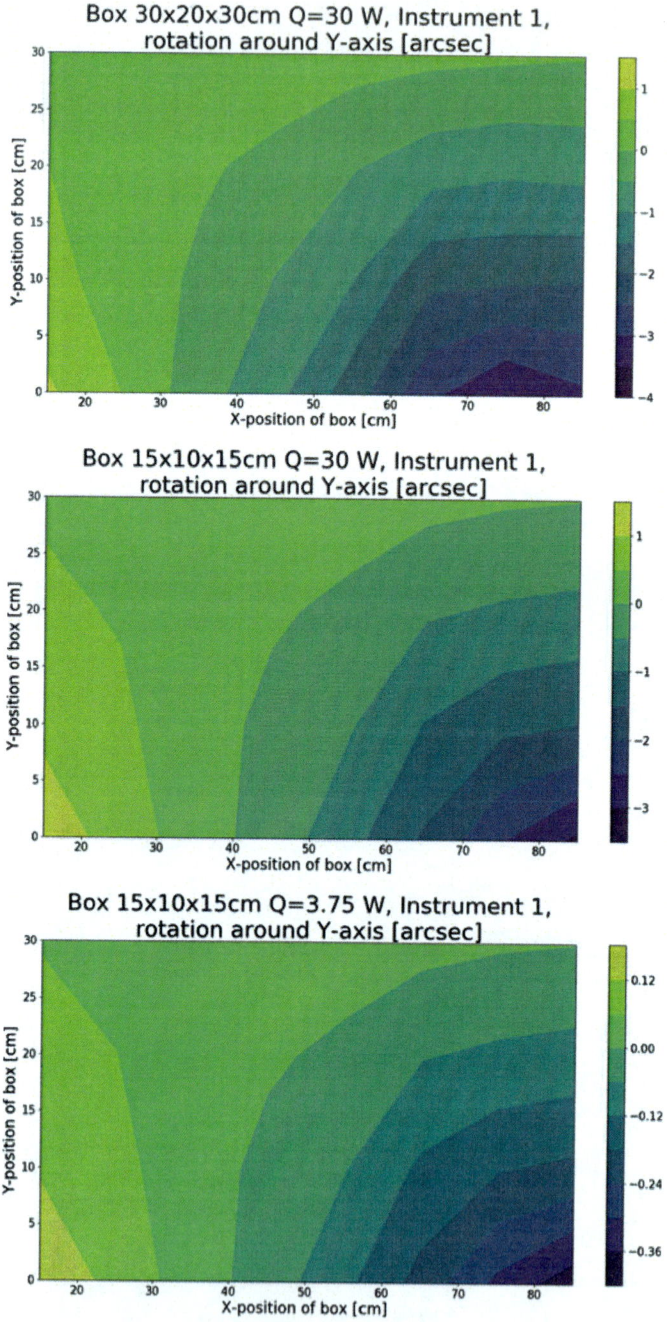

Fig. A.27 Comparing the rotations around the Y-axis for imaginary instrument 1 for half size box with large box

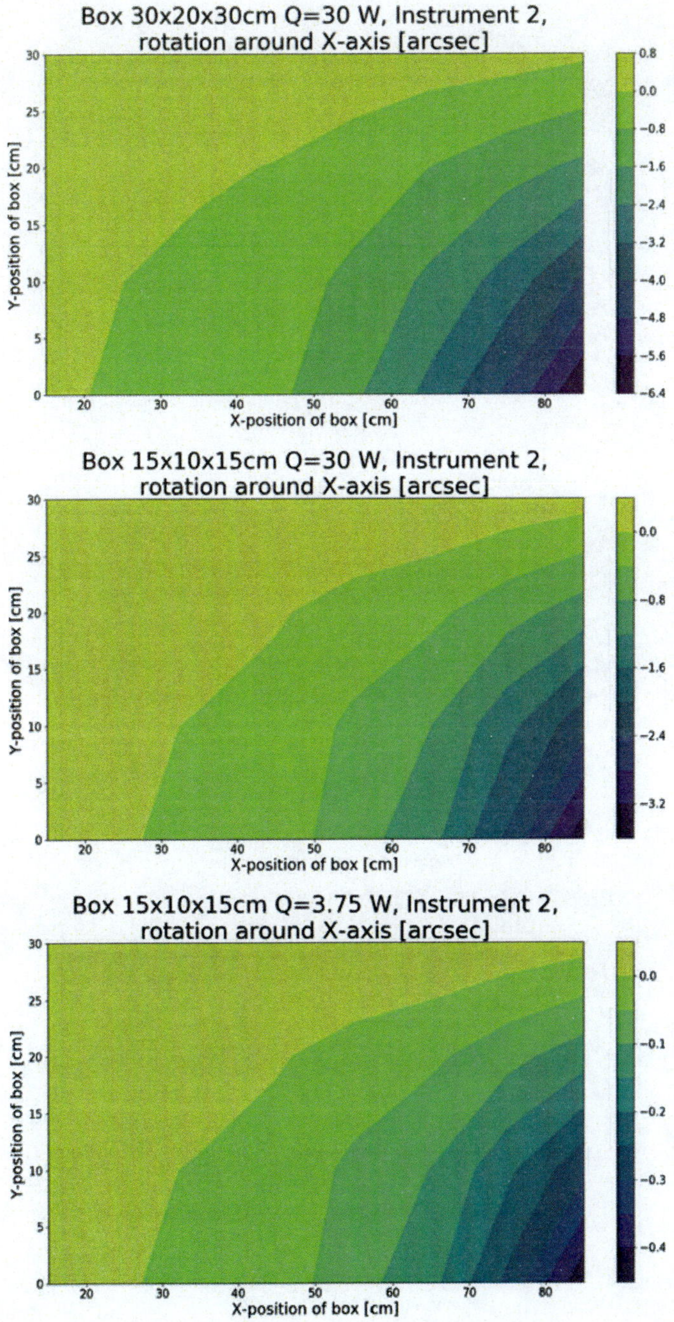

Fig. A.28 Comparing the rotations around the X-axis for imaginary instrument 2 for half size box with large box

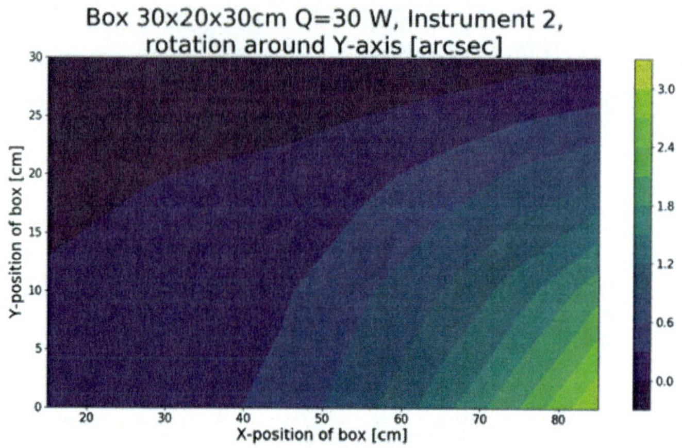

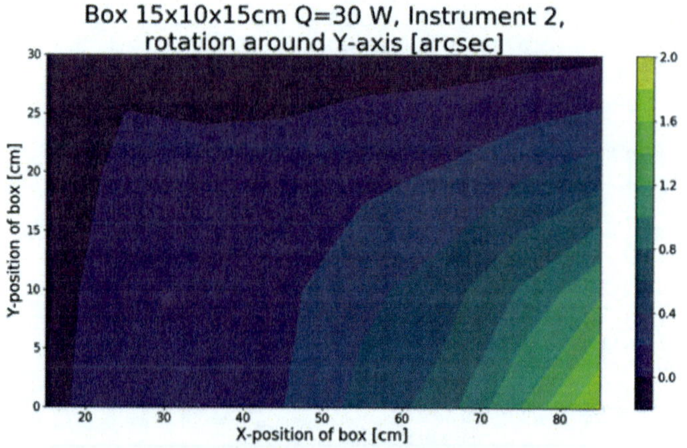

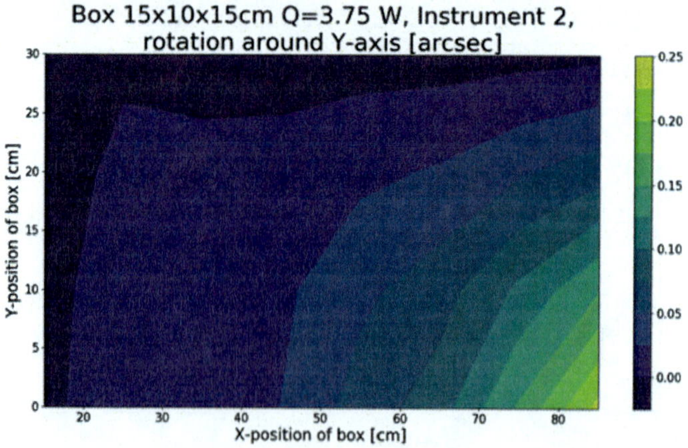

Fig. A.29 Comparing the rotations around the Y-axis for imaginary instrument 2 for half size box with large box

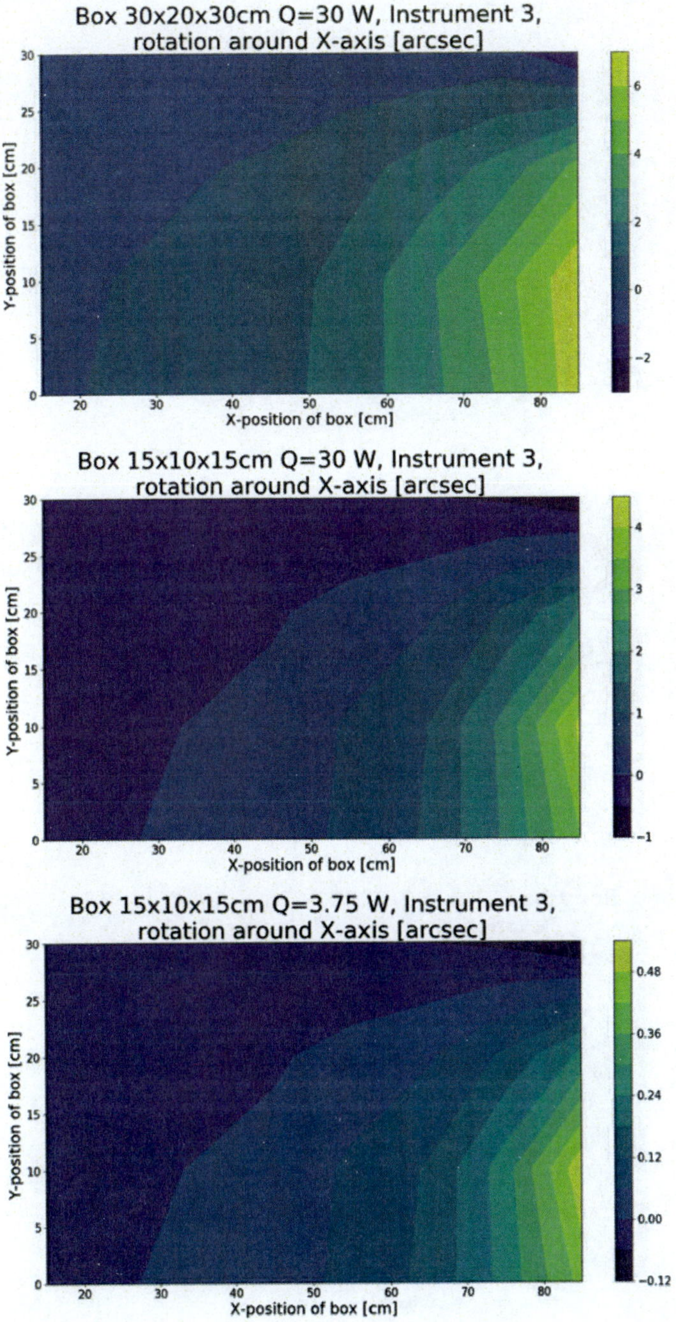

Fig. A.30 Comparing the rotations around the X-axis for imaginary instrument 3 for half size box with large box

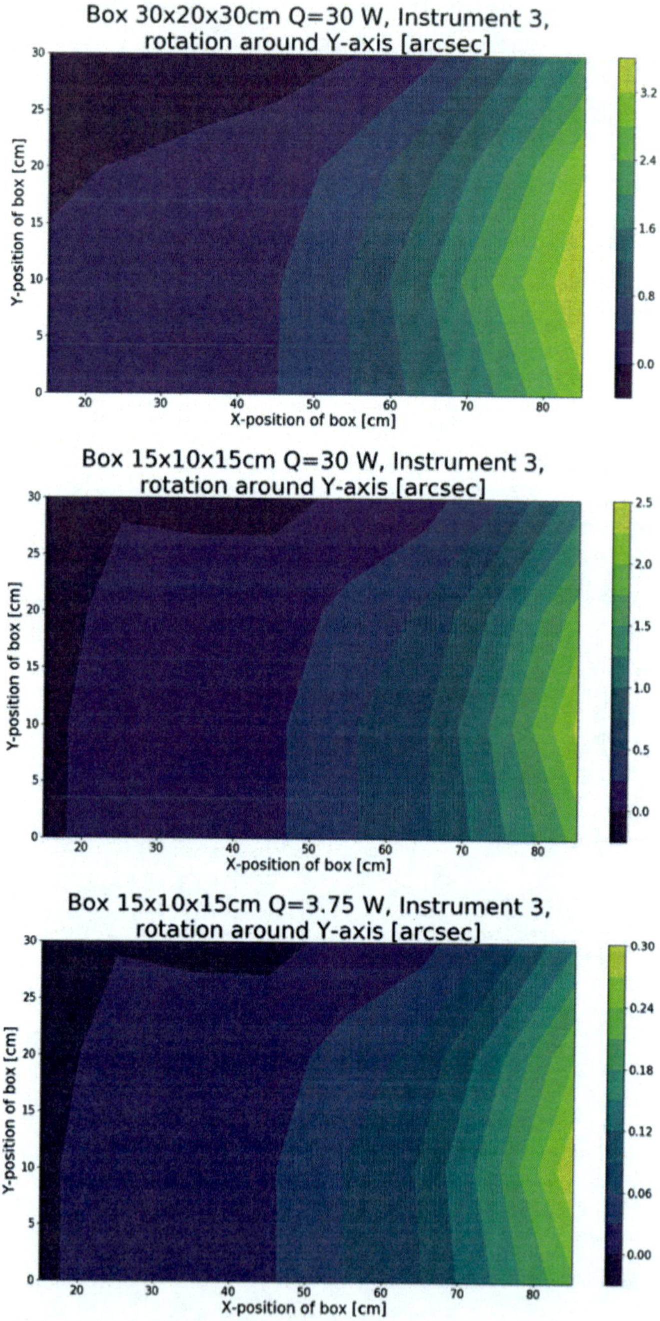

Fig. A.31 Comparing the rotations around the Y-axis for imaginary instrument 3 for half size box with large box

Table A.7 Thermal gradients over the height of three box configurations at position X=35 cm and Y=30 cm with constant heat dissipation density per box

Box dimensions (h × w × d) [cm]	Q total [W]	Top panel Max temperatures [°C]	Base panel Min temperatures [°C]	ΔT [°C]	$\Delta T/\Delta h$ [°C/cm]
30 × 20 × 30	30	36.9	29.6	7.3	0.24
15 × 10 × 15	30	46.1	36.5	9.6	0.64
15 × 10 × 15	3.75	23.3	22.1	1.2	0.08

A.6 Simplified Modelling of an Electronic Box on a Sandwich Plate: Preserve the Physics

In Fig. A.1, a typical modelling of an electronic box in a system-level model is presented. The box is represented by a rigid element and a lumped mass. This way of modelling has proven to be adequate for dynamic simulations. Often, it is assumed that by just adding a CTE value to the RBE element, the simplified "RBE-box" is behaving quite similar to a box modelled in detail.

Similarly, in the system thermal model, a simplified thermal representation for the same box is implemented with only a few thermal nodes.

The mapping of the limited thermal node temperatures to the RBE GRIDs is likely to introduce some ambiguity. The GRID with the lumped mass can be assigned the average temperature of the box. This immediately shows the loss of information about the gradient over the height of the box. The GRIDs at the end of the legs of the RBE represent the feet of the box. When the heat dissipation inside the box is as nicely uniform in the boxes presented above, then the temperatures of these feet are all more or less the same (see for instance Fig. A.8). In real life, the temperatures at each of the feet may differ between each other, but it is expected that the differences in temperature will not be big.

With the RBE leg temperatures being all more or less the same and all sharing the single centre GRID, all the legs have a the same thermal expansion normal to the sandwich surface. As a consequence, the plane of the interface points between sandwich plate and box remains flat. So the RBE with CTE is not at all showing any form of curvature that could be transferred to the sandwich panel. This means that computed thermomechanical deformations of sandwich panels with electronic boxes modelled with RBE elements are wrong. This way of modelling even appears to reduce deformation of the sandwich panel.

It is understandable that a system-level model may become unpractical when all equipment is modelled in full detail. From the previous section, it became clear that the bending stiffness of the box and the ability to capture the gradient, which is essential for generating the curvature in the shape of the box, are the essential ingredients of the physics of the "box on plate" problem.

This requires that both the thermal and structural model, have sufficient resolution to capture the gradient. In addition, it needs to be identified which elements of the box structure are important for the stiffness properties. The modelling of these structural items needs to be present with sufficient resolution in both the structural and thermal models. The development of a simplified, but detailed enough, representation of the structure is an iterative process. During this process, it is recommended to base decisions on which structural items to be included or not, on models that have sufficient resolution in order to prevent that effects are masked by too low mesh resolutions.

A.7 Conclusions

From above exercise, one clear message can be deducted: Temperature levels and even temperature differences are not driving the deformation of the sandwich panel induced by an electronic box. It is the thermal gradient that is driving the deformation of the box and, due to its high stiffness, imposing this deformation on the sandwich panel.

This means that even mild temperature changes relative to the assembly temperature can already cause important deformations in the sandwich panel imposing tilting of instruments on the same panel.

It is a common pitfall that people tend to look only at the temperature levels and sometimes not even at the temperature distribution and claim to be able to make a statement of importance of the temperature field for the deformation of the structure. The above experiment shows clearly that deeper investigation up to the computation of the actual deformation is needed to make such a judgement about the importance or criticality of a temperature field.

Simplified modelling of a box with only an RBE element and a lumped mass gives clearly wrong thermomechanical responses. There may be options to have simple and light models that could possibly reproduce the effect of the detailed model of the same box. This requires thorough tuning of these simplified model before these are implemented in the system-level models with sensitive instruments.

The electronic boxes located closes to the edge of the sandwich panel seem to have a lower influence on the deformation of the panel. This might indicate that the quality of the modelling of the boxes at these locations is less important for the accuracy of the prediction of the deformation of the panel. Nevertheless, it is recommended to have thorough confirmation for the specific situation in which such a box is applied.

Appendix B
One-Dimensional (1-D) Conduction Finite Element

Abstract

This appendix describes the principles of the finite element method applied to heat transfer problems. For illustration purpose, the method is explained on the basis of an 1-D conduction two nodes rod element.

Keywords: Heat transfer equations, finite element method, Galerkin's weighted residual method, conduction rod element.

B.1 Introduction

The general heat transfer equations (conduction, convection and radiation) and associated boundary conditions for the heat transfer analysis using the finite element method are briefly recapitulated. As an example, the characteristics of an 1-D conduction rod element with two nodes are derived. Subsequently, the assembly of conduction finite element matrices are discussed.

B.2 General Heat Transfer Equations

Before introducing the equations of the finite element method, a brief recapitulation of the heat transfer equations is presented in this section.

Considered is an arbitrary-shaped body as is presented in Fig. B.1 with volume V. The material of the body is isotropic and has a density ρ (kg/m^3), a heat capacity c (J/kg/K) and an isotropic conductivity k (W/m^2 K). Inside the body exists a potentially non-uniform temperature field $T(x, y, z, t)$ that is a function of the position (x, y, z) inside the body and time t.

S. Appel and J. Wijker, *Simulation of Thermoelastic Behaviour of Spacecraft Structures*, Springer Aerospace Technology, https://doi.org/10.1007/978-3-030-78999-2

Due to a potentially non-uniform temperature field, heat flow exists at any location in the body. With use of Fourier's law, the heat flow per unit area in the three directions q_x, q_y, q_z (W/m^2) can be expressed as

$$q_x = -k\frac{\partial T}{\partial x}, \quad q_y = -k\frac{\partial T}{\partial y} \text{ and } q_z = -k\frac{\partial T}{\partial z}. \tag{B.1}$$

Also, it is considered that the interior of the body is subjected to an internal heat generation rate per unit volume $Q(x, y, z, t)$ (W/m^3 K). From the equilibrium of an infinitesimal small cube with dimensions dx, dy and dz in the internal of the body, the heat balance at any position in the body can then be written as [63, 71]

$$\frac{\partial}{\partial x}\left(k\frac{\partial T}{\partial x}\right) + \frac{\partial}{\partial y}\left(k\frac{\partial T}{\partial y}\right) + \frac{\partial}{\partial z}\left(k\frac{\partial T}{\partial z}\right) + Q(x, y, z, t) - \rho c\frac{\partial T}{\partial t} = 0. \tag{B.2}$$

With Eqs. (B.1) and (B.2) can be rewritten as

$$\frac{\partial q_x}{\partial x} + \frac{\partial q_y}{\partial y} + \frac{\partial q_z}{\partial z} - Q(x, y, z, t) + \rho c\frac{\partial T}{\partial t} = 0. \tag{B.3}$$

At various parts of the outer surface S of the body, boundary conditions may imposed (see Fig. B.1), such as:

- S_1: Specified temperature: $T_s = T_1(x, y, z, t)$
- S_2: Specified heat flow: $q_x n_x + q_y n_y + q_z n_z = -q_s$
- S_3: Convection boundary conditions: $q_x n_x + q_y n_y + q_z n_z = h(T_s - T_\infty)$
- S_4: Radiation: $q_x n_x + q_y n_y + q_z n_z = (\sigma \epsilon T_s^4 - \alpha q_r)$

where T_s is the prescribed surface temperature on S_1, specified through the function $T_1(x, y, z, t)$, q_s is the specified incoming heat flux, h is the convection coefficient, T_∞ is the convection exchange temperature, σ is the Stefan–Boltzmann constant, ε is the surface emission coefficient, α is the surface absorption coefficient and q_r is the incident radiant heat flow per unit area.

In above boundary conditions, the values of n_x, n_y and n_z are considered to be the components of the normal vector at any location on the outer surface S of the body, combined into the vector (n):

$$(n) = \begin{pmatrix} n_x \\ n_y \\ n_z \end{pmatrix}. \tag{B.4}$$

When q_t is now considered to be the heat flux induced by the prescribed temperature at S_1, the above four boundary loads can be combined into the total applied heat flux $\tilde{q}$ at the boundary S of the body:

$$\tilde{q} = q_t - q_s - h(T - T_\infty) + (\sigma \epsilon T^4 - \alpha q_r). \tag{B.5}$$

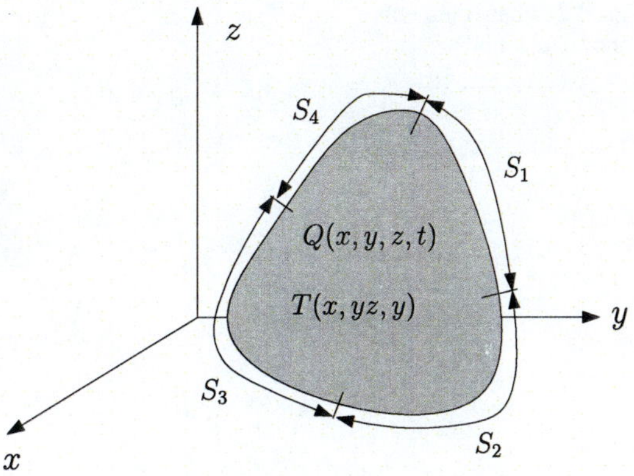

Fig. B.1 Thermal/structural volume and boundaries

At the boundary surface of the body, fluxes coming from inside of the body as defined by Eq. (B.1), projected to the normal (n) of the surface, must be in equilibrium with the applied external fluxes $\tilde{q}$. This results in the following equilibrium equation at the boundary surface [98]:

$$(q)^T (n) = q_x n_x + q_y n_y + q_z n_z = \tilde{q}, \tag{B.6}$$

or

$$(q)^T (n) - \tilde{q} = 0. \tag{B.7}$$

For transient problems, it is needed to specify the initial temperature field for the body at time $t = 0$ as

$$T(x, y, z, 0) = T_o(x, y, z). \tag{B.8}$$

B.2.1 General Finite Element Matrix Derivation

In this section, it is explained how the finite element system matrices for a thermal conduction problem are derived with the use of the equations in Sect. B.2. These matrices form the building blocks of the system to be solved for determining the temperatures at the finite element nodes. The system matrices are assembled from the matrices of the individual finite elements. Therefore, this section focusses on the derivation of the heat transfer relations (equations) of a generic finite element. To illustrate these relations, these are applied in Sect. B.3 to one of the simplest finite elements, being a rod.

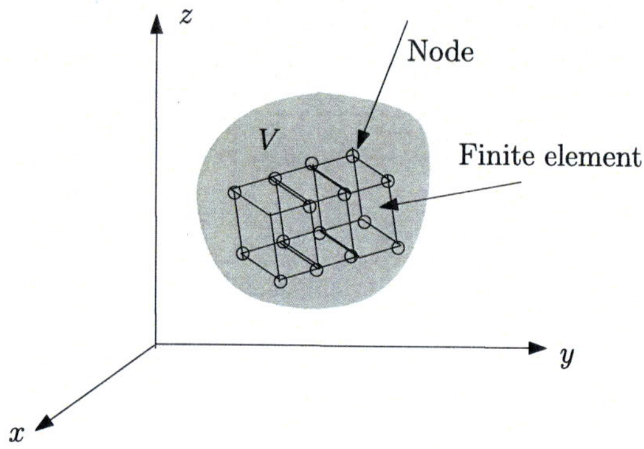

Fig. B.2 Finite elements connected at nodes

Considered is the same body of Fig. B.1 but now with volume V divided into a number of conduction finite elements that are connected at nodes as is illustrated in Fig. B.2. The position of these nodes defines the geometry of each finite element.

The temperatures at the nodes are the unknowns that need to be solved. These unknown temperatures can be presented in a column vector

$$\left(T(t)\right) = \begin{pmatrix} T_1 \\ T_2 \\ \vdots \end{pmatrix} . \tag{B.9}$$

One of the essentials of the finite element method is that a quantity field; in this case, a the temperature field, inside the element, is based on the values of this quantity at the nodes that are in most cases at the outer corners or edges of the element. The temperature field between the nodes is interpolated with interpolation functions, also referred to as shape functions or trial functions. These functions depend on the position inside the element, defined by the coordinates x, y and z. The different functions form the row vector $(\Psi(x, y, z))$.

$$\left(\Psi(x, y, z)\right) = \left(\Psi_1(x, y, z), \Psi_2(x, y, z), \ldots\right) \tag{B.10}$$

Each function Ψ_i corresponds to a nodal temperature T_i, such that with the nodal temperatures from Eq. (B.9) and the interpolation function from Eq. (B.10), the temperature field inside the element can be written as

$$T(x, y, z, t) = \left(\Psi(x, y, z)\right)\left(T(t)\right) . \tag{B.11}$$

Equation (B.11) should also be able to represent the case of a uniform temperature inside the element. In such a case, all nodal temperatures are equal. To assure that this uniform temperature field can be represented by Eq. (B.11), the interpolation functions have to be chosen such at each position in the element the sum of the interpolation function values equals to 1. This condition can be expressed as

$$\sum_i \Psi_i(x, y, z) = 1. \tag{B.12}$$

Differentiation of the temperature field from Eq. (B.11) relative to the spatial coordinates can now be done by differentiation of the interpolation function which results in the interpolation function matrix $[B]$ for the temperature gradients.

$$\begin{pmatrix} \frac{\partial T}{\partial x} \\ \frac{\partial T}{\partial y} \\ \frac{\partial T}{\partial z} \end{pmatrix} = \begin{pmatrix} \frac{\partial}{\partial x}(\Psi) \\ \frac{\partial}{\partial y}(\Psi) \\ \frac{\partial}{\partial z}(\Psi) \end{pmatrix} (T) = [B]\,(T) \tag{B.13}$$

With Eq. (B.13), it is possible to write the directional fluxes from Eq. (B.1), which were derived from Fourier's law, as

$$(q) = \begin{pmatrix} q_x \\ q_y \\ q_z \end{pmatrix} = \begin{pmatrix} -k\frac{\partial T}{\partial x} \\ -k\frac{\partial T}{\partial y} \\ -k\frac{\partial T}{\partial z} \end{pmatrix} = -k\,[B]\,(T). \tag{B.14}$$

The equilibrium condition of Eq. (B.3) has to be fulfilled at every position in the body. At the same time, the boundary condition of Eq. (B.7) at boundary S has to be fulfilled as well. To achieve this, the weighted residual method is applied. This implies that the Eqs. (B.3) and (B.7) are combined and multiplied with weighting functions and integrated over the volume of the body and the boundary surface, respectively. In the approach of Galerkin [14], the weighting functions are chosen to be the same as the interpolation functions (Ψ). With this approach, the basic heat transfer equation from Eq. (B.3) in combination with Eq. (B.7) can be expressed as follows [98]

$$\begin{gathered} \int_V \left(\frac{\partial q_x}{\partial x} + \frac{\partial q_y}{\partial y} + \frac{\partial q_z}{\partial z} - Q + \rho c \frac{\partial T}{\partial t} \right) \Psi_i dV = 0 \\ \text{and} \\ \int_S \left((q)^T (n) - \tilde{q} \right) \Psi_i dS = 0 \quad i = 1, 2, \ldots \end{gathered} \tag{B.15}$$

Product rule gives us

$$\begin{aligned} &\left(\frac{\partial (q_x \Psi_i)}{\partial x} + \frac{\partial (q_y \Psi_i)}{\partial y} + \frac{\partial (q_z \Psi_i)}{\partial z} \right) \\ &= \left(\frac{\partial q_x}{\partial x} + \frac{\partial q_y}{\partial y} + \frac{\partial q_z}{\partial z} \right) \Psi_i + \left(q_x \frac{\partial \Psi_i}{\partial x} + q_y \frac{\partial \Psi_i}{\partial y} + q_z \frac{\partial \Psi_i}{\partial z} \right) \quad i = 1, 2 \cdots, \end{aligned} \tag{B.16}$$

which can be rewritten to

$$\left(\frac{\partial q_x}{\partial x}+\frac{\partial q_y}{\partial y}+\frac{\partial q_z}{\partial z}\right)\Psi_i=\left(\frac{\partial(q_x\Psi_i)}{\partial x}+\frac{\partial(q_y\Psi_i)}{\partial y}+\frac{\partial(q_z\Psi_i)}{\partial z}\right)$$
$$-\left(q_x\frac{\partial\Psi_i}{\partial x}+q_y\frac{\partial\Psi_i}{\partial y}+q_z\frac{\partial\Psi_i}{\partial z}\right)\quad i=1,2\cdots \tag{B.17}$$

With this intermediate step, the first condition from Eq. (B.15) can be rewritten to:

$$\int_V\left(\frac{\partial(q_x\Psi_i)}{\partial x}+\frac{\partial(q_y\Psi_i)}{\partial y}+\frac{\partial(q_z\Psi_i)}{\partial z}\right)dV$$
$$-\int_V\left(q_x\frac{\partial\Psi_i}{\partial x}+q_y\frac{\partial\Psi_i}{\partial y}+q_z\frac{\partial\Psi_i}{\partial z}\right)dV \tag{B.18}$$
$$+\int_V\left(-Q+\rho c\frac{\partial T}{\partial t}\right)\Psi_i dV=0\ \ i=1,2,\ldots.$$

With the divergence theorem of Gauss [54, 87], the first term of Eq. (B.18) above can be transformed into a surface integral, which makes that Eq. (B.18) can be rewritten to

$$\int_S\left(\Psi_i(q)^T(n)\right)dS-\int_V\left(q_x\frac{\partial\Psi_i}{\partial x}+q_y\frac{\partial\Psi_i}{\partial y}+q_z\frac{\partial\Psi_i}{\partial z}\right)dV$$
$$+\int_V\left(-Q+\rho c\frac{\partial T}{\partial t}\right)\Psi_i dV=0\ \ i=1,2,\ldots. \tag{B.19}$$

Equation (B.19) is called the weak form of the heat balance equations (B.2). Making use of Eq. (B.13) allows Eq. (B.19) to be rewritten to

$$\int_S\left(\Psi_i(q)^T(n)\right)dS$$
$$+k\int_V\left(\frac{\partial\Psi_i}{\partial x}\frac{\partial T}{\partial x}+\frac{\partial\Psi_i}{\partial y}\frac{\partial T}{\partial y}+\frac{\partial\Psi_i}{\partial z}\frac{\partial T}{\partial z}\right)dV \tag{B.20}$$
$$+\int_V\left(-Q+\rho c\frac{\partial T}{\partial t}\right)\Psi_i dV=0\ \ i=1,2,\ldots,$$

which can be rearranged into

$$\int_V \rho c \frac{\partial T}{\partial t} \Psi_i dV + k \int_V \left(\frac{\partial \Psi_i}{\partial x} \frac{\partial T}{\partial x} + \frac{\partial \Psi_i}{\partial y} \frac{\partial T}{\partial y} + \frac{\partial \Psi_i}{\partial z} \frac{\partial T}{\partial z} \right) dV$$
$$= \int_V Q \Psi_i dV - \int_S \left(\Psi_i (q)^T (n) \right) dS \quad i = 1, 2 \ldots \tag{B.21}$$

The last term of Eq. (B.21) contains the general boundary conditions which can be written more explicit with the help of Eq. (B.5)

$$\int_S (q)^T (n) \Psi_i dS = \int_{S_1} (q_t)^T (n) \Psi_i dS - \int_{S_2} (q_s)^T (n) \Psi_i dS$$
$$+ \int_{S_3} h (T - T_\infty) \Psi_i dS + \int_{S_4} \left(\sigma \epsilon T^4 - \alpha q_r \right) \Psi_i dS, \quad i = 1, 2, \ldots . \tag{B.22}$$

The finite element equations for the heat transfer problems can now be formulated as follows

$$[C] \left(\dot{T} \right) + ([K_c] + [K_h] + [K_r]) (T) = \left(R_Q \right) + (R_T) + \left(R_q \right) + (R_h) + (R_r) \tag{B.23}$$

where $\left(\dot{T} \right)$ is the nodal vector of temperature time derivatives, (T) is the nodal vector of temperatures and:

The heat capacity matrix

$$[C] = \int_V \rho c (\Psi)^T (\Psi) dV \tag{B.24}$$

The conduction matrix. Note that from the expression can be deducted that the conduction matrix must be symmetric.

$$[K_c] = \int_V k [B]^T [B] dV \tag{B.25}$$

The convection exchange matrix

$$[K_h] = \int_{S_3} h (\Psi)^T (\Psi) dS \tag{B.26}$$

The radiosity matrix

$$[K_r(T)] = \int_{S_4} \sigma \epsilon \left(T^4 \right) (\Psi)^T dS \tag{B.27}$$

The fluxes coming from the internal heat generation rate

$$(R_Q) = \int_V Q\,(\Psi)^T\,dV \tag{B.28}$$

The fluxes coming from the prescribed temperatures at boundary S_1 in Fig. B.1

$$(R_T) = -\int_{S_1} (q_t)^T\,(n)\,(\Psi)^T\,dS \tag{B.29}$$

The fluxes coming from the externally applied heat rate at boundary S_2 in Fig. B.1

$$(R_q) = \int_{S_2} q_s\,(\Psi)^T\,dS \tag{B.30}$$

The fluxes coming from the convective heat input at boundary S_3 in Fig. B.1

$$(R_h) = \int_{S_3} hT_\infty\,(\Psi)^T\,dS \tag{B.31}$$

Finally, the absorbed radiative power received at boundary S_4 in Fig. B.1

$$(R_r) = \int_{S_4} \alpha q_r\,(\Psi)^T\,dS \tag{B.32}$$

The above expressions include all heat transfer mechanisms. In case only a limited set of heat transfer mechanisms are relevant, then the applicable expression can be deducted from the general equation (B.23). For instance, when only the stationary linear heat conduction problem is of interest, then equation (B.23) can be simplified to:

$$([K_c] + [K_h])\,(T) = (R_Q) + (R_q) + (R_h) \tag{B.33}$$

For both the purpose of finite element assisted linear conductor generation, see Sects. 9.2 and 9.3, and conduction-based temperature mapping methods, see Sect. 7.4.2 and Chap. 8, the matrix $[K_c]$ and the vector (R_Q) are frequently used.

B.3 1-D Conduction Rod Finite Element

To illustrate the application of the expressions from previous section, these are applied in this section to one of the simplest finite element, the rod element.

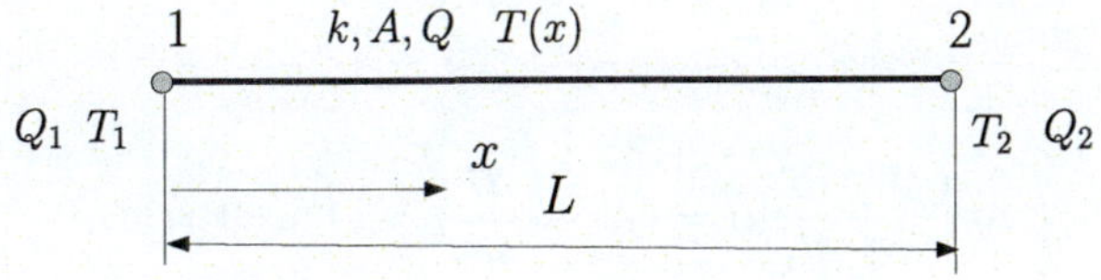

Fig. B.3 Temperature distribution $T(x)$ in rod element

For the rod in Fig. B.3, the general heat transfer problem reduces to a one-dimensional problem. The general heat transfer equation (B.2) can be simplified to [83].

$$k\frac{d^2T(x)}{dx^2} + Q = 0, \tag{B.34}$$

and the boundary conditions at both ends $x = 0, L$ of the rod are:

$$\begin{aligned} Q_1 &= -kA\left.\frac{dT}{dx}\right|_{x=0}, \\ Q_2 &= kA\left.\frac{dT}{dx}\right|_{x=L}, \end{aligned} \tag{B.35}$$

where k is the constant coefficient of thermal conduction (W/mK), $T(x)$ is the temperature distribution in the rod finite element and Q is the internal heat generation rate per unit volume (W/m^3). A is the constant cross section of the rod as shown in Fig. B.3. The application of the weighted residual method to equation (B.34) makes that this expression can be written as

$$\int_0^L (W)\left(k\frac{d^2T}{dx^2} + Q\right)Adx = 0, \tag{B.36}$$

where (W) is the row vector with weighting functions.

Similar to the way Eq. (B.19) is obtained through the application of the divergence theorem of Gauss, the weak form of the differential equation (B.34) is obtained through performing integration by parts on Eq. (B.36). This operation provides

$$\left[(W)Ak\frac{dT}{dx}\right]_0^L - \int_0^L Ak\frac{(dW)}{dx}\frac{dT}{dx}dx + \int_0^L A(W)Qdx = 0, \tag{B.37}$$

or rearranged into

$$\int_0^L Ak\frac{(dW)}{dx}\frac{dT}{dx}dx = (W)\,Ak\frac{dT}{dx}\bigg|_0^L - \int_0^L A\,(W)\,Qdx. \tag{B.38}$$

The temperature distribution $T(x)$ along the length L of the rod can be expressed into the nodal temperatures T_1 and T_2 using the approach of Eq. (B.11) with interpolation or shape functions

$$T(x) = (\Psi_1, \Psi_2)\begin{pmatrix} T_1 \\ T_2 \end{pmatrix} \tag{B.39}$$

with the following functions chosen for Ψ_1 and Ψ_2:

$$\begin{aligned} \Psi_1 &= \left(1 - \frac{x}{L}\right) \\ \Psi_2 &= \left(\frac{x}{L}\right) \end{aligned} \tag{B.40}$$

and their derivatives then become

$$\begin{aligned} \frac{d\Psi_1}{dx} &= -\frac{1}{L}, \\ \frac{d\Psi_2}{dx} &= \frac{1}{L}. \end{aligned} \tag{B.41}$$

In line with the Galerkin method, the weighting functions are chosen to be the same as the interpolations functions:

$$\begin{aligned} W_1 &= \Psi_1, \\ W_2 &= \Psi_2. \end{aligned} \tag{B.42}$$

Performing the integration of Eq. (B.38), with substitution of the weighting function, alias the interpolation functions, from Eq. (B.40) and their derivatives from Eq. (B.41), provides the following matrix equation

$$\frac{Ak}{L}\begin{bmatrix} 1 & -1 \\ -1 & 1 \end{bmatrix}\begin{pmatrix} T_1 \\ T_2 \end{pmatrix} = ALQ\begin{pmatrix} \frac{1}{2} \\ \frac{1}{2} \end{pmatrix} + kA\begin{pmatrix} \frac{-dT}{dx}\Big|_1 \\ \frac{dT}{dx}\Big|_2 \end{pmatrix}, \tag{B.43}$$

or

$$\frac{Ak}{L}\begin{bmatrix} 1 & -1 \\ -1 & 1 \end{bmatrix}\begin{pmatrix} T_1 \\ T_2 \end{pmatrix} = ALQ\begin{pmatrix} \frac{1}{2} \\ \frac{1}{2} \end{pmatrix} + \begin{pmatrix} Q_1 \\ Q_2 \end{pmatrix}, \tag{B.44}$$

where the vector of the second part of the right-hand side of above equations represents the heat fluxes at the two ends of the element. From Eq. B.44, it can be observed that the constant internal heat generation rate Q is equally distributed over both nodes 1 and 2.

This 1-D rod conduction element and its matrices are frequently used in the conduction problem examples to illustrate the theory.

B.4 Assembly of System Matrices

In this section, the process of assembly of system matrices from individual finite element matrices is explained. The explanation uses an example model with two conduction rod elements as is shown in Fig. B.4. The process assembles, or combines, the two individual element conduction matrices from the two elements in the model into a single global conduction matrix. This global conduction matrix can be used to solve simultaneously all unknown nodal temperatures in the model.

The model contains three finite element nodes. The global vector with all nodal temperatures in the model is then therefore as shown in Eq. (B.45).

$$(T_g) = \begin{pmatrix} T_1 \\ T_2 \\ T_3 \end{pmatrix} \tag{B.45}$$

Each of the two rod elements has two nodes connected. The nodal temperature vectors at element level corresponding to each of these two elements follow the convention that was used in Sect. B.3 in which the arbitrary first node was referred as node 1 and the other node as node 2. Since there are here two elements, the node references and the corresponding temperatures need to be distinguished. In Eq. (B.46), this is done through the use of a superscript to indicate to which element the components belong to.

$$(T_{el_1}) = \begin{pmatrix} T_1^1 \\ T_2^1 \end{pmatrix} \qquad (T_{el_2}) = \begin{pmatrix} T_1^2 \\ T_2^2 \end{pmatrix} \tag{B.46}$$

Note that here the superscript "2" is not used to indicate the square operation, but to indicate that the corresponding quantity belongs to element El-2.

The conduction matrix of each of the two elements has the form as was derived in the Sect. B.3 and is presented in Eq. (B.47).

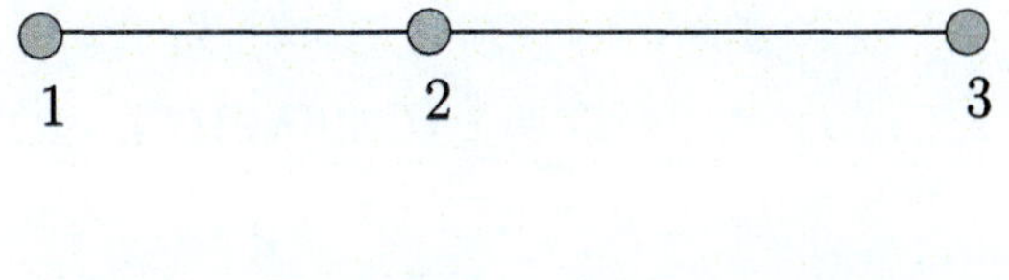

Fig. B.4 Conduction model with two rod elements

$$[K_c^1] = \begin{bmatrix} k_{11}^1 & k_{12}^1 \\ k_{21}^1 & k_{22}^1 \end{bmatrix} \qquad [K_c^2] = \begin{bmatrix} k_{11}^2 & k_{12}^2 \\ k_{21}^2 & k_{22}^2 \end{bmatrix} \tag{B.47}$$

There may be an internal heat generation rate inside each element that is distributed to the nodes of each element like was shown in Sect. B.3. This leads then to a heat flux vector for each element as is indicated in Eq. (B.48).

$$(Q_{el_1}) = \begin{pmatrix} Q_1^1 \\ Q_2^1 \end{pmatrix} \qquad (Q_{el_2}) = \begin{pmatrix} Q_1^2 \\ Q_2^2 \end{pmatrix} \tag{B.48}$$

With Eq. (B.46), the element nodal temperatures can be collected in a single vector (T_{el}). In a similar way, the element nodal heat fluxes from Eq. (B.48) can be combined into a single vector (Q_{el}). Both vectors are indicated in Eq. (B.49).

$$(T_{el}) = \begin{pmatrix} T_1^1 \\ T_2^1 \\ T_1^2 \\ T_2^2 \end{pmatrix} \qquad (Q_{el}) = \begin{pmatrix} Q_1^1 \\ Q_2^1 \\ Q_1^2 \\ Q_2^2 \end{pmatrix} \tag{B.49}$$

The element conduction matrices from Eq. (B.47) can be assembled in a single conduction matrix $[K_c^{el}]$ of which the rows and columns are based on (T_{el}) as is shown in Eq. (B.50).

$$[K_c^{el}] = \begin{bmatrix} k_{11}^1 & k_{12}^1 & 0 & 0 \\ k_{21}^1 & k_{22}^1 & 0 & 0 \\ 0 & 0 & k_{11}^2 & k_{12}^2 \\ 0 & 0 & k_{21}^2 & k_{22}^2 \end{bmatrix} \tag{B.50}$$

The conduction matrix $[K_c^{el}]$ consists of two decoupled sub-matrices. This is due to the fact that, although node 2 of element El-1 is the same node as node 1 of element El-2, these two nodes are till now considered as two distinct entities. Since heat is flowing from element 1 to element 2 and vice versa via node 2, this connectivity has to be reflected in the system matrices. There are various ways to do this. As an example is here a transformation matrix used and is shown in Eq. (B.51), that maps the unknown temperatures that are local to the elements to the unknowns at global model level.

$$(T_{el}) = \begin{pmatrix} T_1^1 \\ T_2^1 \\ T_1^2 \\ T_2^2 \end{pmatrix} = \begin{bmatrix} 1 & 0 & 0 \\ 0 & 1 & 0 \\ 0 & 1 & 0 \\ 0 & 0 & 1 \end{bmatrix} \begin{pmatrix} T_1 \\ T_2 \\ T_3 \end{pmatrix} = [A_g] \left(T_g\right) \tag{B.51}$$

An easy way to show mathematically with Eq. (B.51) how the temperature of the second node of element El-1 is made equal to the first node of element El-2 is through the use of a functional as is shown in Eq. (B.52).

$$J = \frac{1}{2} (T_{el})^T [K_c^{el}] (T_{el}) - (T_{el})^T (Q_{el}) \tag{B.52}$$

Substituting Eq. (B.51) in Eq. (B.52) makes that Eq. (B.52) can be written as

$$J = \frac{1}{2}\left(T_g\right)^T [A_g]^T [K_c^{el}][A_g]\left(T_g\right) - \left(T_g\right)^T [A_g]^T \left(Q_{el}\right). \tag{B.53}$$

This can then be simplified to

$$J = \frac{1}{2}\left(T_g\right)^T [K_c^g]\left(T_g\right) - \left(T_g\right)^T \left(Q_g\right). \tag{B.54}$$

When Eq. (B.54) and Eq. (B.53) are compared and using $[A_g]$ of Eq. (B.51) and $[K_c^{el}]$ of Eq. (B.50), the global model conduction matrix $[K_c^g]$ and the global heat load vector (Q_g) can be written as shown in Eq. (B.55).

$$\begin{aligned} [K_c^g] &= [A_g]^T [K_c^{el}][A_g] = \begin{bmatrix} k_{11}^1 & k_{12}^1 & 0 \\ k_{21}^1 & k_{22}^1 + k_{11}^2 & k_{12}^2 \\ 0 & k_{21}^2 & k_{22}^2 \end{bmatrix} \\ (Q_g) &= [A_g]^T (Q_{el}) = \begin{pmatrix} Q_1^1 \\ Q_2^1 + Q_1^2 \\ Q_2^2 \end{pmatrix} \end{aligned} \tag{B.55}$$

From the expression of Eq. (B.55), it can be noted that rows 2 and 3 as well as columns 2 and 3 of the matrix $[K_c^{el}]$ are combined in row and column 2 of the matrix $[K_c^g]$. Basically, the contributions from element El-1 and element El-2 are combined in the shared node, being node 2 that corresponds to row and column 2 of the global system.

With the expressions of Eq. (B.55), the system conduction matrix $[K_c^g]$ and heat load vector (Q_g) are assembled and all ingredients are available to solve the system equation as presented in Eq. (B.56) for the conduction problem for the complete model.

$$[K_c^g]\left(T_g\right) = \left(Q_g\right) \tag{B.56}$$

Appendix C
One-Dimensional (1-D) Thermoelastic Finite Element

Abstract
The principles of the finite element element method applied to thermoelastic problems are discussed. For illustration purpose, the method is explained on the basis of an 1-D iso-parametric rod element.

Keywords: Finite element method, Galerkin's weighted residual method, principle virtual work, thermoelastic rod element.

C.1 Introduction

This appendix provides as background information the basics of the finite element method explained with a specific focus on the application of thermoelastic analysis. The description of the method is directly applied to one of the most simple finite elements, being the one-dimensional rod element. This element is also used in several examples throughout this book.

The description starts with the differential equations describing equilibrium and compatibility through the applied linear material law. This is followed by the description of two methods that are typically used to solve these differential equations: The Galerkin's weighted residual method and the method based on the principle of virtual work. With the help of shape functions, the discretization of the continuous rod into one-dimensional finite elements is introduced. The description is concluded with the explanation on how to assemble the element matrices into system matrices and system load vectors.

This appendix intends to provide only sufficient background information. Many good textbooks exist that describe in detail many aspects of the finite element methods, such as e.g. [14, 26, 83] that are recommended for further reading.

S. Appel and J. Wijker, *Simulation of Thermoelastic Behaviour of Spacecraft Structures*, Springer Aerospace Technology, https://doi.org/10.1007/978-3-030-78999-2

C.2 Equilibrium Equations for a One-Dimensional Rod

The equilibrium state of an infinitesimal part of a rod is illustrated in Fig. C.1.

The state of equilibrium is given by the following expression [75, 82, 86]

$$A\left(\sigma + \frac{d\sigma}{dx}dx\right) + qdx - A\sigma = 0 \tag{C.1}$$

in which A (m^2) is the area of the cross section, q is the line load with force per unit lenght (N/m) and σ the constant stress in the cross section. Equation (C.1) reworked yields

$$A\frac{d\sigma}{dx} + q = 0. \tag{C.2}$$

The stress strain relation is given by [19]

$$\sigma = E(\varepsilon - \alpha\Delta T). \tag{C.3}$$

where E (Pa) is the Young's modulus, α (m/m °C) the CTE, and ΔT the temperature increase relative to the stress-free condition.

The one-dimensional strain in the rod due to its deformation is defined as

$$\varepsilon = \frac{du}{dx}, \tag{C.4}$$

where $u(x)$ (m) is the axial displacement in the rod as a function of the length coordinate x.

It is good to realise from Eq. (C.3) that the strain ε due to the actual deformation is compensated by the thermal strain $\alpha\Delta T$, to obtain the strain that is responsible for the stress. In case of a rod that is free to expand, the thermal strain is then equal to the strain. This leads then to a zero net strain producing zero stress.

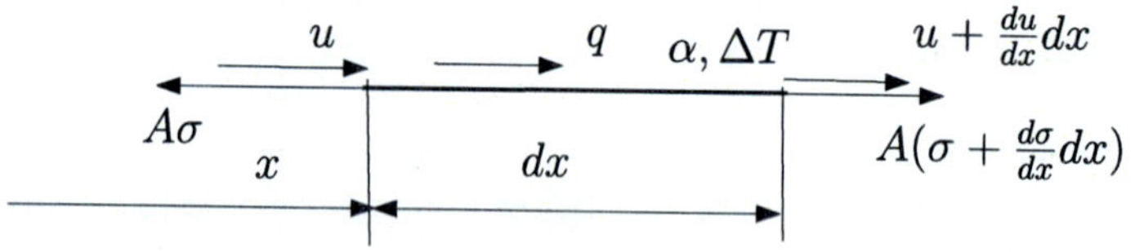

Fig. C.1 Equilibrium state infinitesimal part off rod

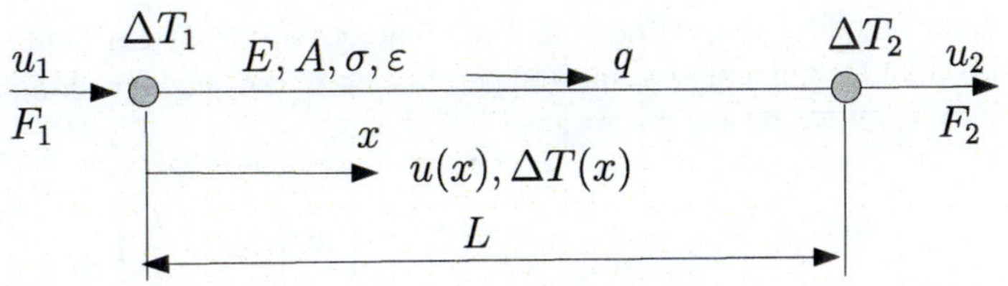

Fig. C.2 1-D rod element with thermal and mechanical loads

Finally, through substitution of Eq. (C.3) and Eq. (C.4) in Eq. (C.2), the following, expression can be derived at x

$$EA\left(\frac{d^2u}{dx^2}-\alpha\frac{d\Delta T}{dx}\right)+q=0 \tag{C.5}$$

This equation can be written in the form

$$\Omega(u)=0 \tag{C.6}$$

Consider now that the rod has a length L and that $x=0$ a force F_1^B is applied, and at $x=L$, an axial force F_2^B is applied (see Fig. C.2). This implies the following boundary conditions are applicable at $x=0$ and $x=L$

$$\begin{aligned} F_1^B &= -\sigma(0)A = -EA\left(\frac{du(0)}{dx}-\alpha\Delta T(0)\right)\\ F_2^B &= \sigma(L)A = EA\left(\frac{du(L)}{dx}-\alpha\Delta T(L)\right). \end{aligned} \tag{C.7}$$

In the following, the parameters E, A, α, q are kept constant along the length of the rod.

In the next sections, the shape functions are introduced which are then in the subsequent sections used to solve Eq (C.5) with two methods that are often used for the finite element method: Galerkin's weighted residual method and the method based on the principle of virtual work.

C.3 Shape Functions for One-Dimensional Element

Considered is the 1-D rod finite element illustrated in Fig. C.2.

In the finite element method, shape functions are used to describe the distribution of a quantity within the space of the element. These quantities are typically displacements or distributed loads for structural applications. Each shape function is associated to a node of the element. The value of the node is used as scaling factor for the corresponding shape function. The complete displacement field within an element is then formulated as a linear superposition of the shape functions. For the rod element with two nodes, this means

$$u(x) = \Psi_1(x)u_1 + \Psi_2(x)u_2 = \begin{pmatrix}\Psi_1(x) & \Psi_2(x)\end{pmatrix}\begin{pmatrix}u_1\\ u_2\end{pmatrix} \tag{C.8}$$

A convenient definition of the individual shape functions assures that shape function Ψ_i is equal to 1.0 at node i and zero at the other node j with $j \neq i$. The following shape function would comply to this convention for a one-dimensional rod with two nodes

$$\begin{pmatrix}\Psi_1(x)\\ \Psi_2(x)\end{pmatrix}^T = \begin{pmatrix}\left(1-\frac{x}{L}\right)\\ \left(\frac{x}{L}\right)\end{pmatrix}. \tag{C.9}$$

Similar to the displacement field, the temperature increase $\Delta T(x)$ (°C, K) with respect to a reference temperature T_{ref}, and also, the distribution of an axial load q (N/m) can be expressed in terms of shape functions

$$\begin{aligned}\Delta T(x) &= \Psi_1(x)\Delta T_1 + \Psi_2(x)\Delta T_2\\ q(x) &= \Psi_1(x)q_1 + \Psi_2(x)q_2\end{aligned} \tag{C.10}$$

Also, the derivative of the displacement field can be expressed in terms of derivatives of the shape functions

$$\frac{du}{dx} = \frac{d\Psi_1}{dx}u_1 + \frac{d\Psi_2}{dx}u_2 = \begin{pmatrix}\frac{d\Psi_1}{dx} & \frac{d\Psi_2}{dx}\end{pmatrix}\begin{pmatrix}u_1\\ u_2\end{pmatrix} \tag{C.11}$$

Using Eq. (C.9), the derivatives of the selected shape functions for the one-dimensional rod can be written as

$$\begin{pmatrix}\frac{d\Psi_1(x)}{dx}\\ \frac{d\Psi_2(x)}{dx}\end{pmatrix}^T = \begin{pmatrix}-\left(\frac{1}{L}\right)\\ \left(\frac{1}{L}\right)\end{pmatrix}. \tag{C.12}$$

C.4 Galerkin's Weighted Residual Method

The principle of the weighted residual method states that when Eq. (C.6) must hold over the full length of the rod, it implies that also the following applies

$$\int_0^L W(x)\Omega(u)dx = 0 \tag{C.13}$$

with $W(x)$ being an arbitrary weighting function over the length of the rod.

Using Eq. (C.5), Eq. (C.13) can be expanded into

$$\int_0^L W(x)\left[EA\left(\frac{d^2u}{dx^2}-\alpha\frac{d\Delta T}{dx}\right)+q\right]dx=0 \tag{C.14}$$

which can be rewritten as

$$EA\int_0^L W(x)\left(\frac{d^2u}{dx^2}\right)-\alpha EA\int_0^L W(x)\frac{d\Delta T}{dx}+q\int_0^L W(x)dx=0 \tag{C.15}$$

Using integration by parts[1], Eq. (C.15) can be transformed into

$$\begin{aligned} &EA\left[W(x)\frac{du}{dx}\bigg|_0^L-\int_0^L\frac{dW}{dx}\frac{du}{dx}dx\right]\\ &-\alpha EA\left[W(x)\Delta T\bigg|_0^L-\int_0^L\frac{dW}{dx}\Delta T dx\right]\\ &+q\int_0^L W(x)dx=0 \end{aligned} \tag{C.16}$$

which could again be reformatted into

$$\begin{aligned} &EA\int_0^L\frac{dW}{dx}\frac{du}{dx}dx\\ &=\alpha EA\int_0^L\frac{dW}{dx}\Delta T dx\\ &+q\int_0^L W(x)dx\\ &+EAW(x)\frac{du}{dx}\bigg|_0^L-\alpha EAW(x)\Delta T\bigg|_0^L \end{aligned} \tag{C.17}$$

[1] $\int f(x)g'(x)dx=f(x)g(x)-\int f'(x)g(x)dx$.

Galerkin proposed to use for the weighting function $W(x)$ the shape functions. For this purpose, two weighting functions are introduced being equal to the corresponding shape functions.

$$\begin{aligned} W_1(x) &= \Psi_1(x) \\ W_2(x) &= \Psi_2(x) \end{aligned} \tag{C.18}$$

It must be noted that Eq. (C.17) is a scalar expression that can be evaluated for each weight function. Using $W_1(x)$ and $W_2(x)$, the left-hand side part of Eq. (C.17) becomes

$$\begin{aligned} EA\int_0^L \frac{dW_1}{dx}\frac{du}{dx}dx &= \frac{EA}{L}(u_1 - u_2) \\ EA\int_0^L \frac{dW_2}{dx}\frac{du}{dx}dx &= \frac{EA}{L}(-u_1 + u_2) \end{aligned} \tag{C.19}$$

and when written in matrix form it becomes

$$\frac{EA}{L}\begin{pmatrix} u_1 - u_2 \\ -u_1 + u_2 \end{pmatrix} = \frac{EA}{L}\begin{bmatrix} 1 & -1 \\ -1 & 1 \end{bmatrix}\begin{pmatrix} u_1 \\ u_2 \end{pmatrix} = [K_e](u) \tag{C.20}$$

with $[K_e]$ being the element stiffness matrix.

The right-hand side part of Eq. (C.17) starts with the a thermoelastic term. This term can be evaluated to

$$\begin{aligned} \alpha EA\int_0^L \frac{dW_1}{dx}\Delta T dx &= \alpha EA\left(-\tfrac{1}{2}\Delta T_1 - \tfrac{1}{2}\Delta T_2\right) \\ \alpha EA\int_0^L \frac{dW_2}{dx}\Delta T dx &= \alpha EA\left(\tfrac{1}{2}\Delta T_1 + \tfrac{1}{2}\Delta T_2\right) \end{aligned} \tag{C.21}$$

As can be noted from Eq. (C.21), the average temperature in the rod element is used as basis for the thermoelastic force. The thermoelastic force vector (F_T) then becomes

$$\tfrac{1}{2}\alpha EA\begin{bmatrix} -1 & -1 \\ 1 & 1 \end{bmatrix}\begin{pmatrix} \Delta T_1 \\ \Delta T_2 \end{pmatrix} = \begin{pmatrix} F_{T1} \\ F_{T2} \end{pmatrix} = (F_T) \tag{C.22}$$

with (F_T) the thermoelastic load vector.

The term with the line load q of Eq. (C.17) evaluates to

$$q\int_0^L W_1(x)dx = \tfrac{1}{2}qL$$
$$q\int_0^L W_2(x)dx = \tfrac{1}{2}qL \qquad \text{(C.23)}$$

and the corresponding load vector can then be written as

$$\begin{pmatrix} \frac{1}{2}qL \\ \frac{1}{2}qL \end{pmatrix} = \begin{pmatrix} F_{Q1} \\ F_{Q2} \end{pmatrix} = (F_Q) \qquad \text{(C.24)}$$

with (F_Q) the vector with the loads applied to the element via the line load.

The terms in the last line of Eq. (C.17) are all related to the boundary conditions as indicated in the expressions of Eq. (C.7). Working out the expression of these last terms of Eq. (C.17) leads to

$$EAW_1(x)\frac{du}{dx}\bigg|_0^L - \alpha EAW_1(x)\Delta T\bigg|_0^L = -\frac{EA}{L}(-u_1 + u_2) + \alpha EA\Delta T_1 = F_1^B$$
$$EAW_2(x)\frac{du}{dx}\bigg|_0^L - \alpha EAW_2(x)\Delta T\bigg|_0^L = \frac{EA}{L}(-u_1 + u_2) - \alpha EA\Delta T_2 = F_2^B \qquad \text{(C.25)}$$

with F_1^B and F_2^B the applied loads as defined in Eq. (C.7) can be written in matrix form:

$$EA\begin{bmatrix} 1 & -1 \\ -1 & 1 \end{bmatrix}\begin{pmatrix} u_1 \\ u_2 \end{pmatrix} - \alpha EA\begin{pmatrix} -\Delta T_1 \\ \Delta T_2 \end{pmatrix} = \begin{pmatrix} F_1^B \\ F_2^B \end{pmatrix} = (F_B) \qquad \text{(C.26)}$$

with (F_B) being the applied loads at the two boundary end nodes of the element.

Now, the expressions of Eqs. (C.20), (C.22), (C.24) and finally Eq. (C.26) can be combined into

$$\frac{EA}{L}\begin{bmatrix} 1 & -1 \\ -1 & 1 \end{bmatrix}\begin{pmatrix} u_1 \\ u_2 \end{pmatrix} = \frac{\alpha EA}{2}\begin{bmatrix} -1 & -1 \\ 1 & 1 \end{bmatrix}\begin{pmatrix} \Delta T_1 \\ \Delta T_2 \end{pmatrix} + \frac{qL}{2}\begin{pmatrix} 1 \\ 1 \end{pmatrix} + \begin{pmatrix} F_1^B \\ F_2^B \end{pmatrix} \qquad \text{(C.27)}$$

or in short

$$[K_e](u) = (F_T) + (F_Q) + (F_B) \qquad \text{(C.28)}$$

Equation (C.28) shows the main elements contributing to the equilibrium of a single rod element. The internal elastic forces due to the deformation and stiffness of the material give the term at the left-hand side of the equation. This has to be in equilibrium with the internal thermoelastic forces (F_T), the line loads applied to the element $\left(F_Q\right)$ and the applied forces at both ends of the element, the boundary forces (F_B).

Equation (C.28) contains the equilibrium equation at element level that can be used for assembling the system equation in Sect. C.8.

C.5 Iso-parametric Formulation

In Sect. C.3, shape functions were introduced to interpolate displacements, temperatures and load. These shape functions are a function of the position in the element in which the position is expressed in physical coordinates. For the one-dimensional only, the single coordinate x was used to indicate the position along the length of the rod element.

Although it may look for a straight rod a bit of overhead, in the formulation of many finite elements, shape functions are also used for the "interpolation" of the position inside the element. For this purpose, the so-called natural coordinates are introduced. For the one-dimensional problem, a single natural coordinate with the symbol ξ is often used. The natural coordinate runs from $\xi = -1$ at node 1 of the element and $\xi = +1$ at node 2, as is illustrated in Fig. C.3. This allows the physical coordinate, being the physical position inside the element, to be expressed as a function of the natural coordinate: $x(\xi)$. For the 1-D rod, the relation between the physical and natural coordinates is defined with the use of shape functions.

$$x(\xi) = \Psi_1(\xi)x_1 + \Psi_2(\xi)x_2 \tag{C.29}$$

where the shape functions are given by

$$\Psi_1(\xi) = \frac{1}{2}(1 - \xi),\ \Psi_2(\xi) = \frac{1}{2}(1 + \xi). \tag{C.30}$$

The choice of the shape functions must be such that

$$\sum_{k=1}^{2} \Psi_k(\xi) = 1 \tag{C.31}$$

which is complied to by choice presented in Eq. (C.30).

Like in Sect. C.3, the shape functions will be used to interpolate the displacements, temperatures and loads. The special feature of iso-parametric element is that also for the conversion from physical coordinates to natural coordinates, shape functions are applied.

Fig. C.3 Iso-parametric rod element

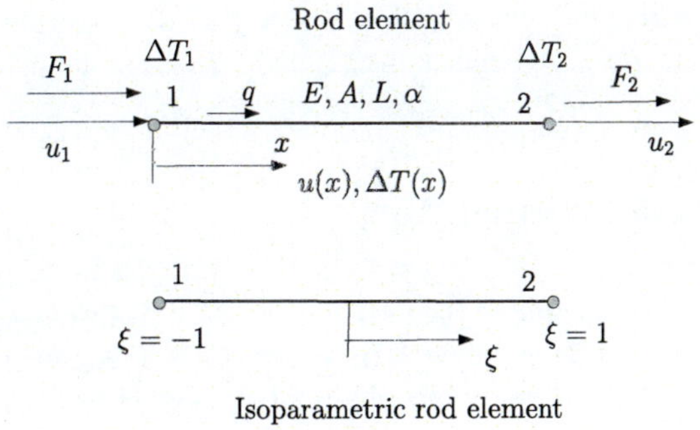

Similar to Eqs. (C.8) and (C.10), the axial displacements in the rod u and the distribution of the temperature increase ΔT can be written as

$$u(\xi) = \Psi_1(\xi)u_1 + \Psi_2(\xi)u_2 \tag{C.32}$$

$$\Delta T(\xi) = \Psi_1(\xi)\Delta T_1 + \Psi_2(\xi)\Delta T_2 \tag{C.33}$$

For the calculation of the strain with the help of the shape functions, the first derivative of the axial displacement u with respect to the physical coordinate x is needed. This property is not immediately available from above expressions. However with the chain rule for differentiation can be written

$$\frac{du}{d\xi} = \frac{du}{dx}\frac{dx}{d\xi} \tag{C.34}$$

which can be reformulated to

$$\frac{du}{dx} = \frac{1}{\frac{dx}{d\xi}}\frac{du}{d\xi} = J^{-1}\frac{du}{d\xi} \tag{C.35}$$

In the case of a one-dimensional element, J is a scalar

$$J = \frac{dx}{d\xi} = -\tfrac{1}{2}x_1 + \tfrac{1}{2}x_2 = \left(-\tfrac{1}{2}, \tfrac{1}{2}\right)\begin{pmatrix} x_1 \\ x_2 \end{pmatrix} = \frac{L}{2}. \tag{C.36}$$

However, for two- and three-dimensional elements, J is a matrix.

The axial strain ε can now be written as

$$\begin{aligned} \varepsilon = \frac{du}{dx} = \frac{du}{d\xi}\frac{d\xi}{dx} = J^{-1}\frac{du}{d\xi} = \\ \left(\tfrac{1}{L}\right)(-u_1 + u_2) = \left(\tfrac{1}{L}\right)(-1, 1)\begin{pmatrix} u_1 \\ u_2 \end{pmatrix} = [B]\begin{pmatrix} u_1 \\ u_2 \end{pmatrix} \end{aligned} \tag{C.37}$$

where the symbol $[B]$ is typically used to indicate the strain matrix that transforms the displacements at the FE nodes to the strain in the element.

C.6 Virtual Work

The principle of the virtual work (VW) is discussed in much detail in [57, 86]. The discussion in the book (in Dutch) of W.T. Koiter[2] is reflected in this section.

Here, the principle of virtual work will be briefly summarised and then in the next sections applied to the one-dimensional rod.

When a deformable (elastic) body with volume V is in an equilibrium state, then is the stress tensor symmetric and fulfils the boundary conditions and equations of equilibrium. The virtual work equation can then be written as follows

$$\begin{aligned} &\int_V [\sigma_{xx}\delta\varepsilon_{xx} + \cdots + \sigma_{xy}\delta\varepsilon_{xy} + \cdots]dV = \\ &\int_V [q_j\delta u + \cdots]dV + \int_A [p_x\delta u + \cdots]dA \end{aligned} \tag{C.38}$$

where

- $\sigma_{ij},\ i,j = x, y, z$ is the symmetric stress tensor,
- $\varepsilon_{ij},\ i,j = x, y, z$ is the strain tensor,
- $f_j, j = x, y, z$ the internal loads per unit volume,
- $p_j, j = x, y, z$ surface loads per unit area on the bounding surface A
- u the displacement at the surface.

The symbol δ in δu or $\delta\varepsilon$ is used to mark the displacement and strain as virtual displacement and virtual strain.

The right-hand side of Eq. (C.38) represents the virtual work of the external loads and left-hand side the internal virtual work in the elastic body.

In accordance with Shanley [84], this method of virtual work is called the "Virtual Displacement Method", which states: Apply a virtual displacement at a point and calculate the virtual work done by the real loads acting through the virtual displacement.

- "Displacements" may be either translational or rotational, "load" may be either a force or a moment.
- During the virtual displacement, the real loads (or moments) remain constant.

[2] Prof. dr. ir. Warner Tjardus Koiter (1914–1997).

C.7 Virtual Work Applied to an Iso-parametric Linear Rod Element

The stiffness matrix $[K]$, the external load vector (F) and the equivalent thermal load vector (F_F) will be derived using the virtual displacement variant of the method of virtual work. The general expression for the virtual work of Eq. (C.38) applied to a rod reads

$$A\int_0^L \sigma\,\delta\varepsilon dx = q\int_0^L \delta u^T(x)dx + F_1^B\delta u_1 + F_2^B\delta u_2 \tag{C.39}$$

This expression can be expanded by using the constitutive relation of Eq. (C.3) and the compatibility relations expresses in the natural coordinates in Eq. (C.29).

$$\begin{aligned} &A\int_{-1}^{1} [\delta\varepsilon(\xi)E\,(\varepsilon(\xi) - \alpha\Delta T(\xi))]\,Jd\xi \\ &= q\int_{-1}^{1} \delta u(\xi)Jd\xi + F_1^B\delta u_1 + F_2^B\delta u_2 \end{aligned} \tag{C.40}$$

With Eq. (C.37), $\delta\varepsilon$ can be written as

$$\delta\varepsilon = [B]\begin{pmatrix}\delta u_1\\ \delta u_2\end{pmatrix} = [B]\,(\delta u) \tag{C.41}$$

and

$$(u) = \begin{pmatrix}u_1\\ u_2\end{pmatrix} \tag{C.42}$$

This allows to expand Eq. (C.40) further to

$$\begin{aligned} &EA\int_{-1}^{1} (\delta u)^T\,[B]^T[B]\,(u)\,Jd\xi \\ &= \int_{-1}^{1} (\delta u)^T\,[B]^T\alpha\Delta T(\xi)Jd\xi + q\int_{-1}^{1} (\delta u)^T\,Jd\xi + (\delta u)^T\begin{pmatrix}F_1^B\\ F_2^B\end{pmatrix} \end{aligned} \tag{C.43}$$

This should hold for every $(\delta u) \neq (0)$ which gives

$$EA \int_{-1}^{1} [B]^T [B] (u) J d\xi = \int_{-1}^{1} [B]^T \alpha \Delta T(\xi) J d\xi + q \int_{-1}^{1} J d\xi + (\delta u)^T \begin{pmatrix} F_1^B \\ F_2^B \end{pmatrix} \tag{C.44}$$

This expression can be written in the same matrix form as Eq. (C.28)

$$[K_e] (u) = (F_T) + (F_Q) + (F_B) \tag{C.45}$$

with

$$[K_e] = EA \int_{-1}^{1} [B]^T [B] J d\xi = \frac{EA}{L} \begin{bmatrix} 1 & -1 \\ -1 & 1 \end{bmatrix} \tag{C.46}$$

$$(F_T) = EA\alpha \int_{-1}^{1} [B]^T (\Psi_i(\xi), \Psi_2(\xi)) \begin{pmatrix} \Delta T_1 \\ \Delta T_2 \end{pmatrix} J d\xi = \frac{EA\alpha}{2} \begin{bmatrix} -1 & -1 \\ 1 & 1 \end{bmatrix} \begin{pmatrix} \Delta T_1 \\ \Delta T_2 \end{pmatrix} \tag{C.47}$$

$$(F_Q) = q \int_{-1}^{1} \delta u^T (\xi) J d\xi = \frac{1}{2} qL \begin{pmatrix} 1 \\ 1 \end{pmatrix} \tag{C.48}$$

$$(F_B) = \begin{pmatrix} F_1^B \\ F_2^B \end{pmatrix} \tag{C.49}$$

This reproduces the expression obtained with the Galerkin method in Eq. (C.28).

C.8 Assembly of the System Equation of the Finite Element Model

In this last section of this Appendix, the principle is explained of assembling element equations of individual one-dimensional rods elements into the system equations of a complete finite element model. For this explanation, a finite element model is considered with two one-dimensional rod elements of which the element equations are derived in the previous sections using two different methods.

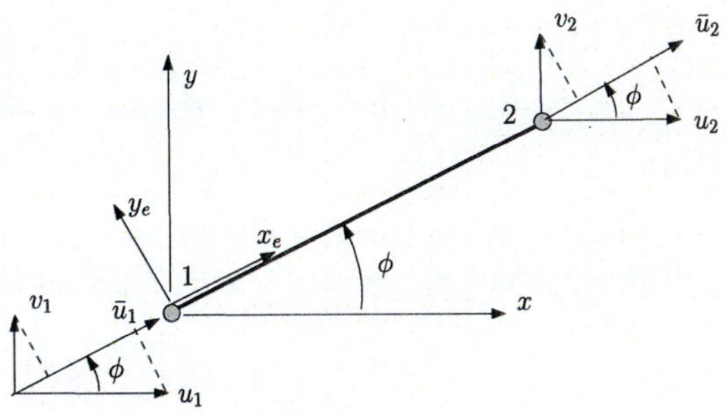

Fig. C.4 Truss element in global coordinate system

For illustration purpose, it is considered that the rod element is placed in a two-dimensional space. This implies that in the discussion below, only two coordinates will be used for defining a position and correspondingly only two translational displacements.

In this two-dimensional space, a position is identified by two coordinates x and y. These coordinates are here referred to as global coordinates and are therefore expressed in the global coordinate system. There is also a coordinate system used to define the position of a point in the element with the coordinates x_e and y_e. Both coordinate systems are indicated in Fig. C.4.

The position of the nodes 1 and 2 of the rod element is indicated in the global coordinate system with the coordinates x_1, y_1 and x_2, y_2. The length of the rod then becomes

$$L = \sqrt{(x_2 - x_1)^2 + y_2 - y_1)^2} \tag{C.50}$$

With the position of the FE nodes expressed in global coordinates, the orientation of the rod element in the two-dimensional space can be expressed in the angle ϕ such that the values of the goniometrical functions become

$$\sin\phi = \frac{(y_2 - y_1)}{L}, \quad \cos\phi = \frac{(x_2 - x_1)}{L} \tag{C.51}$$

The FE nodes of the one-dimensional element only have stiffness in the axial direction of the element. Hence, the equations of the element only consider axial displacements. Consider now that these displacements are referred to as $\bar{u}_1$ and $\bar{u}_2$, which are displacements measured along the local x_e axis of the element. For the two-dimensional finite element model, these displacements in the element coordinate system have to be transformed to the global coordinate system. In this global system, the displacement is in principle expressed in two orthogonal components: u along the global x-axis and v along the global y-axis. The displacements of FE node 1 are then referenced to by u_1 and v_1, and similarly, u_2 and v_2 are the displacements of FE node 2 in the global coordinate system.

With the use of Eq. (C.51), the relation between the displacements in the local system and the global system can be formulated as

$$\begin{pmatrix} \bar{u}_1 \\ \bar{u}_2 \end{pmatrix} = \begin{bmatrix} \cos\phi & \sin\phi & 0 & 0 \\ 0 & 0 & \cos\phi & \sin\phi \end{bmatrix} \begin{pmatrix} u_1 \\ v_1 \\ u_2 \\ v_2 \end{pmatrix} = [R_\phi] \begin{pmatrix} u_1 \\ v_1 \\ u_2 \\ v_2 \end{pmatrix} \tag{C.52}$$

where $[R_\phi]$ is the coordinate transformation matrix.

The expression of Eqs. (C.28) and (C.45) can be simplified to

$$[K_e]\,(\bar{u}) = (\bar{F}) \tag{C.53}$$

with also the components of $(\bar{F})$ being expressed in the element coordinate system

$$(\bar{F}) = (F_T) + (F_Q) + (F_B) = \begin{pmatrix} \bar{F}_1 \\ \bar{F}_2 \end{pmatrix} \tag{C.54}$$

This transformation can be applied to the vector (u) in Eq. (C.53) followed by a premultiplication of all terms with $[R_\phi]^T$, leading to

$$[R_\phi]^T[K_e][R_\phi]\,(\hat{u}) = [R_\phi]^T\,(F) \tag{C.55}$$

with $(\hat{u})$ now being the vector with global displacements

$$(\hat{u}) = \begin{pmatrix} u_1 \\ v_1 \\ u_2 \\ v_2 \end{pmatrix} \tag{C.56}$$

Consider now that based on Eq. (C.46), the matrix $[K_e]$ is of the form

$$[K_e] = \begin{bmatrix} k_e & -k_e \\ -k_e & k_e \end{bmatrix} \tag{C.57}$$

with

$$k_e = \frac{EA}{L} \tag{C.58}$$

The left-hand side term of Eq. (C.55) can then be expanded to

$$\begin{aligned} &[R_\phi]^T[K_e][R_\phi]\,(\hat{u}) \\ &= k_e \begin{bmatrix} cos^2\phi & cos\phi sin\phi & -cos^2\phi & -cos\phi sin\phi \\ cos\phi sin\phi & sin^2\phi & -cos\phi sin\phi & -sin^2\phi \\ -cos^2\phi & -cos\phi sin\phi & cos^2\phi & cos\phi sin\phi \\ -cos\phi sin\phi & -sin^2\phi & cos\phi sin\phi & sin^2\phi \end{bmatrix} \begin{pmatrix} u_1 \\ v_1 \\ u_2 \\ v_2 \end{pmatrix} \\ &= [\hat{K}_e]\,(\hat{u}) \end{aligned} \tag{C.59}$$

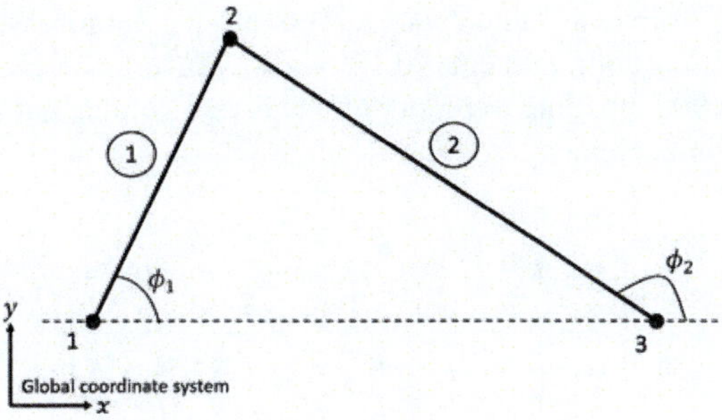

Fig. C.5 FE model with two rod elements

Similarly the right-hand side term of Eq. (C.55) can evaluated to

$$[R_\phi]^T\left(\bar{F}\right) = \begin{pmatrix} \bar{F}_1 cos\phi \\ \bar{F}_1 sin\phi \\ \bar{F}_2 cos\phi \\ \bar{F}_2 sin\phi \end{pmatrix} = \left(\hat{F}\right) \tag{C.60}$$

With above two expressions, the rod is placed in two-dimensional space.

Imagine now a simple finite element model with two rod elements, element 1 and 2, and three FE nodes, 1, 2 and 3, as shown in Fig. C.5. Element 1 connects FE nodes 1 and 2. Element 2 does that for FE nodes 2 and 3. The axis of element 1 has an angle ϕ_1 with the global x-axis. ϕ_2 is the angle between the global x-axis and the axis of element 2.

The total FE model has in total 6 degrees of freedom, being for each FE node 2 displacements along the two axes of the global coordinate system. The system equations for this finite element model shall therefore contain 6 simultaneous linear equations. The stiffness matrix of the finite element model must then have 6 rows and 6 columns. The force vector must then also have 6 rows.

Suppose now that the stiffness matrix of element j in the global coordinate system is written as

$$[K_e^j] = \begin{bmatrix} K_{11}^j & K_{12}^j & K_{13}^j & K_{14}^j \\ K_{12}^j & K_{22}^j & K_{23}^j & K_{24}^j \\ K_{13}^j & K_{23}^j & K_{33}^j & K_{34}^j \\ K_{14}^j & K_{24}^j & K_{34}^j & K_{44}^j \end{bmatrix} \tag{C.61}$$

and the load vector coming from element j is

$$\left(F^j\right) = \begin{pmatrix} F_1^j \\ F_2^j \\ F_3^j \\ F_4^j \end{pmatrix} \tag{C.62}$$

Element 1 and 2 have FE node 2 in common, which implies that the contributions from both elements to FE node 2 have to be combined in the rows and columns corresponding to FE node 2. This results then in the following set of equations of the finite element model

$$\begin{bmatrix} K^1_{11} & K^1_{12} & K^1_{13} & K^1_{14} & 0 & 0 \\ K^1_{12} & K^1_{22} & K^1_{23} & K^1_{24} & 0 & 0 \\ K^1_{13} & K^1_{23} & K^1_{33}+K^2_{11} & K^1_{34}+K^2_{12} & K^2_{13} & K^2_{14} \\ K^1_{14} & K^1_{24} & K^1_{34}+K^2_{12} & K^1_{44}+K^2_{22} & K^2_{23} & K^2_{24} \\ 0 & 0 & K^2_{13} & K^2_{23} & K^2_{33} & K^2_{34} \\ 0 & 0 & K^2_{14} & K^2_{24} & K^2_{34} & K^2_{44} \end{bmatrix} \begin{pmatrix} u_1 \\ v_1 \\ u_2 \\ v_2 \\ u_3 \\ v_3 \end{pmatrix} = \begin{pmatrix} F^1_1 \\ F^1_2 \\ F^1_3+F^2_1 \\ F^1_4+F^2_2 \\ F^2_3 \\ F^2_4 \end{pmatrix} \quad \text{(C.63)}$$

This system allows for implementing displacement boundary conditions, which is not further discussed here. Finally, the solution of this system can be used to determine the local element deformation with Eq. (C.52) with subsequent recovery of the element forces.

Problems

C.1 Derive the stiffness matrix and the equivalent thermal expansion force vector for a three-noded bilinear rod element with aid of the principle of virtual work. The area of the cross section is A (m^2), Young's modulus is E (Pa) and the CTE is given by α (m/m/°C). The physical element is mapped on a iso-parametric (mapped) element. The three-node physical rod element and the corresponding iso-parametric element are shown in Fig. C.6. Only the nodal displacement $u_i,\ i = 1, 2, 3$ are drawn. The nodal forces $F_i,\ i = 1, 2, 3$ and nodal temperature differences $\Delta T_i,\ i = 1, 2, 3$ are not shown. $\Delta T = T - T_{ref}$ is the current temperature T corrected by the reference temperature T_{ref}.

The physical coordinate x, the nodal displacement $u(x)$ and the temperature difference $\Delta T(x)$ can be expressed in the nodal coordinates $x_i,\ i = 1, 2, 3$, nodal displacements $u_i,\ i = 1, 2, 3$ and nodal temperature differences $\Delta T_i,\ i = 1, 2, 3$ using

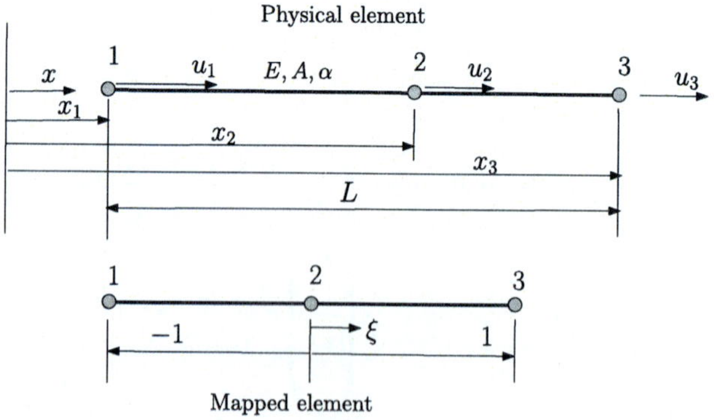

Fig. C.6 Three-noded iso-parametric rod element

shape functions $\Psi_i(\xi),\ i = 1, 2, 3$. For the coordinate x, the following expression is used

$$x(\xi) = \Psi_1(\xi)x_1 + \Psi_2(\xi)x_2 + \Psi_3(\xi)x_3 = \lfloor \Psi_1(\xi), \Psi_2(\xi), \Psi_3(\xi) \rfloor \begin{pmatrix} x_1 \\ x_2 \\ x_3 \end{pmatrix} \tag{C.64}$$

for the discplacement u

$$u(\xi) = \Psi_1(\xi)u_1 + \Psi_2(\xi)u_2 + \Psi_3(\xi)u_3 = \lfloor \Psi_1(\xi), \Psi_2(\xi), \Psi_3(\xi) \rfloor \begin{pmatrix} u_1 \\ u_2 \\ u_3 \end{pmatrix} \tag{C.65}$$

and the temperature difference ΔT

$$\Delta T(\xi) = \Psi_1(\xi)\Delta T_2 + \Psi_2(\xi)\Delta T_2 + \Psi_3(\xi)\Delta T_3 = \lfloor \Psi_1(\xi), \Psi_2(\xi), \Psi_3(\xi) \rfloor \begin{pmatrix} \Delta T_1 \\ \Delta T_2 \\ \Delta T_3 \end{pmatrix} \tag{C.66}$$

where the shape functions are given by

$$\begin{aligned} &\Psi_1(\xi) = \frac{1}{2}(-\xi + \xi^2),\ \Psi_2(\xi) = (1 - \xi^2), \Psi_3(\xi) = \frac{1}{2}(\xi + \xi^2), \\ &\lfloor \Psi(\xi) \rfloor = \left\lfloor \frac{1}{2}(-\xi + \xi^2), (1 - \xi^2), \frac{1}{2}(\xi + \xi^2) \right\rfloor \end{aligned} \tag{C.67}$$

It is required that

$$\sum_{k=1}^{3} \Psi_k(\xi) = 1 \tag{C.68}$$

Appendix D
Theory of Introduction Multipoint Constraint Equations in Thermoelastic Problems

Abstract
The mathematical methods for introduction of the linear multipoint constraints into the finite element system equations are discussed. Both a method based on Lagrangian multipliers and a method based on elimination of degrees of freedom is explained.

Keywords: Finite element method, linear constraint equations, multipoint constraints (MPC), Lagrange multipliers, rigid body element.

D.1 Introduction

Linear constraint relations (MPC) may have a non-physical impact on the thermoelastic responses. This appendix provides the mathematical background on how these relations can be implemented in the finite element system matrix equations. Explained is how the modified stiffness matrix of a FE model is derived for the case in which linear relations, also called multi point constraint (MPC) equations, between degrees of freedom are prescribed.

For structural applications of the FE method, the degrees of freedom of a FE node are the three translations and three rotations, jointly referred to as displacements. When the finite element method is used for thermal problems, the FE nodes have only a single degree of freedom, being the temperature. This appendix is written with the structural application in mind, which makes that the degrees of freedom are mostly referred to as displacements, but the expressions are equally applicable to the thermal domain in which temperatures are used as degrees of freedom. Because of the structural context, reference is made to the stiffness matrix, that can be without any impact of the validity of the method, be read as conduction matrix for thermal applications.

S. Appel and J. Wijker, *Simulation of Thermoelastic Behaviour of Spacecraft Structures*, Springer Aerospace Technology, https://doi.org/10.1007/978-3-030-78999-2

Constraint equations are frequently introduced through the application of rigid body elements (RBE). MSC Nastran provides the possibility simulate thermal expansion of the rigid elements this imposes a relative displacement of the FE nodes that are connected by the rigid elements.

A constraint equation or multipoint constraint (MPC) defines a linear relation between the displacements of one or more FE nodes of the form

$$\sum_{k=1}^{N} a_k u_k = \delta_j, \tag{D.1}$$

where a_k is the weighting factor associated to the displacement u_k of FE node k. δ_j may be a prescribed relative displacement or thermally induced expansion or shrinkage of a leg of a rigid element for constraint relation with sequence number j. In most applications of multipoint constraints, $\delta_j = 0$.

For rigid body elements, the weighting factors a_k are generated from the relative positions of the connected FE nodes. In addition, MSC Nastran allows also for the definition of explicit MPC equations through which the user can manually enter the weighting factors.

Equation (D.1) can be written in matrix form as

$$[U_m](u) = (\Delta), \tag{D.2}$$

in which displacement vector is denoted by (u) and where $[U_m]$ is the matrix with the coefficients a_k of the linear equations. The vector (Δ) contains the prescribed relative displacements δ_j leading to a nonzero vector in case thermal expansion of rigid elements is included.

D.2 Use of Lagrange Multipliers

First the method to implement the linear constraints in the system equations through Lagrange multipliers is explained.

The linear constraint equations from Eq. (D.2) are used to extend the expression for the potential energy, by introducing Lagrange multipliers. This leads to the following expression including (Λ) and can then be written as follows [52]

$$\Pi = \frac{1}{2}(u)^T [K](u) - (u)^T (F) + (\Lambda)^T [[U_m](u) - \Delta], \tag{D.3}$$

with $[K]$ being the stiffness matrix of the finite element model and (F) vector with external load components.

The stationary value of the potential energy (Lagrangian) occurs when the variation of the potential energy $\delta\Pi = 0$ with variation of the displacements and Lagrange multipliers, what means

$$\begin{aligned} \delta\Pi =& \delta\,(u)^T\,[[K]\,(u) - (F) + [U_m]\,(\Lambda)] + \\ & \delta\,(\Lambda)^T\,[[U_m]\,(u) - \Delta] = 0 \end{aligned} \tag{D.4}$$

This equation shall hold with $\delta\,(u) \neq (0)$ and $\delta\,(\Lambda) \neq (0)$, which implies

$$[K]\,(u) + [U_m]\,(\Lambda) = (F)\,, \tag{D.5}$$

and

$$[U_m]^T\,(u) = (\Delta) \tag{D.6}$$

Combining equations (D.5) and (D.6) allows these to be rewritten in

$$\begin{bmatrix} K & U_m \\ U_m^T & 0 \end{bmatrix} \begin{pmatrix} u \\ \Lambda \end{pmatrix} = \begin{pmatrix} F \\ \Delta \end{pmatrix}. \tag{D.7}$$

The solution of this system provides the values of the Lagrange multipliers (Λ) that physically represent the connection forces between degrees of freedom included in the constraint relation.

The method described above is implemented in MSC Nastran and is activated with the CASE control entry RIGID=LAGRAN.

D.3 Elimination of Dependent Degrees of Freedom

Another method to implement the MPC equations into the system equations is based on partitioning the displacement vector into two sets: a set of dependent (slaves) displacements and a set of independent (master) displacements. A transformation matrix is set up to express the dependent displacements as a function of the independent displacements. Because of the use of this transformation matrix, it is not possible to include in this method the relative displacements due to thermal expansion of the FE nodes that are connected to rigid elements. The constraint equation Eq. (D.2) then becomes

$$[U_m]\,(u) = (0) \tag{D.8}$$

As indicated, the displacement vector (u) has to be partitioned in a set of dependent (slaves) displacements (u_d) and a set of independent (master) displacements.

$$(u) = \begin{pmatrix} (u_d) \\ (u_i) \end{pmatrix} \tag{D.9}$$

The matrix of coefficients $[U_m]$ is partitioned in the same manner. The constraint equations Eq. (D.8) can now be written as

$$\begin{bmatrix} U_d & U_i \end{bmatrix} \begin{pmatrix} (u_d) \\ (u_i) \end{pmatrix} = (0)\,. \tag{D.10}$$

When the sub-matrix $[U_d]$ is not singular, the vector with dependent displacements (u_d) can now be expressed in terms of the vector with independent displacements to (u_i) as follows

$$(u_d) = -[U_d]^{-1}[U_i]\,(u_i) \tag{D.11}$$

This is then the basis of the transformation matrix that transforms the full set of displacements into the reduced set of independent displacements

$$(u) = \begin{pmatrix} (u_d) \\ (u_i) \end{pmatrix} = \begin{bmatrix} -[U_d]^{-1}[U_i] \\ [I_d] \end{bmatrix} (u_i) = [\hat{U}_m]\,(u_i) \tag{D.12}$$

with I_d being a unity matrix and $\hat{U}_m$ is

$$[\hat{U}_m] = \begin{bmatrix} -[U_d]^{-1}[U_i] \\ [I_d] \end{bmatrix} = \begin{bmatrix} [\hat{U}_{di}] \\ [I_d] \end{bmatrix} \tag{D.13}$$

This has now to be applied to the system equation of the finite element model

$$[K]\,(u) = (F) \tag{D.14}$$

which can also be partitioned in accordance with Eq. (D.9) leading to

$$\begin{bmatrix} K_{dd} & K_{di} \\ K_{di}^T & K_{ii} \end{bmatrix} \begin{bmatrix} [\hat{U}_{di}] \\ [I_d] \end{bmatrix} (u_i) = \begin{pmatrix} (F_d) \\ (F_i) \end{pmatrix} \tag{D.15}$$

Pre multiplication both sides of this equation with $[\hat{U}_m]^T$ gives

$$\begin{bmatrix} [\hat{U}_{di}]^T & I_d \end{bmatrix} \begin{bmatrix} K_{dd} & K_{di} \\ K_{di}^T & K_{ii} \end{bmatrix} \begin{bmatrix} [\hat{U}_{di}] \\ [I_d] \end{bmatrix} (u_i) = \begin{bmatrix} [\hat{U}_{di}]^T & [I_d] \end{bmatrix} \begin{pmatrix} (F_d) \\ (F_i) \end{pmatrix} \tag{D.16}$$

which can then be summarised into

$$[\hat{K}_{ii}]\,(u_i) = \left(\hat{F}_i\right) \tag{D.17}$$

The expressions for $[\hat{K}_{ii}]$ and $\left(\hat{F}_i\right)$ are

$$\begin{aligned}[\hat{K}_{ii}] &= [\hat{U}_{di}]^T[K_{dd}][\hat{U}_{di}] + [\hat{U}_{di}]^T[K_{di}] + [K_{di}]^T[\hat{U}_{di}] + [K_{ii}] \\ \left(\hat{F}_i\right) &= [\hat{U}_{di}]^T\,(F_d) + (F_i)\end{aligned} \tag{D.18}$$

It may be noticed that quite some operations are required when the elimination method is applied for the implementation of the linear constraints in the system matrices. For that reason, MSC Nastran indicates that the usage of the Lagrange multiplier method is preferred.

Appendix E
Solutions

Problems of Chapter 3

3.1

- The curves for $\beta(T)$ and $\alpha(T)$ are given in Figs. E.1 and E.2.
- The thermoelastic properties of steel had been measured

3.2 Heat pipe

$F = \alpha AE\Delta T_1 \frac{17}{8}\frac{L_o}{L}$

3.3 Euler buckling of bar

- $\Delta T = \frac{\pi^2 I}{A\alpha L^2} - \frac{F}{\alpha EA}$

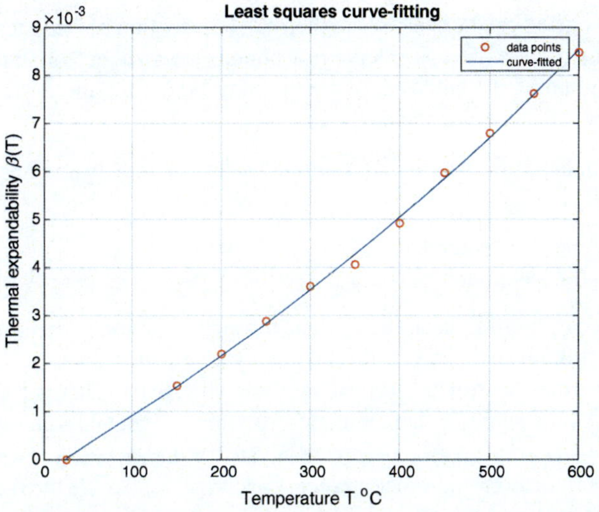

Fig. E.1 Curve-fitted $\beta(T)$

S. Appel and J. Wijker, *Simulation of Thermoelastic Behaviour of Spacecraft Structures*,
Springer Aerospace Technology, https://doi.org/10.1007/978-3-030-78999-2

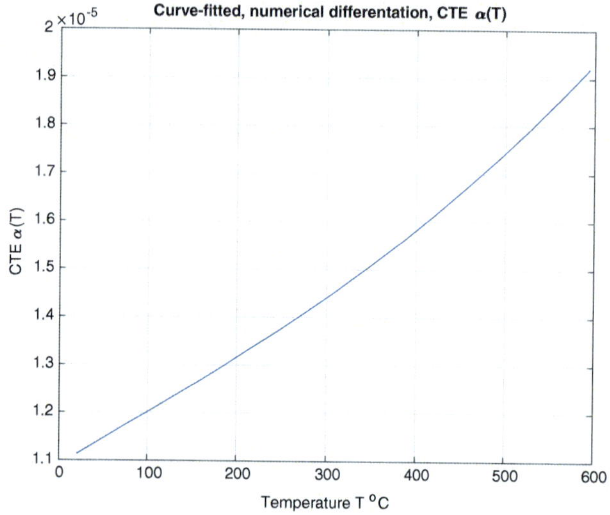

Fig. E.2 CTE $\alpha(T)$

- $\Delta T = 50.6683$ °C.

Problems of Chapter 5

5.1 During the non-operational phases, in which the instruments are in general off, the focus of thermoelastic is on the prediction of thermal stresses in order to verify the strength of the spacecraft. Instruments on spacecraft in many cases contain optical elements of which the images become distorted as soon as these elements move relative to each other due to thermal effects. Also, some types of antennas require accurate pointing and can be sensitive to thermal deformation of the antenna itself or the supporting structure. As soon as the instruments are switched on, the spacecraft is in an operational phase, and also, thermally induced deformations are relevant to be predicted by the thermoelastic analyses.

5.2 The spectral bands for heat transfer through radiation mostly applicable to space are:

- Radiation in the infrared band,
- Radiation in the solar band.

When a radiator is emitting heat, this is done through radiation in the infrared band.
5.3 The main differences between thermal analysis for thermoleastic and thermal analysis for thermal control are driven by their objectives. Thermal analysis for thermal control aims to predict the temperature extremes (hot and cold) of important locations at the spacecraft, while thermal analysis for thermoelastic tries to capture complete temperature fields in order predict as accurate as possible the thermoelastic responses of the structure and possibly the impact on the performance of instruments

due to that. The difference items that are mentioned in Table 5.2 have to be included in the answer.

Problems of Chapter 6

6.1 Spring force $F = 7.9834 \times 10^5$ N

6.2 $R_B = 1.3803 \times 10^4$ N and $R_D = 4.6548 \times 10^4$ N.

6.3 $\Delta T = -0.7340$ °C

Problems of Chapter 7

7.1 See Figs. E.3 and E.4.

Problems of Chapter 8

8.3 $\sigma_{Copper} = 52.3019$ MPa and $\sigma_{Al-alloy} = 39.2264$ MPa.

Problems of Chapter 9

9.1 Conductor: $GL_A(0.5) = 2.71$, $GL_A(1) = 4.0$ W/°C. (10 × 10 Quad4 elements).

9.2

- Conductors: $GL_{12} = 0.5714Ak/L$ and $GL_{23} = 1.7143Ak/L$
- Heat balance

$$\begin{bmatrix} -GL_{12} & GL_{12} & 0 \\ GL_{12} & -(GL_{12} + GL_{23}) & GL_{23} \\ 0 & GL_{23} & -GL_{23} \end{bmatrix} \begin{pmatrix} T_1 \\ T_2 \\ T_3 \end{pmatrix} = \begin{pmatrix} -Q_1 \\ -Q_2 \\ -Q_3 \end{pmatrix}.$$

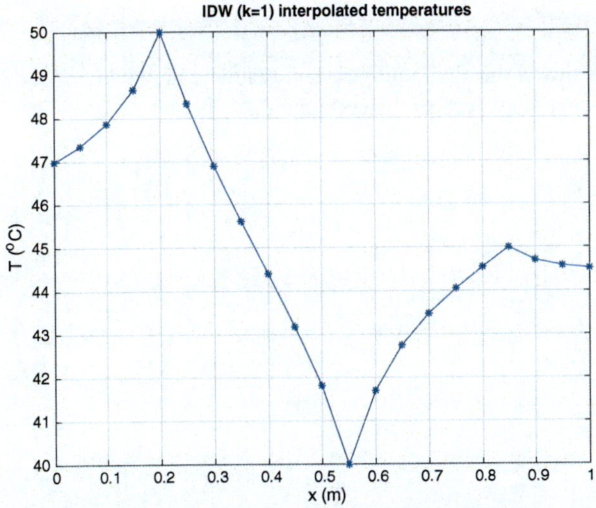

Fig. E.3 IDW interpolated temperatures ($k = 1$)

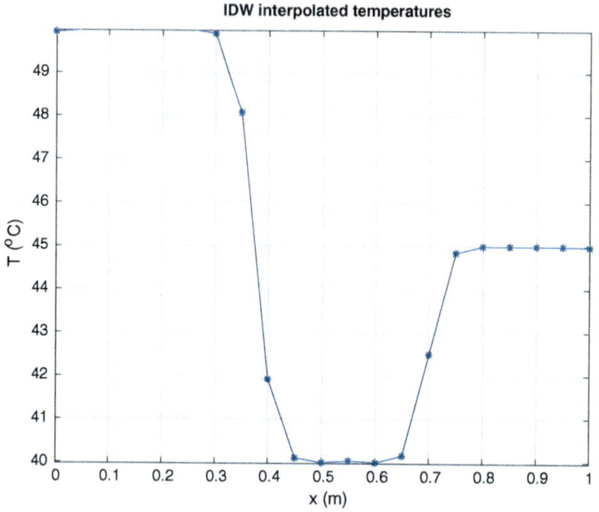

Fig. E.4 IDW interpolated temperatures ($k = 5$)

- Solution:

$$\begin{bmatrix} -(GL_{12}+GL_{23}) & GL_{23} \\ GL_{23} & -GL_{23} \end{bmatrix} \begin{pmatrix} T_2 \\ T_3 \end{pmatrix} = \begin{pmatrix} -Q_2 - GL_{12}T_1 \\ -Q_3 \end{pmatrix}.$$

- Numerical solution: $T_2 = 23.5000$, $T_3 = 24.0833$ °C, $Q_1 = -2.0000$ W.

9.3 $GL_{12} = GL_{13} = 0.2395$, $GL_{23} = 0.0350$ W/°C.

9.4 $GL_{12} = 1.0223$, $GL_{13} = -0.0756$ $GL_{23} = 0.2521$ W/°C (see Fig. 9.22).

9.5 Figure E.5 shows the PAT interpolation results calculated with MATLAB, and Table E.1 presents the values of the conductors.

9.6

$$\begin{bmatrix} GL_{12}+GL_{13} & -GL_{12} & -GL_{13} \\ -GL_{12} & GL_{12}+GL_{23} & -GL_{23} \\ -GL_{13} & -GL_{23} & GL_{13}+GL_{23} \end{bmatrix} \begin{pmatrix} T_1 \\ T_2 \\ T_3 \end{pmatrix} = \begin{pmatrix} Q_1 \\ Q_2 \\ Q_3 \end{pmatrix} \tag{E.1}$$

$$\begin{bmatrix} GL_{12}+GL_{13} & -GL_{12} & -GL_{13} \\ -GL_{12} & GL_{12}+GL_{23} & -GL_{23} \\ -GL_{13} & -GL_{23} & GL_{13}+GL_{23} \end{bmatrix} \begin{pmatrix} T_1 \\ 0 \\ 0 \end{pmatrix} = \begin{pmatrix} Q_0 \\ Q_2 \\ Q_3 \end{pmatrix} \tag{E.2}$$

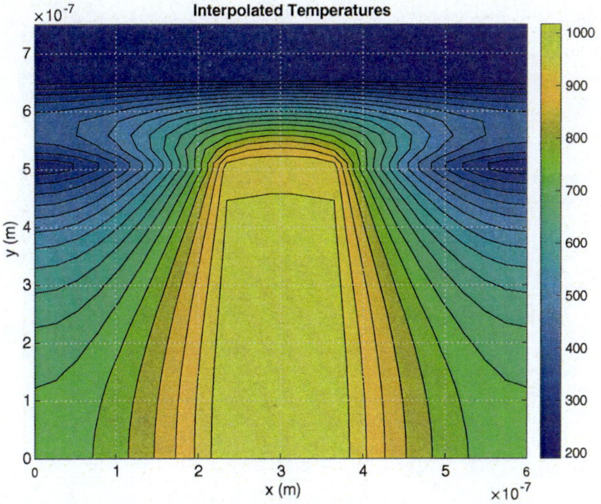

Fig. E.5 PAT interpolated temperatures in PCRAM

Table E.1 Conductors $GL_{ij},\ i \neq j,\ j > i$

	$GL \times 10^6$ (W/°C)					
TNi/TNj	1	2	3	4	5	6
1		0.4072	-0.0477	0.2213	-0.0006	0.0002
2			0.4072	0.5148	-0.0015	0.0005
3				0.2213	-0.0006	0.0002
4					0.6335	-0.2105
5	Symmetric					0.6208

Problems of Chapter 10

10.2 Cracked plate

MCS: $\mu_{K_{ic}} = 1.1787,\ \sigma_{K_{ic}} = 0.2056$ MPAmm$^{1/2}$.
PEM: $\mu_{K_{ic}} = 1.1760,\ \sigma_{K_{ic}} = 0.1613$ MPAmm$^{1/2}$.

10.3 Thermostatic Bimetal strip

- MCS (not unique solution): $\mu_\rho = 1.3501$ m, $\sigma_\rho = 0.0065$ m. The associated PDF is given in Fig. E.6.
- 2k+1 PEM: $k = 5$, $\mu_\rho = 1.3437$ m, $\sigma_\rho = 0.0091$ m.

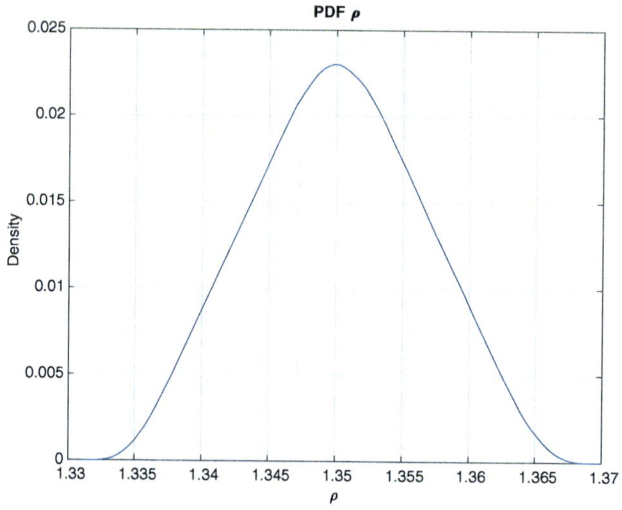

Fig. E.6 PDF MCS analysis

10.4 Thin-walled pipe

Create MATLAB/Octave.m or Python3 py script to repeat all calculations!

References

1. Earth orbit environmental heating. Guidelines GD-AP-2301, NASA Preferred Reliability Practices for Design and Test: Techn. Memo 4322A, Feb. 1999, John F. Kennedy Space Center
2. ECSS. Space engineering, Insert Design Handbook. Noordwijk, The Netherlands, 20 March 2011
3. Sigmadyne, Optomechanical Analysis & Software. Rochester, New York, USA
4. Abramowitz M, Stengun IA (1970) Handbook of mathematical functions with formulas, graphs, and mathematical tables. Applied mathematics series, p 55. National Bureau of Standards
5. Airbus Defence and Space SAS. http://www.systema.airbusdefenceandspace.com/products/thermica.html. Thermica Overview
6. Ang AH-S, Tanf WH (2007) Probability concepts in engineering, 2nd edn. Number ISBN 0-471-72064-X. Wiley, New York
7. Angel GD, Haritos G (2013) An immediate formula for the radius of curvature of a Bimetalic strip. Int J Eng Res Technol 2:1312–1319
8. Anon (2009) Design standard, spacecraft thermal control system. Japan Aerospace Exploration Agency, jerg-2-310 edition, 8 July 2009
9. Appel S (1998) SINAS IV theoretical manual. Technical Report SINAS-IV-R-005, Fokker Space B.V., the Netherlands, 20 April 1998
10. Appel S, Cambicorti L, Etchells J, Blake S, Rieder M, Bandy T, Cessa V, Piazza D (2021) End-to-end STOP analysis on Plato Tou. In: ECSSMET 21, p 11. ESA/DLR/CNES, 23–25 Mar 2021
11. Appel S, Patricio R, de Koning HP, Pin O (2004) Automatic conductor generation for thermal lumped parameter models. In: 34th International conference on environmental systems, number ICES:2004-01-2397
12. Appel S, Peman A, Wijker JJ, van Oostrum A (2021) Uncertainties quantifying. In: Thermoelastic predictions. In: ECSSSMET 21, 23–25 Mar 2021. ESA/ESTEC, DLR/CNES, p 12
13. Appel S, Teichert W (1996) Interpolation of lumped parameter temperatures for thermo-elastic FEM analysis with SINAS. In: European conference on spacecraft structures. Materials & mechanical testing, 27–29 Mar 1996, vol ESA SP-386. ESA/ESTEC, CNES, DLR, pp 277–281
14. Bathe KJ (2014) Finite element procedures, 2nd edn. Number ISBN 978-0-9790049-5-7. Prentice Hall, Watertown
15. Bauchau OA, Graig JI (2009) Structural analysis, with applications to aerospace structures. Number ISBN 978-90-481-2515-9. Springer, Berlin
16. R.V. Baud. 50 Jahre Kirsch'sche Lõsung. Schweizerische Bauzetung, 67(7), 106–107, 1949
17. Blevins RD (1979) Formulas for structural frequency and mode shape. Krieger Publishing Company, Number ISBN 0-89464-894-2
18. Bobba S, Abrar S, Rehman M (2019) Probability Study on the Thermal Stress Distribution in Thick HK40 Stainless Steel Pipe using Finite Element Method. Designs 3(9):24

S. Appel and J. Wijker, *Simulation of Thermoelastic Behaviour of Spacecraft Structures*, Springer Aerospace Technology, https://doi.org/10.1007/978-3-030-78999-2

19. Boley BA, Weiner JH (1997) Theory of thermal stresses. Dover Publications, Inc. Number ISBN 987-0-486-14386-6
20. Bourdeaud'hui M, Ponsy J, Laborde S, Corberand P, Satellites designing, from the main thermoelastic stability contributors quantification-process and tool applied on Juice. In: European conference on spacecraft structures, materials & mechanical testing, ECSSMET, Noordwijk, The Netherlands, 28 May–1 June 2018. ESA/DLR/CNES, ESA
21. Calvi (Conveyor) A (2004) Spacecraft mechanical loads analysis handbook, vol ECSS-E-HB-32-26A of Space engineering. ECSS, Noordwijk
22. Carter JP, Booker JR (1989) Finite Element Analysis of Coupled Thermoelasticity. Computers & Structures 31(1):73–80
23. Cengel YA (2004) Heat transfer: a practical approach, 2nd edn. McGraw-Hill. Number ISBN 978-0072458930
24. Chen WT, Nelson CW (1979) Thermal Stress in Bonded Joints. IBM Journal Resent Developments 23(2):179–188
25. Clough RW (2004) Early history of the finite element method from the viewpoint of a pioneer. Int J Numer Methods Eng 60:283–287
26. Cook RD, Malkus DS, Plesha ME (1989) Concepts and applications of finite element analysis, 3rd edn. Number ISBN 0-471-84788-7. Wiley, New York
27. Craig RR (1981) Structural dynamics, an introduction in computer methods. Number ISBN 0-47187715-8. Wiley, New York
28. Delicata M (2018–2019) The comparison of numerical methods for the calculation of thermoelastic deformation of satellite structures. Master's thesis, Politechnic Di Torino, 2018–2019
29. den Hartog JP (1961) Strength of materials. Dover Publications, Inc, June 1961. Number ISBN 9780486607559
30. Early DA, Halle WB, Turczyn MT (2001) Thermally induced vibrations of the hubble space telescope's solar array 3 in a test simulated space environment. In: 4th International symposium on environmental testing for space programmes, 12 June 2001, vol ESA-SP-467. Liege, Belgium
31. ECSS, Noordwijk, The Netherlands. Space engineering, general structural requirements, ECSS-E-ST-32C, Rev 1, 15 Nov 2008
32. ECSS, Noordwijk, The Netherlands. Space engineering, structural finite element models, ECSS-E-ST-32-03C, 31 July 2008
33. ECSS, Noordwijk, The Netherlands (2008) Space engineering, thermal control general requirements, ECSS-E-ST-31C, 15 Nov 2008
34. ECSS, Noordwijk, The Netherlands (2009) Space engineering, structural factors of safety for spaceflight hardware, ECSS-ST-32-10C, rev. 1 edition, 6 Mar 2009
35. ECSS, Noordwijk, The Netherlands. Space engineering. Thermal analysis handbook, ECSS-E-HB-31-03A, 15 Nov 2016
36. Eslami MR (2014) Finite element methods in mechanics. Springer. Number ISBN 978-319-08036-9. https://doi.org/10.1007/978-3-319-08037-6
37. Prustim T et al (2016) The Gaia Mission (Gaia Data Release 1). Astronomy astrophysics (Special issue), 1–36. https://doi.org/10.1051/0004-6361/201629272
38. Filion R (2017) Thermal network model development for an extended range electric vehicle battery pack with experimental verification through dynamic environment exposure. Faculty of Engineering and Applied Science, University of Ontario Institute of Technology, July 2017
39. Fortescue G, Swinerd P, Stark J (2011) Spacecraft systems engineering, 4th edn. Wiley, New York. Number ISBN 9780470750124
40. Foster CL, Tinker ML, Nurre GS, Till WA (1995) The solar array-induced disturbance of the hubble soace telescope pointing system. Technical Paper 3556, NASA, May 1995
41. Gaspar Venancio LM et al (2014) Euclid payload module: telescope characteristics and technical challeges. In: Conference: SPIE astronomical telescopes + instrumentation, Montreal, vol 9143 of Project: Euclid, p 14. SPIE, 2014
42. Gere JM, Timosshenko SP (1991) Mechanics of materials, 3rd edn. Chapman & Hall. Number ISBN 0-412-36880-3edition

43. Guénégo P, Rivière E, Puillet C, Briet R (2005) Thermal interpolation from a nodal model to a FEM model by using heat fluxes. In: European conference on spacecraft structures, materials & mechanical testing, vol ESA SP-581, Noordwijk, The Netherlands, 10 May 2005. ESA
44. Hetnarsky RB, Eslami MR (2009) Thermal stresses-advanced theory and applications. Solid mechanics and its applications. Springer, Berlin. Number ISBN 978-1-4020-9246-3
45. Holman JP (2010) Heat transfer. Higher education, tenth edition. MacGraw Hill. Number ISBN 978-0-0735936-3
46. Incropera FP, Dewitt DP, Bergman TL, Lavine AS (2007) Fundamentals of heat and mass transfer, 6th edn. Wiley, New York. Number ISBN 978-0-471-45728-2
47. Inman RL, Helton JC (1988) An investigation of Uncertainty and Sensitivity Analysis Techniques for Computer Models. Risk Analysis 8(1):71–90
48. M. Ishizuka, T. Hatakeyama, Y. Funuwatashi, and K. Koizumi. Application of Thermal Network Model to Transient Thermal Analysis of Power Electronic Package Substrate. *Active and Passive Electronic Components*, Volume 2011, Article ID 823654 (doi:1155/2011/823654):8, 2011
49. ITP Engines UK Ltd. (2018) ESATAN-TMS User Manual, Dec 2018
50. Jacques L, Béchet E, Kerschen G (2017) Finite Element Model reduction for Space Thermal Analysis. Finite Elements in Analysis and Design 127:6–15
51. Jain MK (2006) Numerical solution of differential equations. New Age International Limited. Publishers. Number ISBN 0-85226-432-1
52. James G (1993) Advanced modern engineering mathematics. Addison-Wesley Publishing Company. Number ISBN 0-201-56519-6
53. Junker J, Bleicher G, Mottaghibonab A, Nicolay O, Reese C, Appoloni M, Galdo M, Teti D (2021) PLATO spacecraft: thermo-elastic distortion verification concept and demonstration tests. In: ECSSMET 21, p 8. ESA/DLR/CNES
54. Kellogg OD (1967) Poundations of potential theory. Springer, Heidelberg, Reprint 1929 edition
55. Kirtley CJ, Stock NJ, de Koning HP, Appel S (2005) Automatic linear conductor generation for limped parameter models. J Aerosp 114, Sect. 1:546–553
56. Klobučar. Thermal radiation heat transfer between surfaces. Seminar I_b notes, Feb 2016
57. Koiter WT (1972) Stijfheid en Sterkte 1, Grondslagen (Stiffness and strength, basics). In: Mechanica. Scheltema & Holkema, Haarlem. Number ISBN 90-6060-504-7
58. Koot M, Appel S, Simonian S (2018)Temperature mapping for structural thermoelastic analysis; method benchmarking. In: European conference on spacecraft structures, materials & mechanical testing. ESA/ESTEC, CNES, DLR, May–June 2018
59. Koot M, Appel S, Simonian S (2018) Thermal conductor generation for thermal and thermo-elastic analysis using a finite element model and SINAS. In: European conference on spacecraft structures, materials & mechanical testing. ESA/ESTEC, CNES, DLR
60. Korelc J, Wriggers P (2016) Automation of finite element methods. Springer, Berlin. Number ISBN 987-3-319-39005-1. https://doi.org/10.1007/987-3-319-39005-5
61. Kwon YW, Bang H (2000) The finite element method using MATLAB. In: CRC mechanical engineering series. CRC Press, 2nd edn. Number ISBN 0-8493-0096-7
62. Liao L (2011) A study of inertia releaf. AIAA, 52nd AIAA/ASME/ASCE/AHS/ASC structures. structures dynamics and material conference, 4–7 Apr 2011, Denver, Colorado (AIAA-2011-2002), p 10, Apr 2011
63. Lienhard IV JH, Lienhard V JH (2008) A heat transfer textbook, 3rd edn. Phlogiston Press
64. Lim JH (2015) Thermal Pointing Error Analysis of the Observation Satellites with Interpolated Temperature based on PAT method. Journal of Korean Society for Aeronaytical and Space Sciences 44(1):80–87
65. McGinty B (2014) Stress concentration at holes. https://www.fracturemechanics.org/hole.html
66. McMurchy RR (1972) Thermal network modeling handbook. TRW, 14690-h003-r0-00 edition, Jan 1972
67. Meirovitch L (1975) Elements of vibration analysis. McGraw-Hill. Number ISBN 0-07-041340-1

68. Most T (2011) Efficient sensitivity analysis of complex engineering problem. In: 11th International conference on applications and probability on civil engineering, Zurich, Switzerland, p 9. Dynardo GmbH, Germany, Aug 2011. https://doi.org/10.13140/2.1.2587.6162
69. MSC Software. MSC Nastran 2017.1 Linear Static Analysis User's Guide. na:v2017.1:z:z:dc-lin-pdf edition, April–July 2017
70. Nguyen DH, Skladany LM, Prats B (2001) Thermal performance of the Hubble space telescope (HST) solar Array-3 during the disturbance verification Test (DVT). In: 4th International symposium on environmental testing for space programmes, Liege, Belgium, 12–14 June 2001, vol ESA-SP-467, pp 163–170. ESA, Aug 2001
71. Nikishkov GP (2010) Programming finite element in JAVA^{TM}. Springer, Berlin. The University of Aizu, Aizu-Wakamatsu, pp 965–8580, Japan
72. Nowak AS, Collins KR (2000) Reliability of structures. McGraw-Hill. Number ISBN 0-07-116354-9
73. Peery DJ, Azar JJ (1982) Aircraft structures, 2nd edn. McGraw-Hill. Number ISBN 0-07-049196-8
74. NASA preferred reliability practices. Guidelines for thermal analysis of spacecraft hardware. Guideline GD-AP-2303, NASA
75. Prescott J (1961) Applied elasticity. Dover Publications, Inc
76. Rivello RM (1969) Theory and analysis of flight structures. McGraw-Hill, Inc. Number 07-052985-X
77. Rosenblueth E (1981) Two-point estimates in probability. Applied Mathematical Modelling 5:329–335
78. Rubinstein RY, Kroese DP (2008) Simulation and the Monte Carlo method. In: Probability and statistics, 2nd edn. Wiley, New York. Number ISBN 978-0-470-17794-5
79. Saltelli A, Ratto M, Andres T, Campolongo F, Cariboni M, Gatelli D, Saisana M, Tarantola S (2008) Global sensitivity analysis, The primer. Wiley, New York. Number ISBN 978-0-470-05997-5
80. Saltelli A, Tarantola S., Campolongo F, Ratto M (2004) Sensitivity analysis in practice. A guide to assessing scientific models. Wiley, New York. Number ISBN 0-470-87093-1
81. Sarafin TP (ed) (1995) Spacecraft structures and mechanisms-from concept to launch. In: Space technology Series. Kluwer Academic Publishers. Number ISBN 0-7923-3476-0
82. Sechler EE (1968) Elasticity in engineering. Dover Publications, Inc
83. Seshu P (2012) Textbook of finite element analysis. PHI Learning Private Limited, New Delhi, January 2012. Number ISBN-978-81-203-2315-5
84. Shanley FR (1967) Mechanics of Materials. McGraw-Hill, Book Company
85. Sobol IM (1993) Sensitivity Estimates for Nonlinear mathematical Models. MMCE 1(4):407–414
86. Sokolnikoff IS (1956) Mathematical Theory of Elsticity. McGraw-Hill, Book Company, second edition
87. Spiegel MR (1960) Vector aanlysis and an introduction to tensor analysis. Schaum's Outline Series, McGraw-Hill. Number ISBN 07-000228-X
88. Strutt JH, Sock NJ, Kirtley CJ (2014) Lumped parameter thermal conducor generation for 3D geometry. In: 44th International conference on environmental systems, 13–17 July 2014, Arizona, p 8. SAE International. Number ICES-2014-297
89. Thornton EA (1996) Thermal structures for aerospace applications. AIAA Education series, 1996. Number ISBN 1-56347-190-6
90. Timoshenko S, Goodier JN (1951) Theory of Elasticity. McGraw-Hill, Book Company
91. Tse LA, Chang Z, Somawardhana RP Slimko E (2018) Structural, thermal, and optical performance (STOP) modeling and analysis for surface water and ocean topography mission. Albequerque, New Mexico, 8–12 July 2018 2018. 48th International Conference on Environmental Systems. Number ICES-2018-062
92. van Oostrum A, Peman A, Laine B (2021) Modelling guidelines for thermo-elastic analysis. In: ECSSMET 21, p 12. DLR/CNES/ESA

93. Wijker JJ (2005) Mechanical vibrations in spacecraft design. Springer, Berlin. Number ISBN 3-540-40530-5
94. Wijker JJ (2008) Spacecraft structures. Springer, Berlin. Number ISBN 978-3-540-75552-4. DOIurl10.007/978-540-75553-1
95. Wood WL (1990) Practical time-stepping schemes. In: Oxford applied mathematics and computing science series. Clarendon Press, Oxford. Number ISBN 0-19-859677
96. ECSS working group (2016) ECSS-E-HB-31-03A, Space engineering: thermal analysis handbook. ECSS, ESA-ESTEC Noordwijk, The Netherlands, a edition, 15 Nov 15 2016
97. Wyss GD, Jorgensen KH (1998) A user's manual to LHS: Sandia's Latin hypercube sampling software. Manual SANDIA-0210.UC-505, SANDIA National laboratories, Feb 1998
98. Zienkiewicz OC (1985) The finite element method. McGraw-Hill, Book Company, UK Limited, 3rd revised edition 1977 edition

Index

S. Appel and J. Wijker, *Simulation of Thermoelastic Behaviour of Spacecraft Structures*, Springer Aerospace Technology, https://doi.org/10.1007/978-3-030-78999-2